Engineering Materials

This series provides topical information on innovative, structural and functional materials and composites with applications in optical, electrical, mechanical, civil, aeronautical, medical, bio- and nano-engineering. The individual volumes are complete, comprehensive monographs covering the structure, properties, manufacturing process and applications of these materials. This multidisciplinary series is devoted to professionals, students and all those interested in the latest developments in the Materials Science field, that look for a carefully selected collection of high quality review articles on their respective field of expertise.

Indexed at Compendex (2021) and Scopus (2022)

Kuppusamy Thangaraju · Paul Joseph Daniel ·
V. Jayalakshmi · Yogendra Kumar Mishra
Editors

ZnO Tetrapods

Basics, Modeling, Synthesis, and Applications

Springer

Editors
Kuppusamy Thangaraju
Department of Physics
National Institute of Technology Warangal
Warangal, Telangana, India

Paul Joseph Daniel
Department of Physics
National Institute of Technology Warangal
Warangal, Telangana, India

V. Jayalakshmi
Department of Physics
National Institute of Technology Warangal
Warangal, Telangana, India

Yogendra Kumar Mishra
Mads Clausen Institute
University of Southern Denmark
Sønderborg, Denmark

ISSN 1612-1317 ISSN 1868-1212 (electronic)
Engineering Materials
ISBN 978-981-95-1613-1 ISBN 978-981-95-1614-8 (eBook)
https://doi.org/10.1007/978-981-95-1614-8

This Springer imprint is published by the registered company Springer Nature Singapore Pte Ltd.
The registered company address is: 152 Beach Road, #21-01/04 Gateway East, Singapore 189721, Singapore

Foreword

Complex 3D shape of nanostructures could be another dimension which can play a very important role in translational nanotechnology. The role of nanostructure geometry needs to be explored further in context of structure-property relationship and technological applications. Over the last two decades, zinc oxide (ZnO) tetrapods have emerged as a distinctive and multifunctional class of nanostructures, captivating researchers across disciplines. Their unique 3D geometry—comprising four elongated arms interconnected via a central core—confers a high surface-to-volume ratio, structural rigidity, and exceptional optical, electrical, thermal, mechanical, piezoelectrical, optoelectronic and catalytic properties. These characteristics enabled the ZnO tetrapods to transcend conventional applications, finding huge relevance in fields ranging from energy, optical, electronics, catalysis, environmental remediation, healthcare, biosensing, etc., to advanced composites and self-assembling materials.

The importance of this book stems from the growing demand for a comprehensive reference that explores the full potential of ZnO tetrapods, bridging the foundational understanding with advanced technological applications. This volume brings together the multidisciplinary insights into the material properties, synthesis strategies, theoretical modeling, and applied research surrounding ZnO tetrapods. Each chapter has been carefully curated to address a critical area of development in this domain.

The book begins with an in-depth exploration of the fundamental crystallographic, physico-chemical, and optoelectronic characteristics of ZnO material, with an emphasis on the distinct advantages of tetrapod morphology. This establishes the basis for understanding why ZnO tetrapods offer a versatile performance across scientific and industrial domains. The planned chapters delve into specific applications.

Chapter "ZnO Tetrapods: Introduction to Unique Properties, Synthesis Methods and Applications" is devoted to their role in ultraviolet photodetection, where ZnO tetrapods outperform traditional nanostructures due to their high electron mobility and broad surface area. It also examines their photocatalytic efficacy, particularly in degrading organic pollutants, highlighting their promise for environmental sustainability. Further, the biomedical applications of ZnO tetrapods are explored in detail—from antibacterial behavior and drug delivery systems to biosensor design—underscoring their potential in nanomedicine.

Very importantly, the book does not limit itself to experimental studies. It also ventures into theoretical territories, presenting simulation techniques and coarse-grained models that describe the self-assembly and dynamic behavior of ZnO tetrapods. Approaches rooted in statistical physics and augmented by AI/ML methods provide a framework for predictive modeling, supporting the rational design of complex architectures based on tetrapod units.

This book envisions me as both a foundational guide and a springboard for innovation. It is designed to cater to the needs of graduate students, academic researchers, and industrial scientists alike, offering both depth and breadth across the diverse landscape of ZnO tetrapod research. Whether one is interested in synthesis pathways, application-driven engineering, or theoretical modeling, this volume serves as a comprehensive resource.

By consolidating the recent advances and providing forward-looking perspectives, this book will stimulate new interdisciplinary collaborations and paves the way for further technological breakthroughs using ZnO tetrapods. This book on ZnO tetrapods is going to be a foundation book to the scientific community. I wish the editors a very good luck with the book.

July 2025

Prof. Zhong Lin Wang
Regents' Professor and Hightower Chair (Emeritus), School of Materials Science and Engineering
Georgia Institute of Technology
Atlanta, Georgia

Director and Chief Scientist
Beijing Institute of Nanoenergy and Nanosystems
Beijing, China

Preface

The discovery and optimization of zinc oxide (ZnO) tetrapods have unlocked a new frontier in materials science because of its peculiar combination of three-dimensional micro-nanostructure, excellent mechanical strength, porosity, and functional diversity. ZnO tetrapods, in the last twenty years, have become potent building blocks in nanotechnology that can push the boundaries of applications in optoelectronics, photocatalysis, drug delivery, biosensing, and self-assembled soft systems. This book represents a comprehensive collection of chapters that summarizes the key aspects of ZnO tetrapods research, from material properties and syntheses to cutting-edge applications and modeling schemes.

This book consists of five chapters, well selected to present a multi-faceted description of characteristics of ZnO tetrapods and its applications. The book begins with Chapter "ZnO Tetrapods: Introduction to Unique Properties, Synthesis Methods and Applications", which covers the basic crystallographic, optoelectronic, and physico-chemical characteristics of ZnO and highlights the importance of tetrapod morphology. This chapter presents historical views, synthesis routes, and structure-function relationships and emphasizes ZnO's development as a material of preference in both research and industry and also establishes the fact that there is still extendable scope for ZnO tetrapods.

Chapter "A Comprehensive Study on Tetrapod-Structured ZnO for Efficient Photodetector Applications" focuses on ZnO tetrapods for photodetector applications, presenting a detailed account of how ZnO tetrapods enable enhanced ultraviolet detection owing to their large and varying surface area and high electron mobility. The discussion bridges theoretical considerations with experimental outcomes, highlighting the advantages of tetrapods over conventional nanostructures of ZnO in terms of quantum efficiency and responsivity.

Chapter "ZnO Tetrapods for Rapid and Enhanced Photocatalytic Activity: Size, Shape and Geometrical Advantages" discusses the application of ZnO tetrapods in photocatalytic degradation, towards environmental remediation. This chapter outlines the general mechanism involved with ZnO towards photocatalytic degradation and gives a brief outline of the various photocatalytic reactors, working principles, their design and how ZnO is integrated into it. This chapter discusses the contributions of shape, arm geometry, and surface morphology to excellent catalytic performance under UV light, visible light and sunlight, highlighting the feasibilities for wastewater treatment.

Chapter "ZnO Tetrapods in Biosensing, Drug Delivery, and Antimicrobial Strategies" addresses the biomedical and biosensing applications and how the distinctive 3D architecture of ZnO tetrapods increases their therapeutic functionalities in drug delivery, pathogen detection, and antibacterial strategies. This chapter also explains the routes of synthesis, including VLS, CVD, and hydrothermal methods, and how surface modification enhances biocompatibility and therapeutic efficacy of ZnO tetrapods.

Chapter "Statistical Physics View on the Dynamics and Self-organization of Tetrapods" presents various statistical physics-based approaches for coarse-grained simulations and computational models to elucidate the self-assembly and dynamic behavior of ZnO tetrapods and similarly shaped bioactive structures. Simulation techniques such as Langevin dynamics, dissipative particle dynamics, and AI/ML methods are introduced to establish a foundation for predictive modeling and the rational design of hierarchical architectures using tetrapods as modular building blocks.

Since the individual chapters of this book on ZnO tetrapods are drafted for specific applications, to cater the needs of domain specific readers and therefore to offer completeness, every chapter is given briefly with properties of ZnO and its preparation. In short, the chapters of this book seek to fill the gap between basic knowledge and practical applications, providing readers with a consistent overview of the multi-functional contribution that ZnO tetrapods can make to cutting-edge materials research and multitude of applications. We are confident that this book will be a crucial reference for researchers, scientists, scholars and students who aim to create new functional materials, providing a basis for newer interdisciplinary applications. We believe this book will serve as a means to extend further the tuning strategies and applications of ZnO tetrapods with newer and extended perspectives and novel interdisciplinary applications.

Warangal, India — Kuppusamy Thangaraju
Warangal, India — Paul Joseph Daniel
Warangal, India — V. Jayalakshmi
Sønderborg, Denmark — Yogendra Kumar Mishra

Acknowledgements

We take this opportunity to convey our sincere thanks and appreciation to all the contributing authors who have made it possible to successfully complete this book, *ZnO Tetrapods: Basics, Modeling, Synthesis, and Applications.*

We gratefully acknowledge the financial support received from the Scheme for Promotion of Academic and Research Collaboration (SPARC), Government of India, under the project *"Hybrid tetrapod nanomaterials based sustainable optoelectronic technologies"* [SPARC/2019-2020/P2268/SL, dated 28th July 2023]. This funding enabled valuable collaborations, resource sharing, and focused research that has eventually resulted in bringing out this book with enriched information for the benefit of the interdisciplinary scientific community.

We take great pleasure in extending our heartfelt gratitude to Prof. Bidyadhar Subudhi, Director, National Institute of Technology Warangal, for continuous encouragement, infrastructural facilities, and ensuring a rich research ambience. We also thank the Dean, International Relations and Alumni Affairs (IR and AA) and Dean, Research and Consultancy and their office for their support in various aspects. We are also grateful to the faculty members and staff, Department of Physics, NIT Warangal, for intellectually rich environment and access to needed facilities during the course of execution of the SPARC project.

Our sincere gratitude to the reviewers and research scholars for their technical contribution, fine-tuning and proofreading the chapters through their valuable inputs, and tireless efforts that imparted depth and diversity to the content. We also thank the institute post-doctoral fellow Dr. Benjamin Hudson Baby (NIT Warangal) for his efforts in tuning, formatting and proofreading the contents of the book carefully and to bring it out in good format. We also thank the "Springer Nature" and team for publishing this book and taking it to domain specific and larger scientific audience.

Kuppusamy Thangaraju
Paul Joseph Daniel
V. Jayalakshmi
Yogendra Kumar Mishra

Contents

About the Editors

Dr. Kuppusamy Thangaraju is currently working as Professor in Department of Physics, NIT Warangal, India. He obtained his Master of Science (2001) from University of Madras, Chennai, and Ph.D. (2008) from Anna University, Chennai India. After completing his postdoctoral research (2009–2012) in South Korea (Sogang University, Seoul; Gyeongsang National University, Jinju), he worked as Assistant Professor in Department of Physics, Gandhigram Rural University, Gandigram, Dindigul, Tamilnadu, India (2012–2013). Later, he joined as Assistant Professor in Department of Physics, NIT Warangal (2013–2018) and also worked as Associate Professor during 2018–2022 at NIT Warangal. His research area includes organic semiconductors, organic optoelectronic devices such as OLEDs, OPVs, OTFTs, photodetectors, organic/inorganic hybrid nanostructured thin films and their optoelectronic devices, and organic/Inorganic hybrid nanostructures for energy storage devices. He has over 50 publications in peer-reviewed international journals and 04 book chapters published. He has also received research funding worth of Rs. 85 lakhs from Government of India (DST-SERB/ANRF, TEQIP-III, SPARC Government of India) and established research facilities for the fabrication of organic optoelectronic devices. He has also coorganized 1 international conference and 5 Workshops/FDPs. He has also guided 3 PhDs and more than 4 research scholars are pursuing their research under his guidance in his area of research.

Dr. Paul Joseph Daniel is presently serving as an Associate Professor, Department of Physics, NIT Warangal completed M.Sc. (Loyola College, Chennai), B.Ed., M.Phil., and Ph.D. from the University of Madras, India. He was a post-doctoral fellow at National Taiwan University, Taipei (2010–2013). His research areas are alternative transparent conducting thin film electrodes, proximity effect between superconductor/half-metallic ferromagnet, alternative components for dye-sensitized solar cells, nanomaterials for batteries, swift heavy-ion-beam irradiation effects and photocatalytic activity for pollutant degradation. He has 07 contributed book chapters, 80 articles in journals of repute. Five of his X-ray diffraction data have been included in the ICDD database by invitation. As a PI/Co-PI he has completed several funded research projects (SERB/ANRF, TEQIP-III, NPIU, etc.) and a SPARC project is ongoing. He has several academic laurels to his credit; a notable one is 'Young Scientist Award' (2017) by 'Academy of Sciences-Chennai'. He has co-organized 01 international conference, 09 workshops/FDPs, out of which 02 are national level programs each of one-month duration. He has guided 03 Ph.D. scholars and 28 students for their M.Sc. (Tech) and M.Tech. projects. He has designed several indigenous equipment (automated spray pyrolysis unit, standalone dip-coating unit, high-temperature VSM rod adaptor import substitute, bending-twisting unit to study flexible films) and real-time YBCO and bismuth levitation kits as a teaching aid for students.

Dr. V. Jayalakshmi is currently an Associate Professor in the Department of Physics, NIT Warangal. She had postdoc experience from Queen's University, Canada, and Georgia Institute of Technology, USA. She completed Ph.D. (2010) at the Centre for Nano and Soft Matter Sciences, Bengaluru, following her undergraduate and postgraduate at SV University, India, with first-class distinctions. Her research career is distinguished by contributions to the field of materials science, including over 50 international journal papers and a significant presence in conference proceedings. Her research is supported by high-profile grants from SERB, DBT, and SPARC, Government of India, focusing on topics like synthesis of nematic and chiral nematic emulsions, development of technologies for detecting

Alzheimer's biomarkers using liquid crystals and ZnO tetrapods for novel applications. Alongside her research, she is actively involved in academic and administrative roles, significantly contributing to curriculum development, student guidance, member of academic committees, organizing workshops and educational programs, enhancing the learning environment and research capabilities at NIT Warangal.

Prof. Dr. Yogendra Kumar Mishra is Professor (Full) at Mads Clausen Institute, NanoSYD, University of Southern Denmark (SDU), Denmark. Prior joining to SDU, he worked as Privatdozent/Group Leader at Kiel University, Germany. He did Habilitation (Dr. habil.) in Materials Science from Kiel University in 2015 and Ph.D. in Physics in 2008 from Jawaharlal Nehru University (Inter University Accelerator Centre), New Delhi, India. In Kiel, he introduced a new flame-based process for metal oxide tetrapod nanostructuring and their 3D networks which showed many applications in engineering and biomedical fields. At NanoSYD, he is heading 'Smart Materials' group with the focus to develop new multifunctional 3D nanomaterials for green and sustainable technologies. He has published more than 400 publications which are cited more than 25,000 with h-index of over 83. He has served as an Editorial Board Member/Associate Editor for several high impact factor scientific magazines, such as *Advanced Functional Materials*, *Materials Today Electronics*, *Critical Reviews in Biotechnology*, *Energy Materials*, *Nanotechnologies*, *Nano Energy*, etc., He is Humboldtian, honored with FRSC—Fellow of Royal Society of Chemistry (UK) and has received BHJ Young Innovator Award (Denmark). He has attracted significant grants from funding agencies worldwide. He has supervised over 10 Ph.D. students, over 40 M.Sc. students, and over 10 B.Sc. students. He has been teaching solid-state physics, semiconductor optoelectronics, nanoelectronics, etc., at M.Sc. and B.Sc. levels. He is sitting in the advisory board members at funding agencies, institutes, policymakers, and also conferences. He has delivered over 200 speeches (including several plenary) in the conferences worldwide. He is strong advocator for translational nanomaterials research to bridge the gap from lab to daily-life technologies. From last decade, he has put a lot of emphasis on AMIL—Advanced Materials for Industrial Leadership—to address the future Sustainable Development Goals (SDGs).

ZnO Tetrapods: Introduction to Unique Properties, Synthesis Methods and Applications

Yuvashree Jayavelu, D. Paul Joseph, Gouranga Maharana, Benjamin Hudson Baby, and Yogendra Kumar Mishra

Abstract The outstanding popularity of ZnO with reasons to be a very promising, versatile and abundant material is due to its multi-functionality, resulting in applications in physics, chemistry, materials science and engineering, energy, textile, additive manufacturing, cosmetics, and pharmaceutical as well and the ease to grow all kinds of nano-and microstructures. The key tetrapod-shaped ZnO (ZnO-T) structure, with its remarkable properties and multi-functionality has a need for focused attention for emerging versatile applications that are likely to come in the near future. This chapter gives brief introduction on various ZnO properties and multitude of morphologies with emphasis on ZnO-T structure, synthesis techniques covering the scope of unique shape and geometry toward advanced technologies including nano optoelectronics, sensing devices, novel biomedical engineering materials and devices, versatile energy harvesting devices.

1 Introduction

Zinc oxide (ZnO) an exclusive and remarkable material from the view point of nano- and microstructuring is a direct and wide bandgap (about 3.37 eV) metal oxide semiconductor. Several one-dimensional (1D) nano-and microstructures has been grown with ease, thanks to its distinctive hexagonal-wurtzite crystal structure and polar surface. 1D ZnO nanostructures have intriguing potential for usage in nanoelectronics and sensing devices due to its high surface to volume (S/V) ratio [1]. ZnO has special piezoelectric aspects that makes it more appealing for smart technologies, such as energy harvesting, piezotronics, piezophototronics, piezophotocatalysis, self-powered devices, etc., thanks to its non-centrosymmetric lattice with Zn^{2+} and O^{2-} polar facets [2–6].

Y. Jayavelu · D. Paul Joseph (✉) · G. Maharana · B. H. Baby
Department of Physics, National Institute of Technology, Warangal, Telangana, India
e-mail: paul@nitw.ac.in

Y. K. Mishra
NanoSYD, Mads Clausen Institute, University of Southern Denmark, Sønderborg, Denmark

K. Thangaraju et al. (eds.), *ZnO Tetrapods*, Engineering Materials,
https://doi.org/10.1007/978-981-95-1614-8_1

ZnO micro-nanostructures are extremely promising for photodetection and emerging optoelectronics/luminescent applications due to the presence of several intriguing defects [2]. From a crystallographic perspective, flaws are generally undesirable; however, the science underlying them and tunability makes ZnO very intriguing and opens up newer application possibilities, such as defect-mediated ferromagnetism, p-type conductivity by N-doping, etc., in micro-nanostructure form. Defects in ZnO have really been the subject of an ongoing dispute, and extended explorations are necessary to completely understand and utilize its application potential [2, 3].

Oxygen vacancies can be induced on the surface of ZnO using a non-destructive method, by simply shining intense UV light which lead to a significant change in the material's intrinsic characteristics, such as an enhancement in n-type conductivity, switchable wettability (super-hydrophobic to ultra-hydrophilic and the other way around), and various other physico-chemical characteristics that permits for a wide range of utilization in interdisciplinary fields.

Any material's physical robustness is likely the most important pre-requisite for its utility and, in a variety of applications, ZnO is a stubborn material in this regard. For instance its exceptional physical strength, its bulk crystal Young's modulus is approximately 150 GPa and stability at extremely high temperature environment (approximately 1800 °C), holds great promise for use in harsh surroundings. It is readily chemically dissolvable in acidic liquids (below pH $\sim$6) [2, 4] which is advantageous since it provides a means to employ ZnO structures as sacrificial framework to create novel nano- and microstructures from a variety of materials.

Regarding ZnO's biomedical applications, one need to take a closer look at the early days of medicine. The body obtains zinc from food, plants, fruits, and meat as well as from other supplements. Zinc is one of the most vital elements for humans, controlling over 300 metabolic processes. In fact, zinc-fly ash was widely used in ancient medicine to treat a number of deadly diseases, and it was also very effective [4]. Ancient Zn-based medicines were not well understood in terms of scientific advancement (they are still used in some traditional medicine in some parts of the world). Presently, ZnO based materials are being used in biomedical engineering in terms of nanotechnology and nanomedicine [1]

The use of zinc fly ash in ancient medicine was most likely in the form of nanomaterials (the term "nanotechnology" was not yet coined), but the currently optimized ZnO nanostructures are being discussed for their potential applications in biomedicine, especially their strong cytotoxicity on cancer cells at the nanoscale [1]. Although there is a distinct realm for its medicinal significance, extended research is still needed to fully understand ZnO nanostructures. It is believed that even though there is a vast and newer literature available, much more scope for study and applications based on ZnO micro-nanomaterials still persists.

It is significant to note that, according to the global zinc oxide market report 2023, several chemical firms produce several kilo tons (> 1800 Kt) annually, indicating its wide range of utilization and research opportunities. The produced ZnO material is used primarily as additives and fillers in several sectors, including electronics, chemical, textile, pharmaceutical, cosmetics, and other industries [5]. Since most applications depend heavily on the S/V ratio, nanoscale ZnO have recently gained significant attention, which explains the importance of various synthesis techniques.

A thorough literature analysis indicates that, ZnO powder has been used by a number of industries in a variety of ways, and its functions have been independently studied in several domains, including physics, chemistry, biology, engineering, and medicine [6]. Finding a single ZnO structural platform which is useful for many fields is challenging though. Numerous applications have propelled the development of diverse ZnO nano-and microstructures, to which the materials scientists have already given considerable attention, although a number of requirements still remain difficult to be addressed.

Regarding growth, a wide range of ZnO nano-and microstructures with nearly any intricate hypothetical shape are widely found in the literature. The kinetics of the synthesis processes determines the control of dimensions, shapes and properties, and different synthesis procedures have been used for different ZnO structures. The lack of strong chemical reagents is one of the main advantages of physical synthesis techniques over wet chemical and biological approaches; however, the range of nanostructural varieties that can be produced is also limited to a particular method. Another problem is the ease of access to the remarkable micro-nanoscale features for general applications (such as creating a nanodevice based on a single nanowire), which calls for sophisticated integration technologies like advanced clean room processing etc.

Large-scale fabrication is still up for dispute, although a number of standard and unconventional nanostructuring methodologies, such as direct growth on micro-chips/substrates, have been adapted to address the utilization complications. Compared to usual 0D, 1D, and 2D nanostructures, three-dimensional (3D)-architected materials, which are constructed from micro-or nanoscale assembling units like ZnO tetrapods (ZnO-T), multipods, etc. display excellent structural integrity and large free volume to satisfy several of the technological criteria [7]. However, it is extremely difficult to synthesize these 3D-architected materials in appropriate amounts in a facile manner. This chapter summarizes the special potentials of ZnO-T and discusses how the multi-faceted nano- and microstructuring of ZnO could be of significant role in materials science, engineering, and technology.

The macroscopic concrete tetrapods has been employed as a water break to effectively attenuate waves in coastal engineering since about the mid-nineteenth century. Usually, it is composed of four arms that are joined by a central core at angles between approximately in the range between 105° and 110° due to their paddle like structure. Their distinct 3D shape makes it simple to access nanoscale aspects, like the surface area of discrete arms, regardless of their orientation. Indeed, the tetrapod shape is a huge benefit. High stability emerge from the transfer of forces along one arm to the other. Tetrapods autonomously create a distinct macroscopic linked architecture with extremely high porosity, even when a vast number of them are gathered together [1, 8]. A variety of tetrapods and multipods may be readily generated by using relevant synthesis strategies, such as the flame transport approach, ZnO material has special capabilities for micro-nanostructuring in this regard. Crucially, a ZnO micro-nanotetrapod is essentially composed of four 1D ZnO nanorods (which have been extensively explored in the literature); as a result, the tetrapods already have all of the physico-chemical characteristics of 1D nanorods [9]. It is highly distinctive in terms of characteristics and surface accessibility due to its distinct 3D spatial tetrapod shape, which is made up of these four 1D micro-nanorods joined by a central core (if placed together, the nanorods will form an aggregated bundle). Only a few significant uses are highlighted in this chapter, although the tetrapod morphology of ZnO is perhaps the most significant shape from an application standpoint.

2 Physico-chemical Aspects of ZnO

As ZnO has a wide bandgap of 3.3 eV, large exciton binding energy of 60 meV almost similar to that of TiO_2 (3.2 eV), its photocatalytic efficacy is also envisaged to be exceptional to that of TiO_2 [10]. Nevertheless, ZnO is comparatively cost-effective than TiO_2 making its utility more effective for large scale waste water treatment. ZnO has emerged as potential photocatalytic material which could generate reactive oxygen species which in turn enhances the photocatalytic efficiency [6, 11]. Distinct forms of ZnO in nano and microstructures (and mixed dimensions) with a variety of shapes are recorded in literature which depends on the choice of preparation techniques, process and parameters.

ZnO wurtzite structure has a hexagonal unit cell with lattice constants, *a* and *c* in the ratio, $\frac{c}{a} = \sqrt{\frac{8}{3}} = 1.60$ and with the space group *P63mc* (186). The schematics of wurtzite ZnO structure is shown in Fig. 1. In ZnO, the pressure dependent phase transition from the wurtzite (B4) to the rock salt (B1) structure takes place approximately at 10 GPa associated with a 17% reduction in volume. When the bonds in ZnO along the *c*-axis are from cation (Zn) to anion (O), the polarity is indicated as Zn polarity; when it is from anion to cation, the polarity is referred to as O polarity [13]. A number of properties of ZnO such as growth, etching, defects, spontaneous polarization, and piezoelectricity also depend on its polarity. Similarly, ZnO thin films are also crucial for present-day technology and is grown over a variety of substrates such as glass, sapphire, diamond, etc. by various techniques like magnetron sputtering, chemical vapor deposition (CVD), spray pyrolysis, pulsed laser deposition, etc. Due to these exciting properties, nanostructures of ZnO are of great potential for nanodevice applications. Nanorods of ZnO have a significant enhancement in sensitivity due to the high S/V ratio [9]. ZnO is also piezoelectric and could be applied to nanoscale surface acoustic wave devices.

ZnO nanopowders have nearly homogeneous size distribution, although their S/V ratio is higher compared to ZnO-T, the secondary contamination is larger owing to the spherical geometry and nano dimension [7]. Whereas, in the case of ZnO-T, it cannot be solely classified as nano or micro dimensional, since it is having both micro and nano dimensions structure inherently associated in its geometry with the center being micron sized and the arms gradually varying from micron to nano dimensional tip leading to the advantages of both micron and nano sized particles in

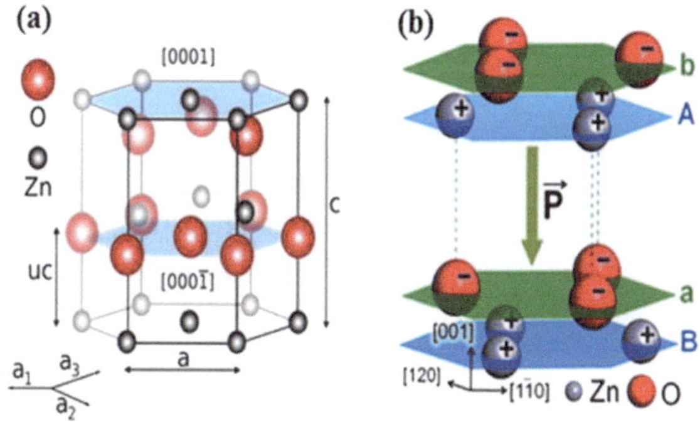

Fig. 1 Schematics of hexagonal ZnO crystal structure. (**a**) With their unit cell, including the tetrahedral coordination and their (**b**) non-centro symmetric crystal structure made with alternate layers of positive and negative ions [12]

a single geometry [7, 14]. The electrical, optical and magnetic characteristics of ZnO can be easily modified or enhanced by doping, incorporating other components into ZnO in nanoscale for better performance. ZnO is non-toxic, environment friendly, compatible with living organisms, and extends itself nicely to a broad range of other daily applications [15]. ZnO has garnered intense attention in the disintegration and complete mineralization of environmental contaminants with the key advantage of its ability to absorb wide range of solar spectrum and several light quanta than most other semiconducting metal oxides.

ZnO providing a plethora of exceptional properties emerges as a robust photocatalyst due to its efficacy in photocatalysis related applications. This is primarily attributed to its diverse range of morphology showcasing versatility from one-dimensional (1D) nanoneedles, nanorods, nanoribbons, nanotubes nano-helixes, nano-rings, nano-belts, nano-wires, nano-springs, and nano-combs, two-dimensional (2D) nano-pellets and nanoplates/nanosheets and three-dimensional (3D) structures like dandelion, flower, snowflakes, tetrapods and coniferous urchin-like shapes [6, 7]. These diverse structural forms of ZnO extend it for multifaceted applications across nanotechnology in various fields including medical, optoelectronic, making it a host material of immense interest and research attention.

Adding to these, the exceptional aspects of ZnO includes excellent physicochemical stability, tunable optoelectronic properties, appropriate bandgap, comprehensive absorption range, photostability, cost-effectiveness, anti-bacterial and environmental compatibility, high electrochemical coupling coefficient with high mobility viable for water remediation and waste water treatment technologies [4, 16, 17]. ZnO encompasses various crystalline forms including hexagonal wurtzite, cubic zinc blende, rock salt structures with the hexagonal wurtzite as the most energetically stable under ambient conditions at the structural level. ZnO is endowed with piezoelectric and spontaneous polarization characteristics, due to its crystal structure significantly influencing formation of defects and chemical interactions [13].

Moreover, ZnO's luminescence characteristics arise from defect-level transitions and excitonic processes with distinct optoelectronic attributes, that can contribute towards defect mediated photocatalytic activity as well [18]. ZnO's interactions with optical phonons vividly impact its light–matter interactions and optical characteristics, emphasizing its key role in optoelectronics and related technologies. Notably, ZnO also exhibits excellent electrical aspects with an exceptional breakdown field strength including high electron mobility, and high breakdown voltages at ambient temperature [2]. In high-power electronic, electro-optical devices, and UV/blue LEDs the electrical aspects further highlight ZnO's functionality and applicability. Additionally, the material's hardness and piezoelectric properties with robust mechanical properties make ZnO as an apex candidate material for a variety of device applications. Further, their induced magnetic properties, particularly classified as diluted magnetic semiconductor (DMS) has generated significant interest, with vivid characteristics for spintronic device applications [16, 19]. Neverthless, ZnO with its thermal expansion coefficient, thermal conductivity, and specific heat is pivotal in leveraging its efficacy in high-power device requirements, signifying its importance in materials science and engineering applications. In addition to these basic characteristics, ZnO demonstrates excellent antibacterial capabilities effectively disrupt micro-organisms [20]. The positively charged ZnO nanoparticles (NPs) can percolate the negatively charged efflux pump modulators (EPMs) resisting multiply of bacteria. Furthermore, the discharge of Zn^{2+} ions from nano dimensional ZnO elevate their toxicity to bacteria, pathogens, and viruses, relatively exceeding the performance of other ions like Cu^{+}, Fe^{2+}, and Al^{3+}. ZnO is also potent for direct interactions with the surface of bacteria and pathogens, emphasising its efficacy as an effective agent in combatting microbial threats [21]. Incorporating ZnO as a photocatalyst into water treatment units can efficiently mitigate the proliferation of harmful bacteria, degrade antibiotics and drugs, and also reduces the risk of waterborne diseases in living organisms [16]. The physico-chemical aspects of ZnO are compiled in Table 1.

Additionally, ZnO is viable to process into a variety of morphologies in aspects of nanostructured synthesis, due to its polymorphic crystal structure and varying growth rates of the diverse crystal planes. ZnO is a promising semiconducting host that has been widely explored in several fields and it was identified that ZnO with (001) facet highly exposed presents excellent performance in gas sensing is, photocatalysis and dye-sensitized solar cells [22]. Nanotechnology enables a promising platform that provides a multitude of novel applicability and technologies for optoelectronic and biomedical applications. One of the key reasons behind the intense interest is that nanotechnology allows the synthesis of materials in various nanostructured forms with at least one dimension less than 100 nm. Although vast literature has been published on the fabrication of ZnO with various synthesis techniques, morphologies and dimensions (0D, 1D, 2D and 3D), still their potential in the form of tetrapods are not extensively explored and exploited for interdisciplinary applications like photocatalysis, photodetectors, and bio-related applications [2, 4, 6] (Fig. 2).

Table 1 Compiled physico-chemical aspects of ZnO

Property	Parameter
Type of compound	Inorganic compound
Molecular formula	ZnO
Lattice constants	a = b = 3.20 Å; c = 5.209 Å
Bulk density	5.6 g/cm^3
Melting point	2248 K
Molecular weight	81.406 g/mol
Boiling point	1974 °C
Thermal conductivity	0.6, 1–1.2 (W/mK)
Relative dielectric constant	8.66
Energy gap and type	3.08 eV, direct
Inherent carrier concentration	$< 10^6$ cm^{-3}
Exciton binding energy	60 meV
Effective mass of electron	0.24 m_e
Mobility of electron	2000 cm^2/V.s
Effective mass of hole	0.59 m_e
Mobility of hole	5–50 cm^2/Vs
Primary diameter	35 nm (spherical geometry)
Specific surface area	31 m^2/g (spherical geometry)

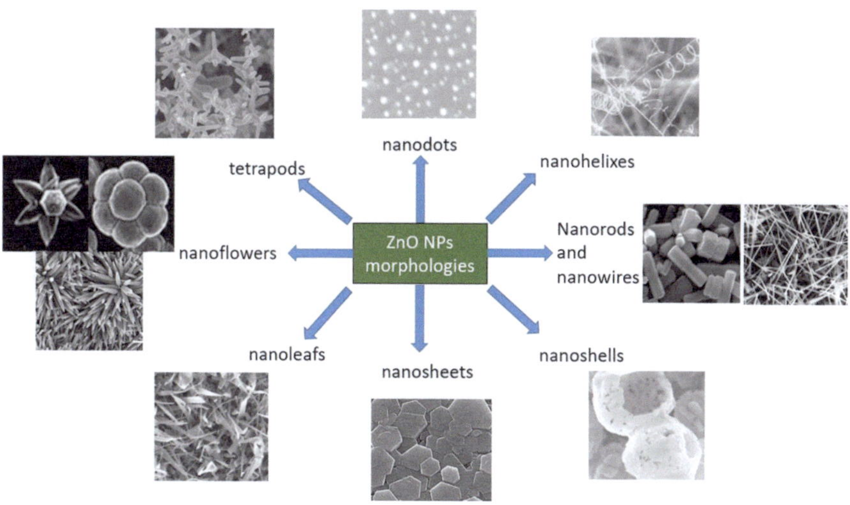

Fig. 2 Multitude of morphologies of ZnO from literature consisting of (but not limited to) nano-flowers, nano-leafs, tetrapods, nanodots, nanohelixes, nanorods, nanowires, nanoshells, nanosheets etc. [23]

3 Multitude of Morphologies of ZnO

Nanostructured ZnO exists in several interesting morphologies due to their ease of fabrication enabling to produce nanostructures in versatile shapes and sizes. Multitude of morphologies of ZnO reported in literature are named as numerous human-perceivable shapes including (but not limited to) nano-spheres, nano-plates, nano-flowers, nano-leafs, tetrapods, nanodots, nanohelixes, nanorods, nanowires, nanoshells, nanosheets. Apart from these much more are achievable in the near future also [16, 19].

ZnO providing a plethora of exceptional properties emerges as a robust photocatalyst due to its efficient performance in photocatalytic related applications which is primarily attributed to the extensive range of morphology showcasing its versatility from one-dimensional (1D) nanoneedles, nano-rods, nanoribbons, nanotubes nano-helixes, nano-rings, nano-belts, nano-wires, nano-springs, and nano-combs, two-dimensional (2D) nanopellets and nanoplates/nanosheets and three-dimensional (3D) variants like dandelion flowers, snowflakes, tetrapods and coniferous urchin-like shapes [8, 19, 24]. These intensity and diverse structural forms of ZnO extends its multifaceted applications across nanotechnology in various fields including medical, optoelectronic, making it a host material of immense interest and research focus.

3.1 1D ZnO Photocatalyst

The ZnO hexagonal crystal structure has an alternating alignment of Zn^{2+} and O^{2-} ions along the c-axis. Typically, during crystal growth, facets of nanomaterials have higher surface energy and smaller surface area. Tunable growth of 1D ZnO nanomaterials including nanoneedles, nano-rods, nano-ribbons, nano-tubes nano-helixes, nano-rings, nano-belts, nano-wires, nano-springs, and nano-combs, have non-polar facets with large surface area along the c-axis making it a promising candidate exclusively not only for photocatalytic activity but also for several other optoelectronic applications [25]. ZnO properties are highly dependent on its crystallographic direction, with the ZnO(00$\bar{0}$1)-O polar surface, polar ZnO(000$\bar{1}$)-Zn and non-polar ZnO(10$\bar{1}$0) surfaces are more effective for photocatalytic applications. Reports show that the crystallographic orientation affect the photocatalytic activity in ZnO in the order of ZnO(10$\bar{1}$0) > ZnO(0001)-Zn > ZnO(000$\bar{1}$)-O surfaces where the non-polar ZnO(10$\bar{1}$0) facet are responsible for higher aspect ratio, higher rate kinetics, photostability, efficiency and photostable photocatalysts [26]. Their exclusive characteristics, including high light absorption aspects, effective e^--h^+ pair separation, and high surface area of ZnO nano structure makes it promising for efficient photocatalytic performance.

1D ZnO are typically nanowires, nanorods, or nanotubes, length much greater than the diameter signifying a high aspect ratio. This signifies their unique shape advantages in photocatalytic applications with its wide-bandgap that can absorb UV light and liberate electron–hole pairs when illuminated. These carriers part take in various redox process, like the disintegration of pollutants, water splitting for hydrogen production, or reduction of CO_2 etc. The 1D structure of ZnO allows efficient charge carrier transport along the length of the nanowires or nanorods. This reduces recombination and enhances photocatalytic efficiency. The 1D ZnO structure, nanowires or nanorods have a relatively higher surface area than 0D structure for catalytic reactions due to their elongated shape [5, 25]. The nanowires and nanorods have a preferred orientation for electron flow, which can facilitate efficient charge carrier separation and which in turn can enhance photocatalytic performance. Especially in photocatalytic reactions that require UV light, the high aspect ratio of 1D structures can improve light trapping and absorption.

Apart from photocatalytic activity, the 1D ZnO photocatalyst are also included in water splitting for hydrogen generation via $H_2O \rightarrow H_2 + O_2$ reaction. The structure of 1D ZnO aids in both the separation of electron–hole pairs and efficient charge carrier transport. Photocatalytic CO_2 reduction, converting CO_2 into value-added products like methane or methanol when irradiated with light. Antibacterial properties make it useful for medical or environmental sanitation applications also [5].

Synthesis Methods: Several methods are adapted to synthesize 1D ZnO nanostructures, including:

Hydrothermal/Solvothermal Synthesis: ZnO nanowires or nanorods by reacting zinc salts with alkalis at high temperatures in aqueous or non-aqueous solvents [27].

Vapor-phase Deposition: CVD or pulsed laser deposition (PLD) are used to form ZnO nanowires directly over appropriate substrates [9, 28].

Electrospinning: Yeilds long ZnO nanofibers that can be used for photocatalysis [9].

Template-assisted Methods: This method use appropriate template to tune the size and morphology of ZnO structures during synthesis [29].

1D ZnO photocatalysts are effective materials for photocatalytic applications due to their high surface area, efficient charge transport characteristics, and potential for improved light absorption, water splitting, pollution degradation, and CO_2 reduction. However, challenges such as charge recombination and limited visible light absorption need to be addressed, often through doping, difficulty in separation by centrifuging or filtration, composite formation, or surface modifications. These characteristics make 1D ZnO a promising morphology in the field of renewable energy, environmental cleanup, and sustainable chemistry.

3.2 2D ZnO Photocatalyst

2D ZnO photocatalysts exist as nanosheets or thin layers, which yields a high surface area for catalytic reactions. The 2D structure enhances light absorption, electron–hole separation, and surface reactivity. The S/V ratio is higher than 1D ZnO which allows for numerous active sites for the photocatalytic process [30]. Thin 2D sheets enhance absorption of light and allow better penetration of UV light through their surface. The 2D layered structure facilitates effective separation and transport of the photoinduced charge carriers and minimal recombination. The characteristics of 2D ZnO can be tuned by controlling the thickness, doping, or by creating heterojunctions with other materials. This also suffers from the swift recombination of charge carriers, which decreases the photocatalytic performace [31]. ZnO primarily absorbs UV-C light, but there are efforts to utilize its absorption into the visible spectrum of electromagnetic radiation through doping or coupling with other materials. To improve the performance of 2D ZnO photocatalyst as an, various strategies can be employed, such as:

Doping with Metallic and Non-Metallic elements: incorporating elements like nitrogen, sulfur, or transition metals can shift the absorption spectrum to the visible range and can also help in charge separation [32, 33].

Composite Formation: 2D ZnO can be easily combined and covered with other material systems like graphene, carbon nanotubes, or other semiconductors to boost its photocatalytic degradation efficiency [30, 34].

Surface Modification: Functionalization of the surface of 2D ZnO is relatively higher than 1D ZnO with various molecules or polymers that can also improve stability and photocatalytic activity. 2D ZnO photocatalysts are a promising set of materials for a diverse type of photocatalytic applications due to their higher S/V ratio, high light absorption aspect, easy surface modification and the ability to facilitate charge separation, making them highly potential for applications like pollution degradation and clean energy production [35].

3.3 3D ZnO Photocatalyst

The 3D ZnO photocatalysts structured in a three-dimensional form, typically in the shape of porous networks, clusters, or other 3D morphologies. In this configuration, ZnO can exhibit unique photocatalytic properties, offering advantages for applications like pollutant degradation, hydrogen generation, and CO_2 reduction. These structures often strengthen the overall photocatalytic performance by promoting light absorption, increasing surface area, and improving charge carrier separation. 3D ZnO photocatalysts include many micro to nanostructures including (but not limited to):

ZnO nano/microspheres: Spherical shaped structures that can increase the surface area with decreasing size and provide multiple active sites for photocatalysis [36].

ZnO nanostructure networks: Interconnected networks or frameworks of ZnO, often formed by nanosheets, nanorods, or nanowires that create a high surface area and improve light scattering [37].

The 3D structures strengthen the photocatalytic properties by increasing the surface area, promoting light scattering, and boosting charge separation and carrier transport [38]. Advantages of 3D ZnO photocatalyst is that they have significantly high surface area with tuneable porosity and viable geometry which provides more active sites with high contact and accessability that is beneficial for photocatalytic reactions, making them much tuneable than the 0D, 1D and 2D ZnO. The 3D ZnO can be extended for applications including photocatalytic water splitting, pollutant degradation, hydrogen generation, CO_2 minimalization, photovoltaic applications, and also for antibacterial and antifungal applications. The 3D morphology can increase the interaction of light with the photocatalyst, enhancing light absorption and photocatalytic efficiency [39]. The complex and tuneable structures of 3D ZnO can minimize the recombination of photoinduced electrons and holes by providing numerous pathways for charge carriers to move in multiple directions, thus improving the photocatalytic efficacy. The porous structure of 3D ZnO allows for easier diffusion of reactants or pollutants to the catalytic sites and can be beneficial for reactions like photocatalytic hydrogen generation or CO_2 reduction.

Although ZnO being mainly active under UV light due to its wide bandgap, the very fact limits its photocatalytic efficiency, as UV light accounts only for a small fraction of the sunlight spectrum [4]. Despite the 3D structure helping to separate charges, ZnO still suffer from the issue of rapid recombination of the photoinduced electrons and holes, which can lower the efficiency of photocatalysis. To improve their photocatalytic performance doping is an efficient strategy to be followed, which can accommodate light absorption in the visible range to improve charge separation by acting as electron ink. ZnO-based photocatalysts can undergo photo-corrosion and disintegration under prolonged UV-light exposure, reducing their long-term stability. Surface of 3D ZnO with organic or inorganic molecules can improve stability, enhance photocatalytic activity, and limit aggregation of NPs along with minimizing their photo-corrosion [17, 40]. Creating heterojunctions between 3D ZnO and other materials can further boost the photocatalytic performance by providing routes for the efficient separation of photoinduced charge carriers. 3D ZnO photocatalysts are a promising class of materials for a diverse range of photocatalytic applications due to their high surface area, enhanced light absorption, and efficient charge transport properties [4, 40]. The 3D structure aid for better light scattering within the medium, thereby leading to increased access to active sites for photocatalytic process, and improved charge carrier separation, making them effective for pollutant degradation, hydrogen production, and CO_2 reduction. However, challenges like moderate visible light absorption and fast carrier recombination remain to be fully addressed, and efforts to modify or combine 3D ZnO with other materials are actively ongoing to enhance its efficiency and stability [17].

Among the various 3D ZnO morphologies, tetrapods are a specific type of 3D ZnO photocatalyst that possess a unique four-legged projected structure, that consists of a central trunk with four branching arms extending outward in a tetrahedral configuration [1, 8]. The comparison of ZnO-T vs other ZnO morphologies for the photocatalytic activity is briefed in Table 2. This highly distinct geometry provides several advantages for photocatalytic applications, making them a practically viable choice.

The tetrapod shape consists of a central trunk and four arms that extend outward in a highly branched configuration, resembling a "starfish" like or a "tetrahedron". This 3D morphology contributes significantly to high and varying surface area of the material providing numerous active sites for effective photocatalytic process [17].

Table 2 Comparison of ZnO tetrapods versus other ZnO morphologies as an effective photocatalyst

ZnO property/feature	3D tetrapods	0D NPs	1D nanorods	2D nanosheets	1D nanowires
Surface area	Very high (3D structure)	High (smaller than tetrapods)	Moderate (1D structure)	High (2D structure)	Moderate (1D structure)
Light absorption	High (light trapping and scattering)	Moderate (poor light scattering)	Moderate (better than NPs)	Moderate (UV light absorption only)	Moderate (better than NPs)
Charge carrier separation	Excellent (Effective separation due to length arms	Poor (fast recombination)	Moderate (better than NPs due to their length)	Moderate (Limits transportation along Z-direction)	Moderate (better than NPs)
Charge transport	Excellent (continuous path along four arms)	Poor (limited transport) due to smaller and near spherical shape	Good (due to path available for transport)	Moderate (2D transport limitation)	Good (rapid transport due to path available)
Complexity of synthesis	Facile and moderate	Simple and cost-effective	Moderate (depend on method)	Costliest (depend on method)	Moderate (depend on method)
Photocatalytic efficiency	High (due to structure and charge dynamics)	Low to moderate (due to high recombination)	Moderate (better than NPs)	Moderate (but limit the carrier transport)	Moderate (better than NPs)
Geometrical advantages	Excellent (Feasibility for self-assembly)	Less (Advantageous for membrane filtration)	High (Advantageous due to their different chemical reactivity along middle and tip of rods)	Moderate (Advantageous due to their mechanical stability)	Low (Advantageous can be used for drug delivery application than NPs)
Chemical stability	Relatively higher than other metal-oxide photocatalysts, but has considerable photocorrosion which can be overcome by suitable modification of ZnO's surface				

4 Preparation Techniques of ZnO Tetrapods

ZnO-T can be produced by various techniques like thermal evaporation, controlled oxidation in specific atmosphere, catalytic oxidation, microemulsion and microwave irradiation. These synthesis methods can be predominantly divided into two major categories (i) starting from a solid phase and then proceeding via a solid–vapour transport–condensation step and (ii) is a liquid medium reaction process [41].

The fabrication of ZnO-T involves a range of methods that aim to create this unique, three-dimensional nanostructure. ZnO-T are highly attractive for various applications due to their distinct protruding arms geometry. This 3D structure not only enhances the material's S/V ratio but also offers a variety of potential applications in optics, sensing, photovoltaics, and nanocomposites [1]. The fabrication techniques used to synthesize ZnO-T need to consider factors such as degree of crystallinity, morphology, and scalability to achieve high-performance. Here, various methods for fabricating ZnO, including vapor-phase growth, hydrothermal synthesis, solution-based methods, and template-assisted approaches are discussed in brief including how each fabrication technique influences the structural and optical characteristics of the formed ZnO-T.

4.1 *Vapour Transport Process*

The vapour transport method is the simplest synthesis technique to produce ZnO-T because of its cost-effective production nature, high yield, and minimal aggregate. The vapour transport method systematically involves sublimation of Zn metal powder, oxidation, nucleation and growth, all of these taking place at a temperature above 600 °C (Fig. 3). The nucleated tetrapods grow swiftly at high-temperature and connect into ZnO-T of arm diameter higher than 100 nm and arms longer than 2 μm. The key parameters in vapour transport process are the partial pressures of reactant gases, growth temperature and initial size of the nuclei. In spite of commonality between solid–gas-solid, the vapour transport method also consist of various modifications, due to the specific experimental parameters and the choice of precursors [42].

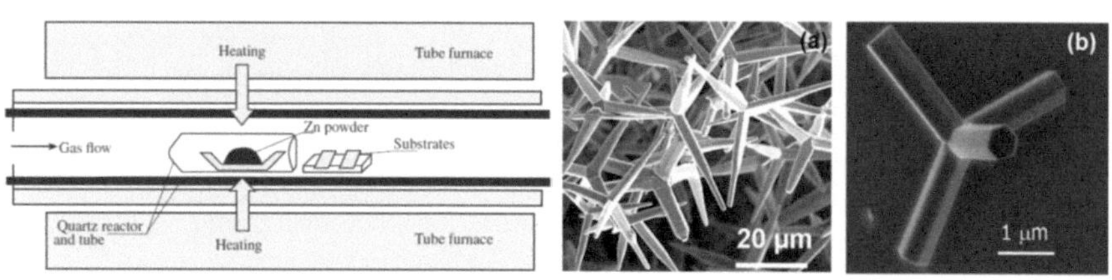

Fig. 3 Schematics of vapour transport process of ZnO-T and the SEM images of the fabricated tetrapodal morphology [43]

4.2 Thermal Evaporation

In thermal evaporation technique, zinc metal powder is taken in the reactor crucible for thermal sublimation and is directly oxidized. No catalyst is necessary for this method, as the reaction atmosphere in the setup is uncontrolled. The quantity of zinc metal powder is taken as per the reactor volume and necessity, while taking some oxygen deficiency in the reactor. The heating procedure is classified into three steps with a pre-heating step to 700 °C, continued by a heating step to the reaction temperature, with a heat preservation step at the reaction temperature for required duration [44]. The experimental observation confirms that at a low pre-heating rate of 3 °C/min, the morphology of ZnO-T is optimal and the product output is about 95%. With higher pre-heating rate, the size of the tetrapod produced increases and tend to be uneven. With process temperature above 1000 °C, the whisker size, yield and homogeneity decline (Fig. 4) [44].

In the thermal evaporation method, tetrapods, upon nucleation, grows swiftly in the high-temperature regime. This growth can be restricted in a controlled fashion so as to alter the size and dimensions of the tetrapod arms which can be tuned by swiftly transporting the growing tetrapods with gas purging to the low-temperature region [1].

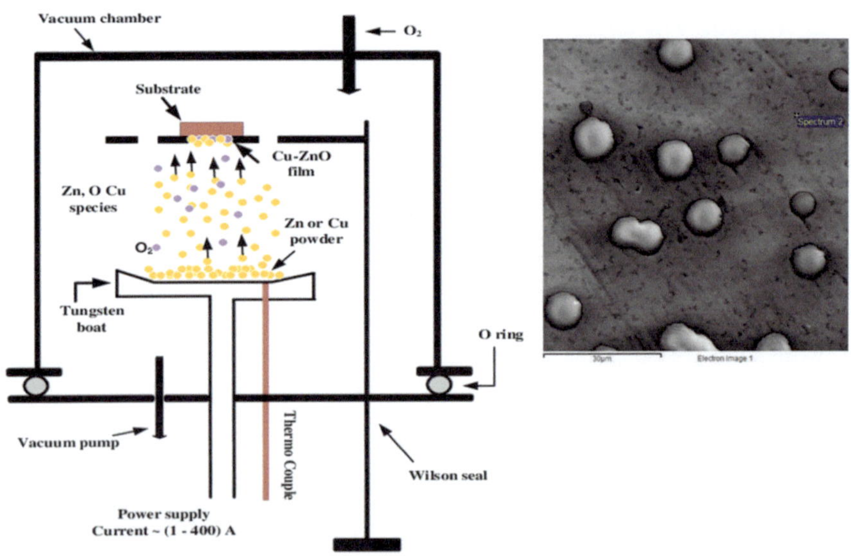

Fig. 4 Schematics of fabrication of ZnO nanostructures by thermal evaporation technique shown along with the spherical morphology by SEM [45]

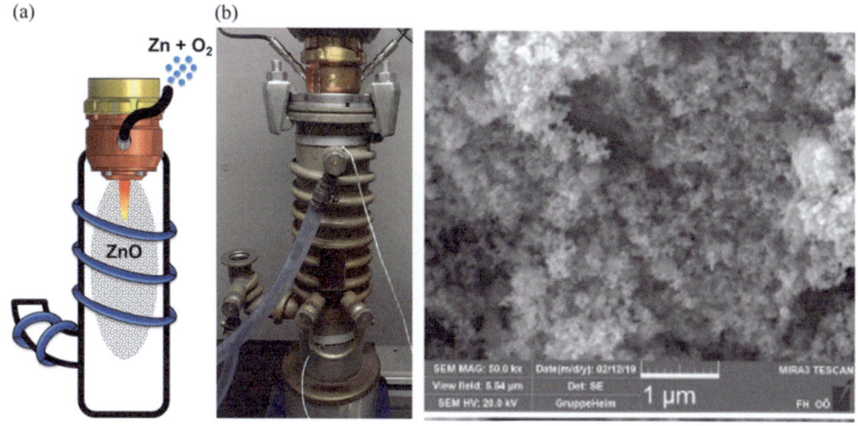

Fig. 5 (**a**) Schematics of ZnO synthesis by atmosphere-controlled oxidation technique, (**b**) and picture of plasma jet reactor with the hot gas plasma nozzle and, **c** with their fabricated nanostructured morphology [46]

4.3 Atmosphere-Controlled Oxidation

Controlling the oxidation rate of zinc by stabilizing uniform, thin, fine ZnO layer on the surface of zinc metal powders to minimize zinc vapours from spreading out rapidly and O_2 from penetrating the particles is the key factor in the method of synthesizing ZnO-T (Fig. 5) [46, 47]. ZnO-T can be synthesized in a combined atmosphere of inert and oxygen gases and the diluted content of oxygen can result in conditions needed for growth of whiskers. The inclusion of water vapour to the reactive gas has a significant effect on the morphology and size control of ZnO-T. Tetrapods synthesized with the mixing of water vapour have resulted in longer arms and a lower average diameter (8–20 nm) than those made without it. The process of Zn powder with water vapour yields hydrogen gas that may aid the generation of the suboxides of zinc, which is identified as the highly probable intermediate form in the creation of ZnO-T [48]. However, the reaction of zinc powder with excess of water vapour result in the passivation of the surface limiting the transportation eventually decreasing the yield of ZnO. Therefore, it is identified that traces of water vapour may be aiding the creation of ZnO-T with controllable dimension, whereas higher water vapour content reduce the yield of ZnO significantly.

4.4 Carbothermal Reduction

ZnO is blended with graphite powder as a precursor and treated at about 800–1100 °C to lower the decomposition temperature. The graphite reduces ZnO to result into CO/ CO_2 vapours and eventually react to result into ZnO nanocrystals. A mixture of equal

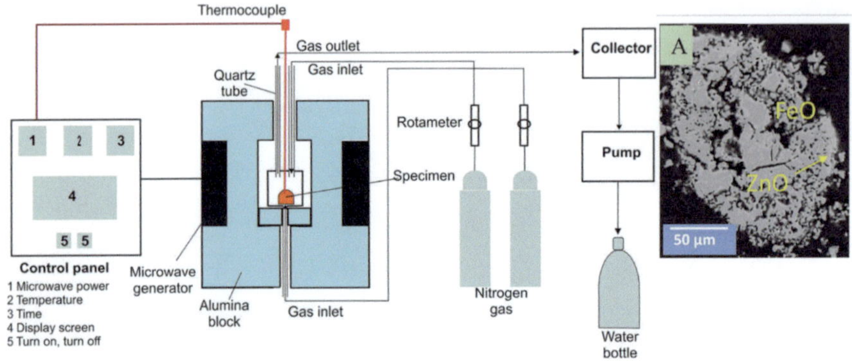

Fig. 6 Schematics of carbothermal reduction of ZnO working mechanism along with their SEM image representing nanostructured morphology [50]

ratio of pure ZnO and graphite powders were kept as the precursor material in the middle of a quartz tube. The mixture was treated at a temperature around 1100 °C under purging of high-pure N_2 carrier gas [49]. Various products formed in different locations on the inner wall of the quartz tube when the rate of flow was modified. With a higher carrier gas flow rate of carrier gas homogeneous tetrapod-shaped ZnO resulted on the inner side of the quartz tube positioned downstream. The temperature of growth was assessed to be in between 955 °C and 990 °C [49]. Thus, the rate of flow of carrier gas is a main aspect for tuning the growth of ZnO-T by carbothermal reduction. An appropriate gas flow rate leads to a considerable supersaturation of Zn vapour and oxygen, which result in the synthesis of ZnO-T (Fig. 6).

4.5 *Catalytic Oxidation*

Catalysts like Pb, Cd, Sn, Al, Ag and Sb substituted into pure Zn matrix, ZnO-T were grown after the oxidation step. The ZnO nucleation and growth phenomena of whiskers with deviations in the morphological features of ZnO is due to the influence of the catalyst elements. Gold catalyst yielded a bulk amount of tetrapod-like ZnO, which had morphological aspects akin to those formed by a catalyst-free method [51, 52] (Fig. 7).

Nanostructured ZnO were produced by the thermal sublimation of Zn and carbon nanotubes over on a silicon wafer with a sputter coated Ag catalyst. Ag catalysts substituted ZnO yield high (> 95%) polypods, with the majority in tripod morphology and the rest of structures mainly in tetrapod form. In the case of catalyst free layer, the most of the polypods are in tetrapod form and the size is comparatively small with the density and yield of polypods were significantly much lower (< 50%) with respect to Ag catalyst layer [52]. ZnO NPs nucleated on the Ag catalyst have varying crystallographic directions from ZnO particulates randomly nucleated on plain Si

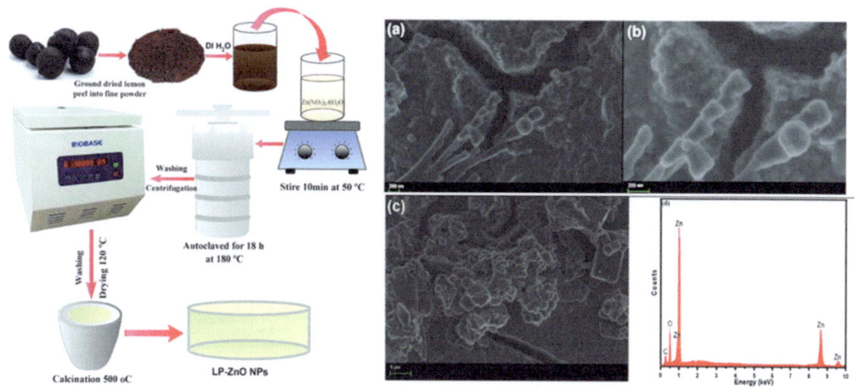

Fig. 7 Pictorial representation of catalytic oxidation of ZnO with their respective SEM and EDS data showcasing the morphology [53]

wafer, such that ZnO tripods are prominently grown. In contrast, nanotetrapods are basically grown from arbitarily nucleated ZnO-T on plain Si wafer.

4.6 Thermal Oxidation in Combination with Solution

ZnO-T were synthesized by the thermal oxidation method using Zn metal dust blended with various solutions like methanol, ethanol and H_2O_2 in a 10:1 ratio. The mixture was treated at 1000 °C in ambient atmosphere with ZnO formed by heating Zn and H_2O_2 resulted in tetrapod-like structures, and that H_2O_2 served as an intense oxidizer that delivered highly reactive oxygen species to Zn to convert into ZnO-T (Fig. 8) [3, 41].

4.7 Microwave Plasma Process

In a regular microwave irradiation method, Zn metal powder taken in a quartz crucible was set at the middle of the plasma heating zone of a horizontal quartz tube. Pure Ar and O_2 gases purged inside the tube and a microwave source was combined to a rectangular wave-conducting pipe to the tube's mid region for creating stable plasma. An additional tubular furnace was utilized to maintain the required temperature zone and the products were gathered on the inner wall at the downstream edge of the tube. The formed ZnO-T had predominantly four linear and uniformly sized arms with diameter in between 10 and 25 nm, and length up to 160 nm [55].

Among the vapour-transport preparation techniques, thermal evaporation at atmospheric condition is the simplest, and ZnO-T produced by thermal evaporation have

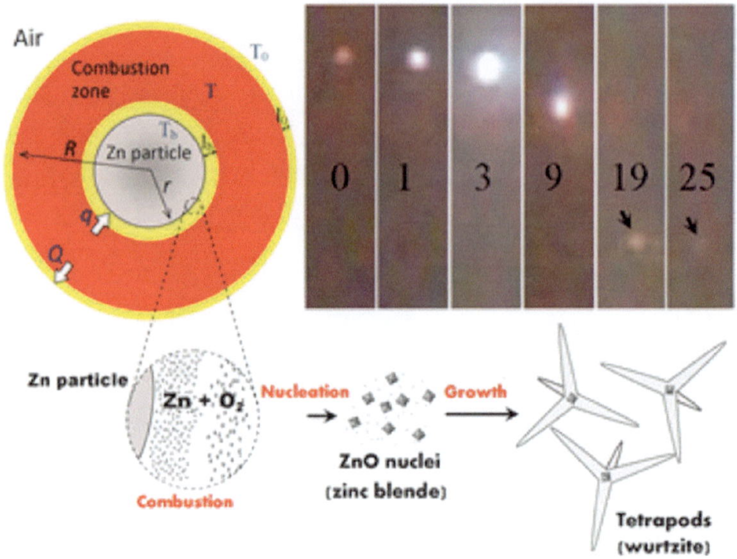

Fig. 8 ZnO-T emerging from Zn particles with combustion zone formed around a Zn particle with high-speed camera photograph progression of a burning Zn particle in the reactor, with time representation in milliseconds along with their growth process [54]

desired morphology with high crystalline quality. The size of ZnO-T formed using this process is in the micro range. The smaller size aids the creation of thin polymer layers that are needed for efficient charge transport in polymer-nanocrystal solar cells [56]. Even though the vapour transport method is a prominent synthesis procedure for producing ZnO-T, liquid reaction methods have become highly significant in formation of ZnO-T with arm diameter lower than 100 nm and with controllable arm lengths. These processes can essentially fill the size gaps and significantly extend to various applications of this distinct nanostructure (Fig. 9).

4.8 Wet Chemical Process

Neverthless the vapour transport method is the predominantly used synthesis process for forming ZnO-T, other growth process like wet chemical method has also been optimized. These techniques aid the synthesis of ZnO nanostructures at relatively low temperatures, which leads to the generation of ZnO-T [47, 56] (Fig. 10).

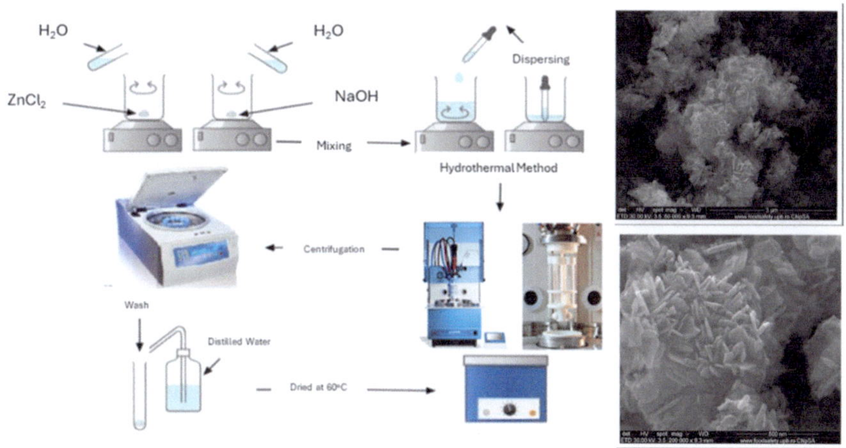

Fig. 9 Pictorial representation of microwave plasma synthesis technique of nanostructured ZnO platy rod like morphology [57]

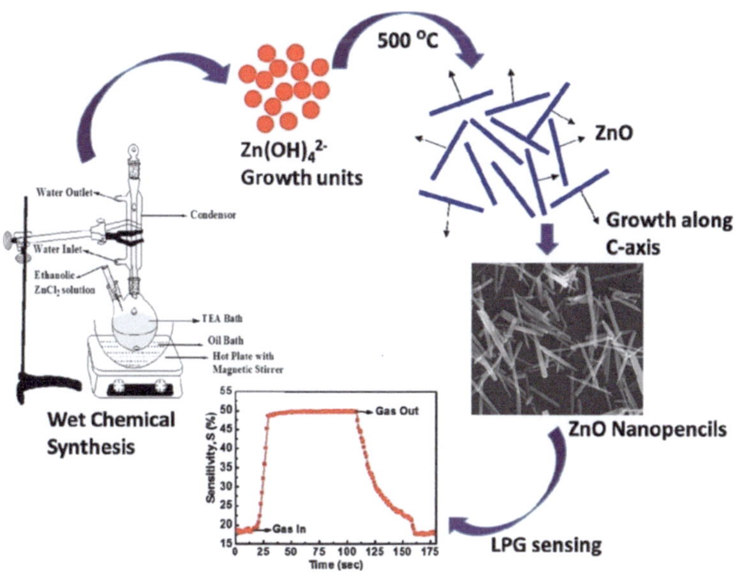

Fig. 10 Schematics of the steps involved in the wet chemical synthesis process to prepare ZnO nano-pencils [58]

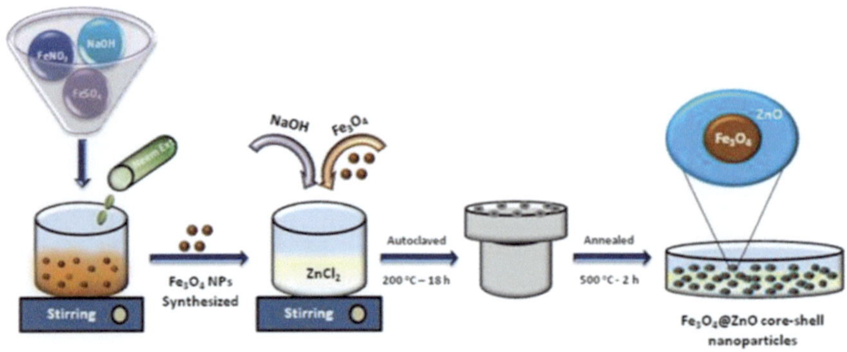

Fig. 11 Steps involved in the microwave emulsion combined with hydrothermal process method to prepare ZnO-tetrapods [60]

4.9 Micro-emulsion in Combination with Hydrothermal Process

Micro-emulsion combination with thermal hydration can be utilized for the synthesis of ZnO nanotetrapods due to its facileness. This process is relied on the benefits of microemulsion in forming fine particles, and the hydrothermal step in modifying morphology and quality. The hydrothermal synthesis duration and cyclohexane concentration were explored to study their influence on morphology and size [59]

The length and diameter of tetrapods get modified slightly with duration of ZnO whiskers formed within a short growth-duration had tetrapod framework along with several smaller particles attached over the arms [55]. With growth duration elevated to four hours, the edges of the arms tend to be clear with considerably good growth. The key morphology was nanotetrapod, apart from the sheet- and flower-shaped morphologies. With a increased growth period, more number of flower-like whiskers were formed and the surface of the arm being rough. With a further extended growth period, secondary growth occurred and several tiny whiskers were formed on the arms. This process indicates the commencement of polar and secondary growth phenomena, and that the growth period significantly influences the morphology of ZnO (Fig. 11).

4.10 Non-Hydrolytic Alcoholysis Process

To further minimize the size of ZnO-T, the alcoholysis of zinc stearate was adapted to form ZnO-T about 100 nm in size. Growth of nanostructured ZnO nanocrystals by the reaction of zinc stearate with alcohol in hydrocarbon solvents (non-coordinating medium) at higher temperatures was selected due to the nanocrystals formed in this system were stable at ambient temperature, and no aggregation or preferred features

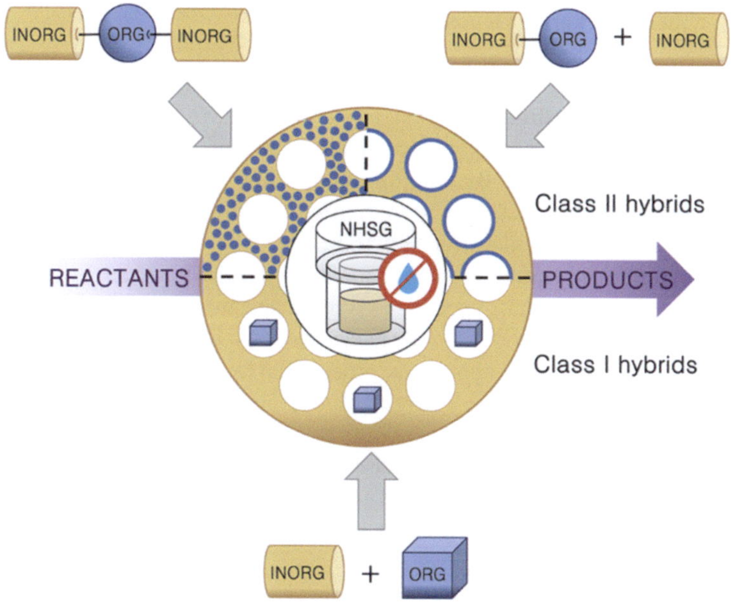

Fig. 12 Pictorial representation of non-hydrolytic alcoholysis synthesis route of ZnO-T [61]

was noticed. The incorporation of Mg dopants considerably influenced the growth of the host lattice by tuning the crystallographic phase of ZnO seeds, which led to a significant shape evolution. A small quantity of Mg precursor significantly changed the morphology of the formed nanocrystals, yeilding tetrapods [49]. Nanocrystals with tetrapod geometry were also identified in the resultant samples had a higher yield of high-quality ZnO nanotetrapods (Fig. 12).

Although several of the above-described techniques enable controllable growth of ZnO-T, a facile, scalable and cost-effective synthesis process which can yield large quantity of size-controlled ZnO-T is still on high requirement. Compared to vapour-transport process, ZnO-T formed from liquid reaction process have smaller sizes; nevertheless, their yield and purity both are limited due to the tedious washing procedure.

4.11 Vapor-Phase Growth Techniques

Vapor-phase growth techniques are widely utilized to synthesize high-quality ZnO nanostructures due to their ability to produce well-defined morphologies and high-purity materials. Two common vapor-phase techniques are CVD and hydrothermal synthesis, which can be adapted to fabricate ZnO-T.

4.11.1 Chemical Vapor Deposition

Chemical Vapor Deposition (CVD) is utilized frequently for the fabrication of high-quality ZnO nanostructures, including ZnO-T. In this process, a precursor gas is decomposed in a high-temperature reaction chamber, and the ZnO material is deposited onto a substrate. By carefully controlling the growth parameters, like temperature, precursor concentration, and carrier gas flow rate, it is possible to induce the formation of tetrapod-like structures [62].

The substrate temperature is ramped to where the ZnO precursor vaporizes and decomposes. The vapor-phase species then condense onto the surface of the substrate, where they nucleate and grow into ZnO nanostructures. By adjusting the temperature and pressure conditions, the vapor-phase process can favor the formation of tetrapods, with the central body and arms emerging due to anisotropic growth of ZnO micro crystals (Fig. 13).

The CVD process permits for the precise tuning of the growth process, enabling the production of high-purity ZnO-T [62]. This technique can also be adapted for large-scale production, making it suitable for industrial applications. The CVD process

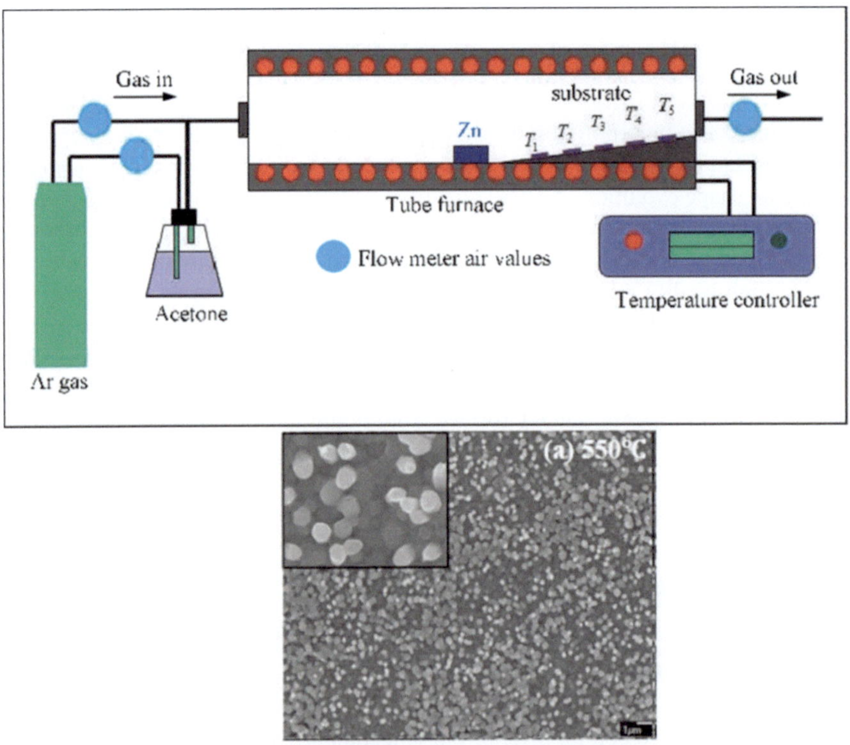

Fig. 13 Schematics of synthesis of vertically oriented ZnO nanowires by CVD method with the SEM images showing the morphology [63]

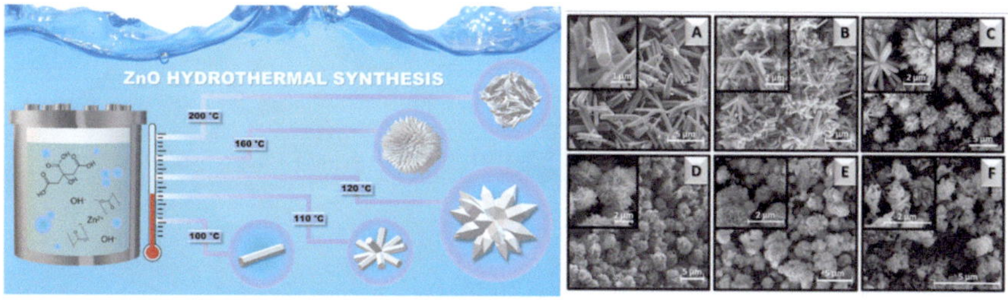

Fig. 14 Pictorial representation of hydrothermal process with various SEM images of nanostructured ZnO morphologies [65]

requires high temperatures and a controlled environment, which can limit scalability and make the process energy-intensive.

4.11.2 Hydrothermal Synthesis

Hydrothermal technique is a viable method used to grow a variety of ZnO nanostructures, including tetrapods, by reacting zinc salts (zinc nitrate or zinc acetate) with a strong base (like NaOH or ammonium hydroxide) in a relatively high-temperature, high-pressure aqueous environment. This method is advantageous since it can be performed at relatively lower temperatures than CVD and thermal evaporation.

The hydrothermal method involves placing the zinc precursor and base in a sealed autoclave, which is then heated to temperatures around 180–200 °C. Over time, the solution undergoes a series of chemical reactions that leads to the formation of ZnO. The process can be tailored to create multiple morphologies, including the tetrapod structure, by controlling the concentration of the precursor, pH of the solution, choice of precursor and temperature (Fig. 14) [64].

Hydrothermal synthesis is simple, cost-effective, and scalable. It also allows for the preparation of high-quality ZnO-T with high crystallinity and uniformity. Additionally, it does not require high vacuum conditions, making it easier to implement. The process is limited by the choice of substrates and typically results in smaller quantities compared to vapor-phase methods. Also, the reaction time and pressure conditions must be carefully tuned to achieve the desired morphology.

4.11.3 Solution-Based Process

Solution-based methods are another approach for producing ZnO-T. These methods involve the use of liquid-phase reactions, typically in the presence of surfactants or templates, to guide the growth of ZnO nanostructures (Fig. 15). The benefit of solution-based techniques is that they are very facile to implement, cost-effective, and can produce large quantities of material.

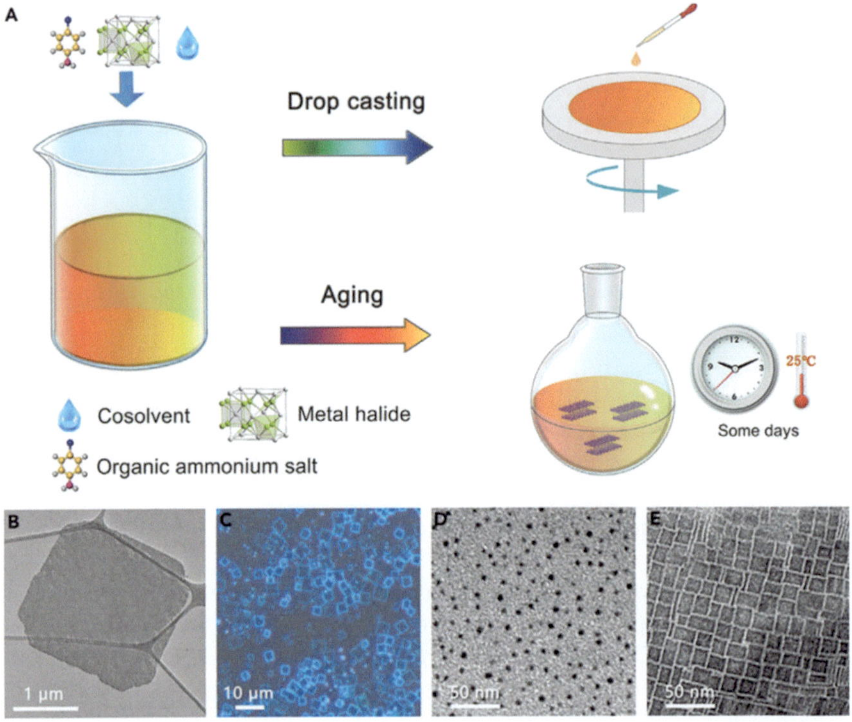

Fig. 15 Pictorial representation of fabrication of nanostructured ZnO synthesis by solution-based methods with their morphology from SEM and TEM images [66]

4.11.4 Solvothermal Method

The solvothermal method is a solution-based technique that uses a solvent other than water, such as ethanol, methanol, or other organic solvents, to facilitate the growth of ZnO-T. This process is similar to the hydrothermal process except the high pressure and temperature condition but is performed with a different solvent to alter the growth dynamics and achieve different morphologies. In the solvothermal method, a solution of zinc precursors (such as zinc acetate) is prepared in an organic solvent. The solution is then enclosed in an autoclave and heated to temperatures between 150 and 200 °C. This process promotes the growth of ZnO-T with good control over size, aspect ratio, and crystallinity [67] (Fig. 16).

The solvothermal method offers flexibility in terms of solvent choice, which can influence the crystal growth direction and final resultant morphology. It also allows for better control over the nucleation process and can result in high-quality, well-defined ZnO-T. Similar to other solution-based methods, the solvothermal process may not be easily scalable for industrial production. Additionally, the nature of the solvent and precursors must be carefully controlled to avoid unintentional impurities in the final product.

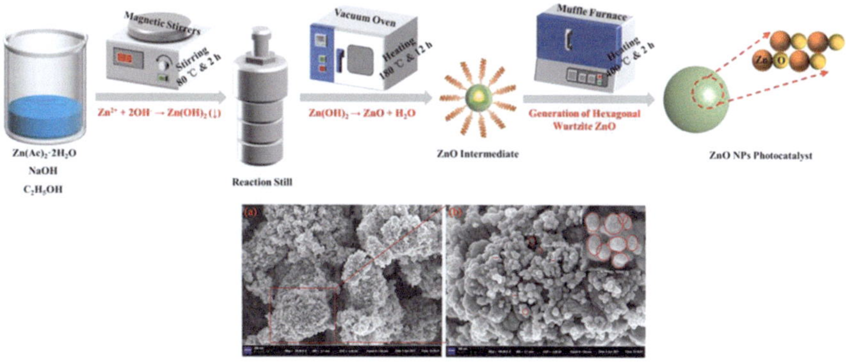

Fig. 16 Pictorial representation of solvothermal synthesis of ZnO NPs along with their morphology [68]

4.11.5 Electrochemical Deposition

In electrochemical deposition process, electrical current is applied to a solution of zinc salts to induce the reduction of zinc ions and their subsequent deposition as ZnO on a substrate [69]. This process can be used to create various ZnO nanostructures, including tetrapods. In electrochemical process, a conductive substrate is immersed in an electrolytic medium containing a zinc precursor, such as zinc sulfate. An electric current is applied from an external source to reduce the zinc ions onto the surface of the substrate, where they nucleate and grow into ZnO nanostructures. By carefully controlling the deposition conditions (e.g., current density, temperature, and precursor concentration etc.), ZnO-T can also be formed (Fig. 17). Electrochemical coating is a relatively cost-effective process that can be performed at room temperature, making it a suitable option for scalable production. It also allows for fine control of the deposition rate and material properties. The process may require additional steps, such as post-deposition annealing, to improve the crystallinity and morphology of the ZnO-T.

4.12 Template-Assisted Methods

Template-assisted methods use a pre-existing structure, or template skeletal network, to guide the growth of ZnO-T. This method allows for the creation of highly ordered ZnO-T with controlled size and shape.

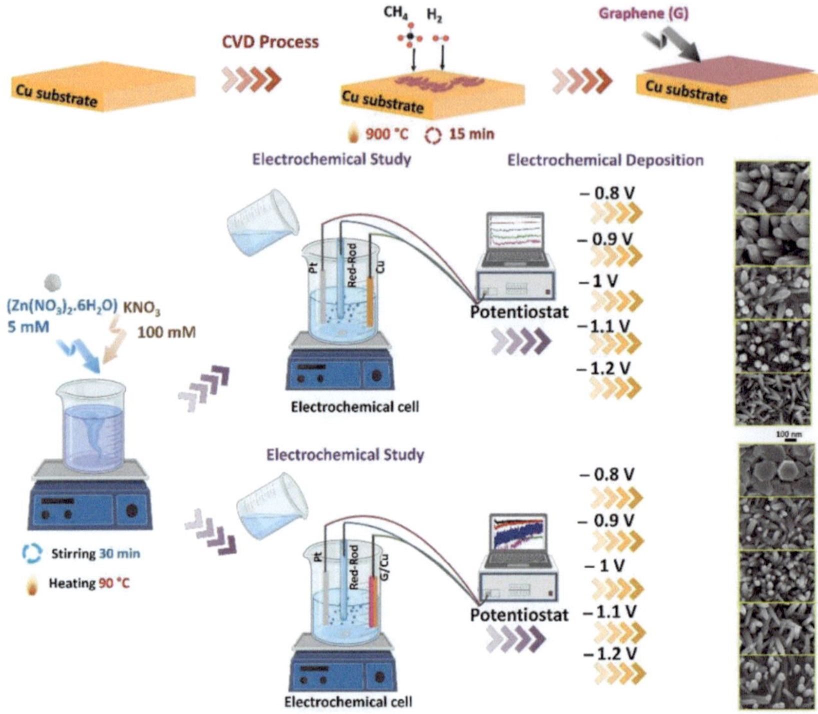

Fig. 17 Schematics of electrochemical deposition technique to prepare ZnO nanostructures with the SEM images of the fabricated samples [70]

4.12.1 Anodized Aluminium Oxide Templates

One common approach is using anodized aluminium oxide (AAO) templates to grow ZnO-T. The AAO template has a highly ordered array of nanopores, which can act as a mould to prepare nanostructured ZnO. The ZnO precursor is deposited within the nanopores of the AAO template through methods like vapor deposition or solution-phase reactions. After ZnO is deposited, the template is removed, leaving behind ZnO-T that are well-aligned and have uniform size (Fig. 18) [72]. Template-assisted methods permit for the accurate tuning of the size, shape, and alignment of ZnO-T. This is particularly useful for applications in photonics, sensing, and optoelectronics. The major limitation is the need for template removal, which can introduce defects or impurities in the final product. Additionally, the scalability of template-based methods is limited due to cost and complex procedure.

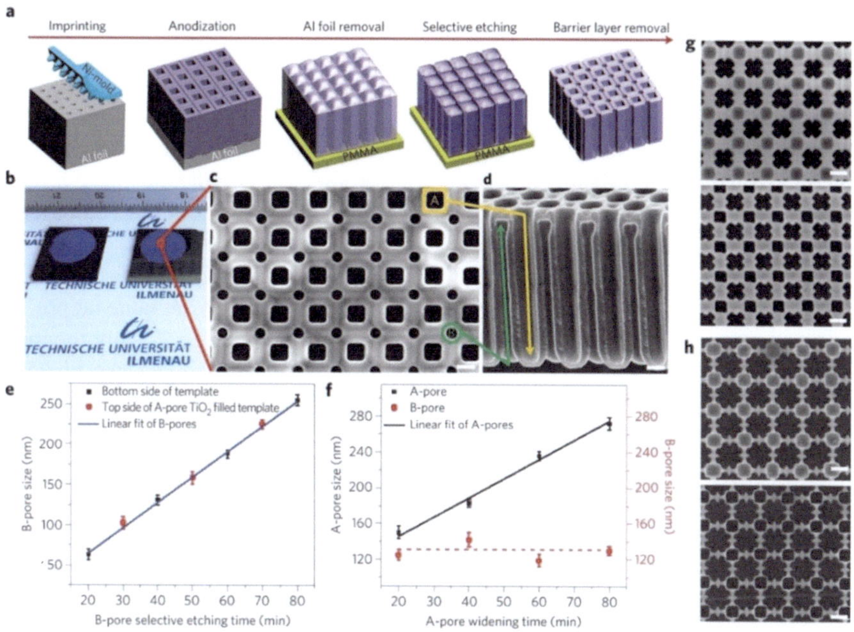

Fig. 18 Schematics of the fabrication of imprinting on an Al foil anodization on the imprinted Al foil to form square A-pores with photograph and enlarged, cross-sectional SEM image of binary-pore templates with the error bars representing standard deviations of pore sizes (**a–h**) [71]

4.12.2 Influence of Fabrication Method on the Properties of ZnO Tetrapods

The choice of fabrication method has a prominent role on the properties of the ZnO-T, including their size, crystallinity, optical behavior, and surface morphology. Here are some key factors influenced by the fabrication process:

- *Crystallinity:* Vapor-phase methods like CVD and hydrothermal synthesis often produce ZnO-T with high crystallinity, which is critical for optoelectronic applications [6, 73].
- *Size and Aspect Ratio:* The size of the tetrapods and the length of the arms can be controlled by tuning the reaction conditions. Solution-based and template-assisted methods provide better control over the arm dimensions of the tetrapods [1, 2, 73].
- *Surface Area and Defects:* The surface area of ZnO-T can be increased by their 3D structure, which is beneficial for applications in catalysis and sensing. However, defects such as oxygen vacancies can also be formed during the synthesis step, which may influence the material's optical and electronic characteristics [1, 73].

Although tetrapods have garnered a lot of interest due to their applicability, their scalable production was a significant technical hurdle which was overcome with the flame transport synthesis (FTS) method. However, physical methods (vapor deposition, vapor–liquid–solid, microwave, etc.) have disadvantages like the complexity in making nanostructures with required size, shape, and yield (for instance, exclusive ZnO-T is complicated to synthesize, low yield), as well as high in price. Wet-chemistry approaches to synthesis typically involve strong chemical reagents that undoubtedly leave back their residue. Since the production of high amounts of ZnO-T will be dubious (multistep procedure, high in cost, etc.), synthesis options employing catalytic NPs are already out of reach.

Furthermore, the catalytic particles function as flaws that some applications would find undesirable. In this situation, self-catalytic processes are quite helpful. A summary of some of these synthesis techniques is given in Table 3. Only the most pertinent procedures have been compared in Table 3 out of the extensive range of ZnO-T manufacturing techniques. The prime factors for utilizing these tetrapods are likely process simplicity, tetrapod size, size distribution, yield, cost-effectiveness, etc. A close comparison indicates that both combustion and microwave methods at least show promise for producing ZnO-T on a large scale and may expand their applicability to industrial process.

The solid–vapor-solid (SVS) preparation phase in the FTS method immediately transforms the microsized metallic Zn particles into ZnO-T in the flame (within the muffle furnace) at a elevated temperature [23]. In earlier publications, the growth mechanism for ZnO-T structures in the FTS method has been thoroughly examined [1, 6, 24]. Tetrapod arm morphologies can be adapted from low to extremely high aspect ratios and from nano to microscale by simply changing the synthesis parameters in the FTS technique [23]. The FTS-prepared ZnO-T and nanostructures provide access of ZnO surface characteristics, which makes it highly intriguing for a variety of applications since no expensive or complicated chemicals are used [1].

The fabrication of ZnO-T requires careful control over growth conditions to achieve the desired morphology and material properties. Vapor-phase techniques, solution-based methods, and template-assisted approaches each offer unique advantages for fabricating ZnO-T. CVD and hydrothermal methods are often preferred for producing high-quality, well-defined tetrapods, while solution-based techniques such as solvothermal synthesis and electrochemical deposition offer flexibility and scalability. Template-assisted methods, such as AAO, can provide highly ordered ZnO-T with controlled dimensions. By understanding and controlling the synthesis conditions, it is possible to stabilize the characteristics of ZnO-T for specific needs in photonics, sensing, optoelectronics, and photocatalysis, bio-medical energy conversion.

Table 3 Outline of catalyst-free synthesis methods for upscaling ZnO tetrapods

Methods	Conditions	Size, quality, yield	References
Vapor-phase	Sublimation of Zn in air O_2 mixture; pressurized electric furnace (~920 °C); multistep process	ZnO-T with long arms (30–200 μm); low yield	[1, 74]
Wet chemical process	Growth in a chemical vessel (200–300 °C); $Zn(OH)_2$, NH_4OH and chloride solutions; particularly designed Ag lined tube	Tetrapod arms; wide size variation; low yield	[1, 75]
Hydrothermal	Aqueous medium, $ZnSO_4$ and NH_3 in reactor @100 °C; washing required (150 °C hot air)	Uniform tetrapods; multistep process; low yield	[1, 76]
Thermal evaporation	Zn sublimation in the tubular furnace (800–1000 °C); Zn filled quartz boat in tube	Mixed tetrapods; wide size variation; low yield	[1]
Evaporation & recondensation	Zn evaporation; tubular furnace (1100 °C); in air	Primarily tetrapods; total size range 300 nm–15 μm; arm dia: 30–650 nm; poor yield	[1, 77]
CVD	Zn sublimation in tubular furnace (630 °C); on Si substrates; air + O_2 environment	Asymmetrical tetrapods; very poor yield	[1]
DC thermal plasma	DC plasma reactor; N_2 gas; fast process (10 ms); facile process; direct conversion of Zn dust into ZnO nanostructures	Mixed structures including tetrapods; broad distribution; very low yield	[1, 78]
Vapor phase growth	Zn evaporation; tubular furnace (700 °C); carrier gases (Ar, O_2); gravimetric isolation; ~ longer processing duration	Arm length ~ up to 2 μm, dia 100–300 nm; aggregates; Zn traces; high yield; broad size variation	[1, 79]
Microwave	Zn oxidation; tube furnace (4 Torr, up to 600 °C), microwave (400 W, 2.45 kMHz, ~1000 °C), vacuum, carrier gasses (Ar, O_2); processing time ~ 30 min	Tetrapod arm (length ~ 0.5 μm, dia ~ 50 nm); crystalline; yield ~ 75%; broad size distribution	[1, 80]
Combustion process	Combustion of Zn powders in air in a reactor in air; continuous process and multistep synthesis process	Homogeneous nanotetrapods; comparatively wide distribution (size); poor yield	[1]

(continued)

Table 3 (continued)

Methods	Conditions	Size, quality, yield	References
Flame transport method	Conversion of Zn particles into tetrapods; sublimation and regrowth (solid–vapor–solid process); muffle/ backing furnace (~900 °C); air environment; single-step process; (i) using sacrificial polymer (PVB), (ii) direct growth. PVB process: micro-nanotetrapods; arm dia 200 nm–2 μm; arm length; 1–15 μm; (4 gm Zn: 1 g ZnO-T); (ii) direct growth; arm dia: 100 nm–1 μm; length: 2–20 μm; 10 g Zn wire: 3 g ZnO-T	Tetrapods, high quantity; narrow size distribution; high crystallinity; PVB process: yield ~ 25%; direct growth: yield ~ 30%	[1, 81]
		(Entirely tetrapods in both methods)	

5 Properties of ZnO Tetrapods

5.1 Thermal Properties of ZnO Tetrapods

The thermal characteristics of ZnO-T are an important aspect of its performance in various applications, as it affects their stability, heat dissipation capabilities, and overall functionality in high-temperature environments. ZnO, a wide-bandgap semiconductor with a wide range of nanostructured morphologies, including tetrapods, exhibits unique thermal characteristics that differ from those of bulk ZnO because of its high surface area and reduced dimensionality.

Understanding the thermal behavior of ZnO-T is critical for their use in fields like sensors, energy harvesting, optoelectronics, and high-temperature applications. In this section, the various thermal properties of ZnO-T, including their thermal conductivity, thermal stability, specific heat capacity, and thermal expansion are elucidated.

5.1.1 Thermal Conductivity

Thermal conductivity is a measure of a material's ability to conduct heat. In nanomaterials, including ZnO-T, the thermal conductivity is significantly impacted by the size, shape, and crystallinity of the structures. ZnO-T, like many other nanostructured materials, typically have lower thermal conductivity compared to bulk ZnO of around 50 W/mK at room temperature. The 3D nature of the tetrapod morphology, characterized by a central body and four arms, increases the S/V ratio leading lead to enhanced phonon scattering at the surface and grain boundaries, thus reducing the overall thermal conductivity [82].

- *Influence of Nanostructure:* In nanostructured ZnO, the lower thermal conductivity is often attributed to the fact that phonon transport is limited due to the smaller dimensions of the structure. Phonons, the primary carriers of heat in non-metallic materials, are scattered more frequently in nanomaterials, leading to lower heat conduction. For ZnO-T, the unique 3D morphology may also cause additional scattering due to surface defects and the structural complexity of the arms, resulting in high thermal resistance [73, 82].

- *Comparison with Bulk ZnO:* Bulk ZnO has a relatively high thermal conductivity (approximately 200 W/mK), but in the case of ZnO-T, thermal conductivity is generally lower, around 50 W/(mK) due to the increased surface area and nanometer-scale dimensions [6].

5.1.2 Thermal Stability

Thermal stability implies to the capability of a material to retain its structural integrity and related characteristics when subjected to high temperatures. For ZnO-T, thermal stability is critical, especially in high-temperature environments where devices may be exposed to thermal stress. ZnO-T being single crystalline in nature with no grain boundaries, the stability is way better than ZnO powders. ZnO-T demonstrate good thermal stability in oxidizing environments and at elevated temperatures, making them apt for use in sensors, photocatalysis, and energy harvesting where thermal conditions can vary [83, 84].

- *Decomposition and Oxidation:* ZnO is familiar for its high thermal stability, with a melting point about 1975 °C and a boiling point of approximately 2360 °C. ZnO-T, owing to their crystalline nature, exhibit good stability up to moderate temperatures, generally in the range of 500–700 °C. Beyond this range, ZnO may start to decompose or undergo oxidation, especially in the presence of oxygen, potentially altering its morphology and associated properties [84].
- *Sintering Effects:* At high temperatures, ZnO-T may undergo sintering, a process in which individual particles or nanostructures fuse together to form larger clusters. This can result in a loss of the desired tetrapod morphology and may negatively impact the material's functional properties. To avoid sintering, ZnO-T are often stabilized with additives or processed at comparatively lower temperatures [83].

5.1.3 Specific Heat Capacity

Specific heat capacity refers to the quantity of heat energy necessary to increase the temperature of a unit mass of a material by 1 °C. In nanostructured materials like ZnO-T, the specific heat capacity can be influenced by factors such as size, surface area, and phase transitions 0.494 J/g °C.

- *Size and Surface Effects:* The specific heat capacity of ZnO-T is generally higher than that of bulk ZnO due to the enhanced surface area and the contribution of surface atoms, which can behave differently from atoms in the bulk phase. In nanostructures, the specific heat is typically influenced by the phonon behaviour, where lower dimensionality (as in tetrapods) can lead to higher heat capacity compared to bulk materials [85].

- *Anomalies in Nanomaterials:* In some nanomaterials, the specific heat capacity shows deviation from bulk values due to quantum size effects and surface phonon modes that are prominent at the nanoscale. For ZnO-T, these effects may cause their specific heat capacity to differ from that of bulk ZnO, particularly at low temperatures [83].

5.1.4 Thermal Expansion

Thermal expansion describes how the size of a material changes with temperature. This property is significant when considering the integration of ZnO-T into devices and systems, where dimensional stability is required during temperature fluctuations.

- *Coefficient of Thermal Expansion (CTE):* ZnO has a relatively low CTE compared to other materials, with a value of about 4.2×10^{-6}/K at room temperature. This low CTE helps ZnO-T to maintain dimensional stability upon thermal cycling [83].
- *Effect of Nanostructure:* The thermal expansion of ZnO-T may be influenced by their unique morphology. The arms of the tetrapods, being thinner and more exposed to surface effects, may exhibit slightly different thermal expansion behavior compared to the central body of the tetrapods. However, overall, ZnO-T are expected to exhibit minimal thermal expansion due to their relatively low CTE, enabling them relevant for applications that need thermal stability, such as in electronics or sensors that undergo temperature fluctuations [73].
- *Thermal Stress:* As with other nanomaterials, ZnO-T could experience thermal stress when subjected to rapid temperature changes, potentially leading to cracking or fracture. This makes them more sensitive to thermal cycling compared to bulk ZnO, although their 3D structure may help mitigate some of these effects by distributing the stress more evenly [83].

5.1.5 Thermal Conductivity Applications

The low thermal conductivity of ZnO-T can be both an advantage and a limitation, depending on the application.

- *Advantage in Sensing and Energy Harvesting:* The reduced thermal conductivity can be beneficial for energy harvesting applications, such as nanogenerators or thermoelectric devices, where heat is converted into electricity. The lower thermal conductivity improves the thermoelectric performance by increasing the temperature gradient across the material, enhancing the energy conversion efficiency [85, 86].
- *Limitation in Heat Dissipation:* For applications where effective heat elimination is essential, like in high-power optoelectronic devices or LEDs, the low thermal conductivity of ZnO-T might pose a challenge. In such cases, hybrid structures

that incorporate ZnO-T with high thermal conductivity materials may be used to balance the performance [1].

5.2 Electronic Properties of ZnO Tetrapods

Wide-bandgap ZnO semiconductor has a range of remarkable electronic characteristics, like high electron mobility, piezoelectric behaviour, and optoelectronic activity. When ZnO is fabricated into nanostructures like tetrapods, these electronic properties are often enhanced due to the enhanced surface area, reduced dimensionality, and unique tetrapodal structure. The electronic aspects of ZnO-T are crucial for their performance in several applications that includes optoelectronics, sensors, piezoelectric devices, and energy harvesting devices.

This section describes the key electronic properties of ZnO-T, including their band structure, electrical conductivity, carrier mobility, piezoelectric properties, and surface effects.

5.2.1 Band Structure and Electronic Configuration

ZnO is a wide-bandgap semiconductor (3.37 eV at ambient temperature) which allows it to exhibit strong ultraviolet (UV) absorption and emission properties. The electronic structure of ZnO possess a valence band (VB) and conduction band (CB), with the energy gap between them determining its semiconductor behaviour.

- *Effect of Nano structuring:* The electronic band structure of ZnO-T can be influenced by their nanometer-scale dimensions and unique 3D morphology. In some cases, the bandgap may be slightly altered compared to bulk ZnO, depending on the size and geometry of the tetrapods, but typically, it remains in the UV region [82].
- *Surface States:* The surface of ZnO-T plays a significant role in their electronic properties. Since tetrapods have a high S/V ratio, the presence of surface states may influence the dispersal of charge carriers and affect the overall electronic behaviour. Surface states can trap electrons or holes, modifying the material's conductivity and transport properties [86].

5.2.2 Electrical Conductivity and Carrier Transport

The electrical conductivity of ZnO-T is governed by the concentration and mobility of charge carriers (electrons and holes). The electronic properties of ZnO, including its conductivity, can be modulated by factors such as doping, crystallinity, temperature, and morphology.

Conductivity in Pure and Doped ZnO Tetrapods

- *Intrinsic Conductivity:* Pure ZnO typically exhibits n-type conductivity because of the presence of O_2 vacancies and Zn interstitials, which act as electron donors. This makes ZnO-T moderately conductive, though their conductivity may be lower compared to metals or other low bandgap semiconductors. The free electrons in the conduction band can move through the material when subjected to an external electric field [87].
- *Doping for Improved Conductivity:* To enhance the conductivity of ZnO-T, dopants can be introduced to modify the carrier concentration and increase electrical conductivity. Common dopants include aluminium (Al), gallium (Ga), and indium (In), which act as donor impurities, enhancing the free electron concentration and improving the electrical conductivity [87, 88]. Al-doped ZnO is one of the most promising materials for transparent conductive oxide (TCO) applications due to its combination of high electrical conductivity with sheet resistance of 32 Ω/□ to 8.2 Ω/□ and mobility of 36.8 $cm^2\ V^{-1}\ s^{-1}$ and optical transparency in the visible light spectrum. The addition of aluminium into ZnO helps improve its electrical properties while maintaining the transparency of ZnO, making it suitable for a several of optoelectronic applications, such as in solar cells, displays, touchscreens, and light-emitting diodes (LEDs).

5.2.3 Carrier Mobility

- *Electron Mobility:* ZnO has a relatively high electron mobility (150–200 cm^2/V s), especially in its bulk form. In nanostructures like ZnO-T, electron mobility can be influenced by the size of the structure, the surface states, and the crystallinity of the material. The 3D architecture of ZnO-T may hinder carrier transport along certain directions, especially along the arms of the tetrapods, but the electron mobility (10–50 cm^2/V s), is typically still sufficient for a variety of applications [89].
- *Surface Effects on Mobility:* High surface area of ZnO-T can cause increased surface scattering, which can reduce the mobility of charge carriers compared to bulk materials. However, this effect can be mitigated by controlling the synthesis to minimize defects and by functionalizing the surface to passivate surface states [89]

5.2.4 Piezoelectric Properties

One of the most distinctive electronic properties of ZnO, particularly in its nanostructured form, is its piezoelectric behaviour. This property arises from the non-centre of symmetry of the crystal structure, which implies that when ZnO is subjected to mechanical stress, electric charge is generated on the surface of the material.

Piezoelectric response in ZnO Tetrapods

- *Enhanced Piezoelectricity:* ZnO-T, due to its 3D geometry, exhibit enhanced piezoelectric properties compared to 1D structures like nanowires or nanorods. The tetrapod arms are aligned along specific crystallographic axes (e.g., [0001] and [10–10]), which maximizes the piezoelectric response [19, 90]. This is because each arm of the tetrapod can generate a piezoelectric potential when subjected to stress, and the configuration allows for efficient coupling between mechanical strain and the resulting electric field.
- *Applications in Energy Harvesting:* The piezoelectric properties of ZnO-T make them ideal candidates for use in nanogenerators that convert mechanical motion (like vibrations or pressure) into electrical energy that is particularly useful in applications such as self-powered sensors, wearable devices, and energy harvesting systems where mechanical energy from the environment is converted into usable electrical power [19].
- *Flexibility and Efficiency:* The flexibility of ZnO-T allows for a higher degree of deformation under mechanical stress, which can boost the efficiency of piezoelectric energy translation. This makes ZnO-T especially suitable for harvesting energy from low-frequency mechanical vibrations [35].

5.2.5 Surface Effects and Charge Carrier Distribution

The electronic characteristics of ZnO-T are strongly impacted by their high and varying surface area along the arms. Surface effects can modify the distribution of charge carriers and contribute to various phenomena, such as surface conductivity, trap states, and electrostatic properties.

Surface States and Defects

- *Surface Defects:* ZnO-T, like other nanomaterials, are highly sensitive to surface defects, like O_2 vacancies or Zn interstitials, which can lead to the formation of surface states. These surface states can trap free charges (electrons or holes), reducing the effective conductivity and altering the electronic properties [2].
- *Surface Passivation:* To mitigate the effects of surface states, surface passivation techniques can be employed, such as coating on the ZnO-T with a thin layer of insulating material or functionalizing the surface with specific molecules. This can reduce the density of trap states and improve carrier mobility and overall electronic performance [2, 91].

Carrier Distribution and Screening Effects

The high surface area and nanoscale dimensions of ZnO-T lead to enhanced carrier screening effects, where free carriers near the surface screen the internal electric fields, affecting the charge distribution. This can influence the electrostatic potential within the material and alter its electronic behavior [2]. In applications such as field-effect transistors (FETs) or sensors, the control of carrier distribution is critical for ensuring reliable performance.

5.2.6 Device Applications Based on Electronic Properties

The unique electronic properties of ZnO-T enable them to be utilized in a several of electronic devices and systems:

Field-Effect Transistors (FETs)

ZnO-T can be used in field-effect transistors (FETs), where the tetrapod structure allows for enhanced charge carrier transport and modulation of conductivity in response to an external electric field. ZnO-T-based FETs could be employed in flexible electronics, sensors, and optoelectronic devices [4].

Photodetectors

Due to their high electron mobility, piezoelectric properties, and UV absorption capabilities, ZnO-T can be utilized in photodetectors, particularly for detecting UV light. The ability to modulate the conductivity of ZnO-T in response to light makes them relevant for a wide range of optical sensing applications [1, 4]. ZnO photodetectors have fast response rise time as fast as 20 ns and fall time of 350 ns. Typically, the range is between some micro to milliseconds which can be further increased by surface modification using other materials including Al_2O_3 or SnO_2 etc. ZnO photodetectors exhibit high sensitivity to ultraviolet (UV) light, with sensitivities ranging from 10 to 4750 depending on the specific material and fabrication techniques used [92]

Photovoltaics and Solar Cells

The electronic properties of ZnO-T, especially their ability to effectively separate and transport photogenerated charge carriers, make them promising candidates for use in multi-junction solar cells and 3rd generation photovoltaics. ZnO-T can be used as the electron transport layer in hybrid organic–inorganic solar cells to host the efficiency of light absorption and charge collection [1, 47].

5.3 Mechanical Properties of ZnO Tetrapods

Zinc oxide exhibits a variety of unique mechanical properties when structured at the nanoscale, particularly in 3D tetrapod like nanostructures. ZnO-T have gained considerable interest because of its impressive mechanical performance, including their high strength, flexibility, and characteristics properties, which make them viable for a wide range of applications, that includes sensors, energy harvesting devices, and flexible electronics. In this section, we will brief the key mechanical aspects of ZnO-T, including their strength, flexibility, hardness, tensile behavior, and piezoelectric mechanical response.

5.3.1 Strength and Stiffness

Mechanical strength of ZnO-T is primarily influenced by their crystallinity, size, and morphology. While bulk ZnO exhibits significant mechanical strength of Vicker's Hardness of 1200–1400 HV, nanostructures like ZnO-T can demonstrate enhanced or altered mechanical properties due to their reduced dimensions and high S/V ratio with a scale of 300–800 HV.

High Strength

- *Intrinsic Strength*: ZnO-T typically exhibit high mechanical strength due to their strong covalent bonds and single crystalline structure with negligible grain boundary. The wurtzite crystal structure of ZnO, which is a hexagonal arrangement of atoms, contributes to the material's inherent strength and stability. The tetrapod morphology, with its arms and central core, enhances the mechanical stability and strength by distributing mechanical stress across multiple directions [93].
- *Size-Dependent Strength*: At the nanoscale, ZnO-T may experience size-dependent strengthening effects. This phenomenon, often referred to as the "size effect", occurs since the strength of nanomaterials can be higher than that of the bulk materials, mainly due to the minimal number of point and surface defects and the increased surface area relative to the volume. Smaller structures can resist deformation more effectively compared to their bulk counterparts [4, 94].

Stiffness

ZnO-T are relatively stiff due to their crystalline nature, which makes them resistant to deformation under mechanical stress. The stiffness of a material is typically characterized by its Young's Modulus, a measure of how much it deforms under an applied force. The bulk ZnO has 100–200 GPa at room temperature along the c-axis and a-axis making them stiff. For ZnO-T, the Young's Modulus varies between 20 and 60 GPa depending on their size and the quality of the material, but it is typically high due to the strong atomic bonds in the wurtzite crystal structure. The mechanical rigidity of the tetrapod arms is essential for maintaining their shape during the service period, particularly in applications involving mechanical stress and strain [95].

5.3.2 Flexibility

Despite their high strength and stiffness, ZnO-T can exhibit some degree of flexibility, especially under low-stress conditions. This flexibility is crucial for their application in flexible electronics and energy harvesting devices, where mechanical deformation is often required.

Deformability Under Stress

- *Mechanical Deformation:* The ability of ZnO-T to deform under stress is influenced by their nanostructure. Tetrapod arms, being thin and relatively long, allow for limited deformation in response to mechanical forces without breakdown. When subjected to external stress, the arms of the tetrapods can bend or stretch to a limited extent before they fracture. This makes them suitable for use in applications such as nanogenerators or flexible sensors, where mechanical stress is often applied to the material [95].
- *Elasticity:* The elastic properties of ZnO-T, characterized by their Young's Modulus, allow them to return to their original shape after the withdrawl of stress. While the overall elasticity of ZnO-T may be lower than that of some softer materials, they still exhibit enough elasticity to withstand repeated mechanical deformation without permanent damage [95].

5.3.3 Hardness

Hardness refers to the resistance offered by a material to deformation, in particular permanent indentation or scratching. ZnO is known for its high hardness due to its strong ionic and covalent bonding in the wurtzite crystal structure, and this property is maintained when ZnO is fabricated into nanostructures like tetrapods.

Nanoindentation Hardness

- *High Hardness at the Nanoscale:* The nanoindentation hardness of ZnO-T has been reported to be higher than that of bulk ZnO, mostly due to the reduced size and the strong S/V interactions that dominate at the nanoscale. Nanoindentation tests, which involve pressing a sharp indenter tip upon the surface of the material to measure resistance to deformation, have shown that ZnO-T are relatively hard, making them suitable for applications that require wear resistance or durability, such as coatings on mechanical components [96]. The Vickers hardness of bulk ZnO typically ranges from 6.5 to 8 GPa, which indicates that it is relatively hard compared to many other materials.
- *Size-Dependent Hardness:* As with many other nanomaterials, the hardness of ZnO-T is often size-dependent, meaning that the smaller the tetrapods, the greater the hardness relative to their size. This behavior is due to the fact that smaller structures tend to exhibit fewer dislocations and defects, contributing to enhanced resistance to indentation and deformation [95, 96]. Arger Tetrapods (Micrometer Scale): For ZnO-T with larger arm sizes (e.g., several micrometers in length), the Vickers hardness tends to be around 8–10 GPa. These larger structures may still benefit from the nanostructured properties but are also subject to lower defect densities and the influence of larger grain boundaries, meaning their hardness will be relatively stable but not maximized. Smaller tetrapods in nanoscale range as the

size of the ZnO-T decline (i.e., reducing the length of the arms to the nanometer scale), the hardness may increase slightly due to the reduction in grain size and the Hall–Petch effect, which strengthens the material. In very small tetrapods (less than 100 nm in arm length), the hardness can approach 10–12 GPa. Smaller nanostructures often have a higher proportion of atoms at or near the surface, where they experience different bonding environments. If the tetrapods are well-formed and passivated, the increased S/V ratio and reduced defect density in small tetrapods can lead to an increase in hardness.

5.3.4 Tensile Strength and Fracture Behavior

Tensile strength implies to the highest stress a material can withstand upon stretching or pulled prior to breaking. The fracture behavior of ZnO-T is critical in determining its suitability for use in applications that involve significant mechanical stress.

Tensile Strength

- *Brittle Behavior:* ZnO is generally considered a brittle material, meaning it does not exhibit significant ductility or plastic deformation before fracture. This characteristic is preserved in ZnO-T, where the arms of the tetrapod can break under high tensile stress. However, since the tetrapods have a 3D structure, the forces acting on each individual arm are distributed, which can enhance their overall strength and allow the structure to withstand greater tensile forces before failure commences [97].
- *Influence of Nanostructure:* The fracture behavior of ZnO-T can differ from that of bulk ZnO due to their nanostructured nature. When the tetrapod arms are subjected to tensile stress, they are more likely to undergo brittle fracture compared to more ductile materials. However, the high aspect ratio (length-to-width ratio) of the arms allows them to dissipate mechanical forces in multiple directions before breaking, providing some additional resilience [90, 97].

Fracture Toughness

- *Crack Propagation:* ZnO-T may experience crack propagation when subjected to high stress, but the unique 3D structure of the tetrapods can help slow down the propagation of cracks due to the arms' geometry and the ability of the material to distribute stress. Nonetheless, as with many brittle materials, ZnO-T are susceptible to crack formation under high mechanical loading, and understanding their fracture toughness is crucial for preventing failure in practical applications [97].

5.3.5 Piezoelectric and Triboelectric Nanogenerators

One of the most unique mechanical properties of ZnO-T is their piezoelectric behaviour. ZnO exhibits piezoelectricity, meaning that it generates an electrical charge when applied with mechanical stress. This property is especially pronounced in the nanostructured form of ZnO, including tetrapods. ZnO-T have an unusually high S/V ratio due to its branched structure. This large surface area enhances contact with other triboelectric materials, which is crucial for generating high amounts of charge during the contact-separation process in Triboelectric nanogenerators (TENGs). The tetrapod shape helps to improve mechanical stability and facilitates better mechanical interlocking with other surfaces, which leads to increased charge transfer and improved energy conversion efficiency. This geometry also contributes to effective stress distribution under mechanical deformation. ZnO-T maintain the piezoelectric properties of ZnO, meaning they can induce an electrical charge when given a mechanical deformation. These properties add to the triboelectric effect, resulting in enhanced electrical output when the TENG is subjected to motion or vibration. ZnO-T can be synthesized on flexible substrates, making them well-suited for applications in flexible or wearable electronics. The flexibility is important for many TENG applications that involve dynamic movement. ZnO-T are integrated into one of the triboelectric layers (often paired with other materials like PTFE, PDMS, or elastomers) that exhibit a strong triboelectric effect. When two materials come into contact and then separate (through mechanical motion, vibration, or bending), electrons are transferred from one material to another. The ZnO-T help increase the contact area during the contact-separation cycle. As a result, the charge transfer is enhanced, producing a higher voltage and current from the mechanical motion. When the ZnO-T are subjected to mechanical stress or strain (due to the motion in TENGs), they exhibit piezoelectric behavior and generate additional charge. This piezoelectric contribution complements the triboelectric effect, leading to a more efficient energy conversion process. The charges generated from the triboelectric and piezoelectric effects accumulate on the electrodes of the TENG, which can then be stored in capacitors or used directly to power small electronic devices or sensors.

The high surface area and 3D structure of ZnO-T lead to a higher density of charges, which enhances the efficiency of the TENG. ZnO-T are mechanically stable and resilient, even under repeated mechanical stress. This durability is particularly beneficial in applications that require long-term energy harvesting without significant degradation in performance. The ZnO-T can be integrated into flexible substrates, making them suitable for wearable, portable, or stretchable electronics. This adds a significant level of versatility in the design and use of TENGs. ZnO is a relatively low-cost and environmentally friendly host, and it can be prepared using facile and scalable techniques. This makes ZnO-T-based TENGs an promising option for large-scale energy harvesting applications. ZnO-T-based TENGs are used in smart textiles, energy-harvesting shoes, or wearable health-monitoring devices, where energy can be harvested from human movement. These TENGs can power sensors used for environmental monitoring, health tracking, or motion detection without the need for external power sources. ZnO-T-based TENGs can be used to capture energy from vibrations in industrial settings, mechanical systems, or even from environmental sources like ocean waves or wind. These TENGs can power small, low-energy devices such as LED lights, small sensors, or portable electronics by harvesting ambient mechanical energy from their surroundings (Fig. 19).

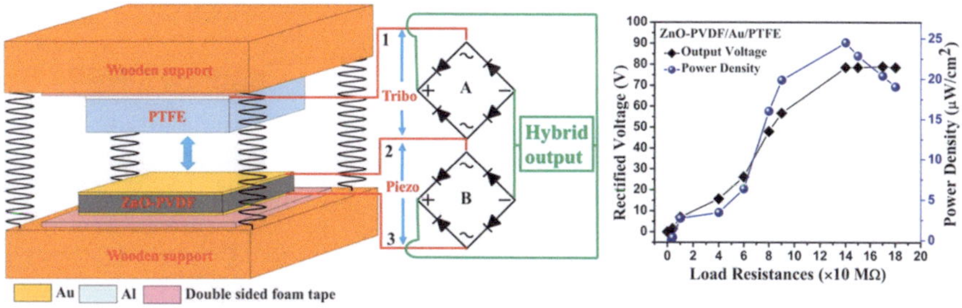

Fig. 19 Schematics of triboelectric nanogenerator along with their experimental data showing variation in rectified voltage [98]

Fabrication of ZnO Tetrapods for TENGs

Solution-Based Methods: ZnO-T can be synthesized via techniques such as hydrothermal synthesis, which involves the formation of ZnO nanostructures in an aqueous medium under controlled conditions.

CVD: ZnO-T can also be fabricated using vapor-phase deposition methods, which allow for precise control over the nanostructure morphology.

Electrospinning: In some cases, ZnO-T can be fabricated by electrospinning ZnO fibers and manipulating them into the desired tetrapod shape.

ZnO-T is a promising material for the development of highly efficient triboelectric nanogenerators. Their distinct 3D morphology, enhanced surface area, piezoelectric properties, and flexibility make them outstanding candidates for improving the performance of energy-harvesting devices. As research continues to progress, ZnO-T-based TENGs possess the capability to power a wide range of gadgets like wearable electronics, self-powered sensors, and environmental energy harvesting systems.

Piezoelectric Effect in ZnO Tetrapods

- *Mechanical Stress to Electrical Energy Conversion:* Due to 3D shape, ZnO-T can generate a piezoelectric potential when subjected to mechanical deformation. The tetrapod arms act as flexible beams that bend or stretch under stress, producing an electrical charge that can be harvested for use in nanogenerators or sensors. This piezoelectric response is highly beneficial for energy harvesting applications, like converting mechanical energy from vibrations or movement into usable electrical energy [90].
- *Enhanced Response:* The 3D structure of ZnO-T enhances their piezoelectric response since each arm of the tetrapod contributes to the overall electrical output. The geometry of the tetrapod maximizes the amount of deformation that occurs in response to mechanical forces, thus improving the overall efficiency of piezoelectric energy harvesting device components [90].

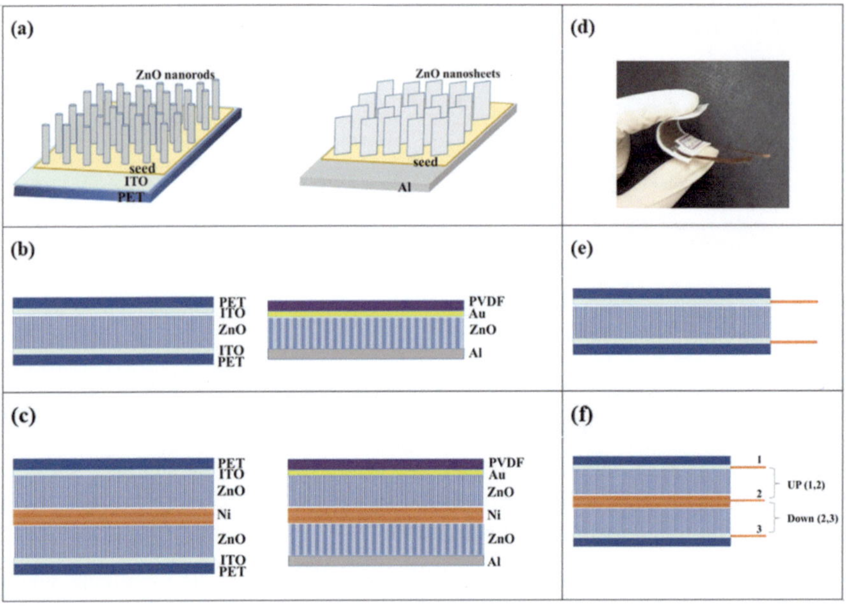

Fig. 20 Schematics of functioning of piezoelectric TENG devices (**a–f**) fabricated using nanostructured ZnO [99]

The mechanical aspects of ZnO-T, including their strength, flexibility, hardness, and tensile behavior, make them highly suitable for several applications, particularly in the fields of sensors, energy harvesting, and flexible electronics. The association of high strength and stiffness with some degree of flexibility makes ZnO-T well-suited for applications that involve mechanical stress, while their piezoelectric properties enable them to efficiently convert mechanical energy into electrical energy (Fig. 20).

Their nanoindentation hardness and tensile strength ensure durability in harsh conditions, while their piezoelectric mechanical response opens up exciting possibilities for self-powered systems. However, the brittle nature of ZnO-T remains a limitation, and further research into improving their fracture toughness or enhancing their flexibility by suitably doping ZnO-T, although it is a challenging process it may help broaden their use in more demanding applications.

5.4 Optical Properties of ZnO Tetrapods

Wide-bandgap ZnO semiconductor exhibits interesting optical properties, such as high transparency in the visible region and strong ultraviolet (UV) absorption and emission. These properties are further enhanced when ZnO is structured at the nanoscale, especially in 3D tetrapod like nanostructures like tetrapods. The branched-out arms of ZnO-T influences how light interacts with the material. ZnO-T showcase a wide variety of optical phenomena that make them perfect contender for several applications, including optoelectronics, photodetectors, photocatalysis, light-emitting devices, and UV sensors. In this section, we will discuss the key optical properties of ZnO-T, including their bandgap, optical absorption, photoluminescence, light scattering, and surface plasmon resonance.

5.4.1 Bandgap and Optical Absorption

The optical bandgap of ZnO-T can be influenced by its size and morphology. In nanostructured ZnO materials, including tetrapods, the quantum confinement effect may modify the electronic structure and lead to a slight increase in the effective bandgap compared to the bulk ZnO. This effect is particularly pronounced in quantum dots or NPs, but the tetrapod structure also shows a similar trend due to its reduced dimensions and unique 3D morphology [11], especially the fine tip leading to concentrating the electric fields for applications like flat panel display, electron microscopes, higher sensitivity in gas and chemical sensors, increased catalytic activity due to enhanced active sites, higher mechanical stability and can be fuctionalized for targeted drug delivery.

ZnO-T generally exhibit strong absorption in the UV regime, with the absorption edge corresponding to the energy bandgap (~3.37 eV). This makes them suitable for UV photodetector and UV light-emitting applications. The absorption efficiency of ZnO-T can also be enhanced by their high surface area, which increases the number of active sites for photon interaction [100].

5.4.2 Optical Absorption Properties

- *UV Absorption:* ZnO is a popular material for UV light absorption and emission. ZnO-T exhibit enhanced UV absorption efficiency compared to their bulk counterparts due to their high S/V ratio and potential for higher surface defect density. This can lead the material absorbing a wide range of UV wavelengths, making it valuable for UV protection coatings, UV sensors, and photodetectors [101, 102].
- *Enhanced Absorption in Nanostructures*: The optical absorption of ZnO-T can be enhanced relative to bulk ZnO due to of their high surface area. Nanostructured ZnO typically has a larger number of surface states, which can contribute to enhanced light-matter interaction at the surface. The tetrapod structure can further increase this interaction due to its 3D architecture, where the arms of the tetrapod allow for multiple light-scattering events [14, 15].

5.4.3 Photoluminescence Emission Characteristics

In ZnO-T, PL emission is a key optical property due to the material's ability to emit light when excited by photons, particularly in the UV and visible ranges.

UV and Visible Emission

- *Exciton Emission*: ZnO is known for its strong excitonic emission in the UV region, which occurs due to recombining of electrons and holes bound together and release of light energy in an exciton. This emission is typically formed around 380 nm (the near-UV region). ZnO-T, due to their high surface area (range of 50–200 m^2/g), and crystallinity, often exhibit stronger exciton-related emission compared to bulk ZnO (0.1–5 m^2/g) [14]. The S/V ratio of ZnO-T morphology is higher consequently which can be exploited for various applications.
- *Defect-related Emission*: In addition to exciton emissions, ZnO-T can exhibit defect-related photoluminescence, which occurs due to surface defects or oxygen vacancies in the material. This emission is generally observed in the visible region, with peaks around 500–600 nm (green to orange light), which is often referred to as the green band or deep-level emission. The intensity of this emission can be significantly influenced by the quality of the ZnO-T, with lower-quality material exhibiting stronger defect-related emissions [14, 103].
- *Influence of Nanostructure*: The PL emission spectrum of ZnO-T can be modified based on the size, surface defects, and crystallinity of the tetrapods. The increased S/V ratio in tetrapods can lead to higher rates of surface recombination, which may contribute to the defect-related PL emission in the visible range. On the other hand, the high crystallinity of the tetrapod structure can enhance the intensity of the UV exciton emission [18].

Excitation Dependence

The PL characteristics of ZnO-T are also highly based on the excitation wavelength. Under UV excitation (e.g., at 325 nm), ZnO-T often show intense UV emission, while visible emissions (such as the green band) are more prominent when the material is excited with lower-energy photons. The ability to tune the excitation wavelength allows for the tuning of ZnO-T for specific applications, like optical sensors and light-emitting gadgets.

5.4.4 Light Scattering and Surface Plasmon Resonance

ZnO-T have the potential to exhibit light scattering properties and surface plasmon resonance (SPR), both of which are highly beneficial for certain optical applications, including sensing and energy harvesting.

Light Scattering

- *Enhanced Scattering*: Due to their 3D structure, ZnO-T exhibit enhanced light scattering properties. The arms of the tetrapod can scatter incident light in different directions, which increases the overall interaction between light and the material. This scattering effect can be particularly useful for applications where light trapping is desirable, such as in solar cells or photodetectors [55].
- *Nanostructure Effects*: The high aspect ratio of the tetrapod arms further enhances light scattering and absorption, contributing to greater light-matter interactions. This aspect can be leveraged in photonic crystals-based devices, where light needs to be efficiently trapped and guided in a specific path or direction [55].

Surface Plasmon Resonance

- *SPR in ZnO-T*: Surface plasmon resonance (SPR) occurs when light interacts with free electrons on the surface of a material, leading to collective oscillations. While SPR is most commonly observed in noble metals like gold and silver, ZnO-T, due to their high surface area and nanostructured geometry, can also exhibit considerable plasmonic behaviour under specific conditions. This is typically observed in the visible and infrared regions [104]. The SPR peak value of ZnO with metal doping like gold NPs typically have around 520–550 nm and silver NPs exhibit around 400–450 nm.
- *Potential for Plasmonic Applications*: The ability of ZnO-T to support SPR could enhance their performance in sensing applications, particularly for biosensors and chemical sensors, where plasmonic effects can help detect small changes in the local environment by enhancing the scattering or absorption of light [104, 105].

5.4.5 Non-Linear Optical Properties

ZnO-T, like other semiconducting nanostructures, can also exhibit non-linear optical properties when subjected to high-intensity light. These properties include phenomena like second-harmonic generation (SHG) and optical rectification, which can be useful for optical switching, signal processing, and laser generation.

- *Second-Harmonic Generation*: The non-linear optical properties of ZnO-T can make them suitable for SHG applications, where the material generates light at twice the frequency of the incident light. This is particularly valuable in the generation of coherent UV light ($\sim$110 nm), where ZnO-T can be used in femtosecond lasers or ultraviolet light sources [105].
- *Non-linear Optics in Sensors and Communications*: The non-linear optical properties of ZnO-T could also be utilized in communication systems and optical sensors, where they can be used to detect specific wavelengths or improve the efficiency of signal transmission and detection [106].

5.4.6 Applications Based on Optical Properties

The exceptional optical properties of ZnO-T make them excellent candidates for a wide spectrum of applications in various fields. In view of their strong UV absorption, high surface area, and variable optical characteristics, ZnO-T are widely used in photodetectors for UV and visible light detection. Their ability to respond to both UV and visible light makes them beneficial in applications such as UV sensors, light sensors, and environmental monitoring [1]. ZnO-T can be enplaned in UV–Visible light-emitting diodes (LEDs) and laser diodes, specifically for UV applications. Their strong excitonic emission in the UV range makes them ideal for UV LEDs, which are used in applications such as disinfection, medical treatments, and security markings [107]. ZnO-T, with their enhanced absorption and scattering properties, can be utilized in solar cells to improve light trapping and energy conversion efficiency. Their high surface area permits for more efficient interaction with light, making them a promising material for thin-film solar cells or photovoltaic devices [47]. The power conversion efficiency of the solar cell where in ZnO-T were utilized as single electron acceptors was mere 0.002%.

ZnO-T exhibit a range of remarkable optical properties, including strong UV absorption, photoluminescence (both exciton and defect-related emissions), light scattering, and potential surface plasmon resonance behavior. These properties are influenced by their unique nanostructure, which enhances light-matter interaction due to the high S/V ratio and the 3D architecture of the tetrapods. As a result, ZnO-T show great promise in numerous optoelectronic devices, including photodetectors, LEDs, solar cells, and sensors. Their ability to respond to both UV and visible light, combined with their non-linear optical properties, opens the door to exciting possibilities in non-linear optical devices, plasmonics, and light-based sensing applications.

5.5 *Whispering Gallery Mode of ZnO Tetrapods*

The concept of Whispering Gallery Mode (WGM) refers to a phenomenon in which light or sound waves propagate along a curved surface, such as the inner walls of a circular or spherical cavity, in a manner similar to whispers traveling along the perimeter of a dome. The light or sound waves repeatedly reflect off the surface, maintaining their intensity over large distances due to constructive interference, without significant energy loss. This principle, first observed in architecture (in the famous St. Paul's Cathedral in London), has been extended to various fields, including optics and nanotechnology, to describe certain modes of light confinement within spherical or cylindrical structures (Fig. 21) [1].

In the context of ZnO-T, the Whispering Gallery Mode (WGM) effect is an intriguing optical phenomenon where light is confined and guided along the surface of the tetrapod arms. This is particularly potential for applications in sensing, lasing, and cavity optoelectronics. The unique 3D nanostructure of ZnO-T makes them an

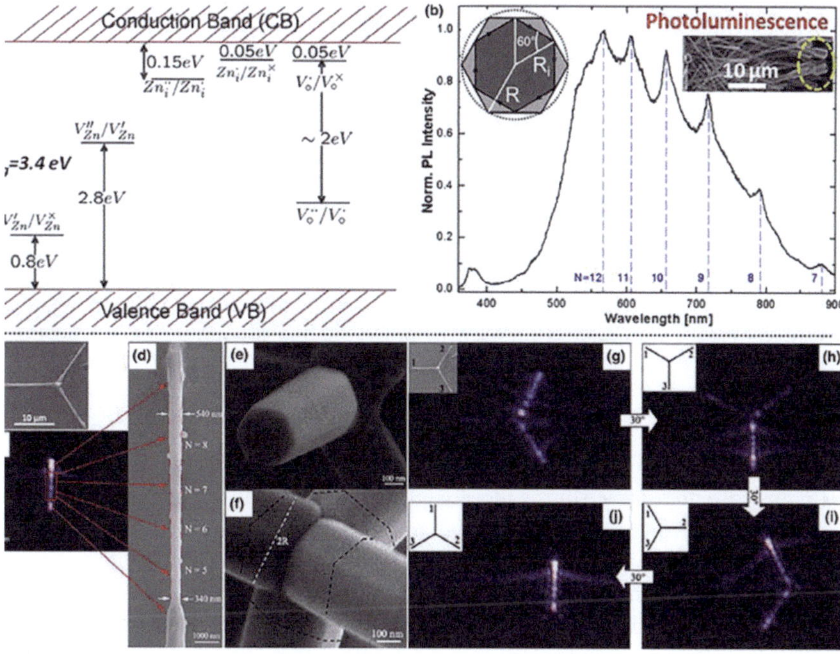

Fig. 21 Schematics of (**a**) Defect levels in ZnO with their (**b**) Whispering gallery mode and (**c–e**) SEM images of ZnO tetrapod with their (**g–i**), and (**j**) CCD images of single harmonic-whispering gallery mode [1]

ideal candidate for exhibiting WGM behavior due to their high S/V ratio and the nanoscale geometry that supports resonant light trapping [108].

In nanoscale materials, such as ZnO-T, WGM can occur when light interacts with the nanostructure's surface and is trapped within the material because of total internal reflection at the interfaces. This phenomenon can be utilized in various ways, including optical resonators and biosensing applications. Nanostructures like ZnO-T, which have sharp edges and curved geometries, are able to support WGM resonances even at the nanoscale [108]. ZnO-T have a tetrahedral structure that includes four arms extending from a central core, giving them a geometry that may support WGM when illuminated with light at certain angles. The arms of the tetrapod can serve as waveguides for the light, and the light propagates along the surfaces of the arms with multiple reflections [108].

5.5.1 Mechanism of WGM in ZnO Tetrapods

Whispering Gallery Modes in ZnO-T arise from the interaction of light with the nanostructured surface of the tetrapods. The morphology of ZnO-T, characterized by sharp edges and angles, can lead to multiple light reflections, which results in constructive interference at certain resonant frequencies. These resonant modes can be divided based on the various geometrical characteristics of the tetrapod arms and the wavelength of light.

Surface Plasmon Coupling

- ZnO-T can exhibit plasmonic effects due to their nanostructure, where incident light excites collective electron oscillations (plasmon modes) at the ZnO surface. These plasmonic effects can further enhance the WGM resonance by amplifying the light-matter interaction at the surface of the tetrapod arms [108, 109].
- The surface plasmon resonance (SPR) phenomenon in ZnO-T can lead to enhanced light trapping, allowing for more efficient coupling of light into the Whispering Gallery Modes [108].

Resonant Condition

- The WGM condition is satisfied when the circumference of the curved surface (in this case, the surface of the faceted tetrapod arms) is an integer multiple of the wavelength of light inside the material. In ZnO-T, this condition can be controlled based on the size and geometry of the branching arms [110].
- As light is incident on the tetrapod arms at specific angles, the total internal reflection occurs along the surface of the arms, with light circulating around the perimeter. This results in resonant optical modes that can be selectively excited at specific wavelengths, often in the UV or visible ranges due to the optical characteristics of ZnO.

Influence of the Tetrapod Geometry

- The sharp edges and angular structure of ZnO-T allow for the light to travel along the arms, potentially enhancing the WGM effect. These structures create optical cavities where light is confined for longer periods before it decays [1, 109].
- The geometry of the tetrapod arms influences the resonance conditions, where the length, thickness, and curvature of the arms play a critical role in determining the resonant wavelength and the intensity of the confined modes.

5.5.2 Applications of the Whispering Gallery Mode of ZnO Tetrapods

The ability of ZnO-T to support WGM has potential applications in various fields due to their unique optical and physical characteristics. Given below are some of the key applications of WGM in ZnO-T:

Sensing and Biosensing

One of the most significant applications of WGM is in sensing, particularly for detecting small changes in the surrounding environment. Since WGM are highly sensitive to variation in the material's refractive index close to the surface, ZnO-T can be used in biosensors and chemical sensors.

- *Biosensors*: The surface of ZnO-T can be sensitized with specific recognition molecules (like antibodies or DNA strands), allowing them to detect biological or chemical substances. The WGM resonance can shift as the refractive index near the tetrapod surface changes due to the binding of target molecules, providing a sensitive platform for biosensing [100].
- *Gas Sensing*: ZnO-T can also be used for gas sensing applications, where changes in the surrounding gas molecules alter the optical properties of the material, influencing the WGM resonance. This can be used for environmental monitoring, pollution detection, and medical diagnostics [16].

Light Emitting Devices and Lasers Using ZnO Tetrapods

ZnO-T exhibit strong photoluminescence emission, and is aspect along with the WGM phenomenon can be leveraged in light-emitting devices and lasers. The confined light within the WGM can be efficiently coupled out of the system, creating a resonant cavity that produces coherent light. This is particularly relevant for applications in solid-state lasers, LEDs, and optical communication systems.

- *Lasing Action*: The WGM effect can enable laser oscillations in ZnO-T, where the tetrapod arms act as optical cavities that confine and amplify light. This property is promising for micro-lasers and nano-lasers, especially for UV light generation, which can be utilized in applications such as sensing, disinfection, and high-resolution imaging [108].

Optical and Photonic Devices

ZnO-T can be used in photonic devices, such as filters, resonators, and optical switches. The WGM is a powerful means for optical waveguiding and light manipulation in nanoscale devices. This allows ZnO-T to be integrated into nanophotonics for applications requiring precise control of light, including integrated photonic circuits and optical communications.

- *Optical Filters*: By tuning the WGM resonances, ZnO-T may be utilized in filter specific wavelengths of light, making them ideal for spectral filtering applications in telecommunications and sensing (Fig. 22I) [110].
- *Non-linear Optics*: The WGM phenomenon can also enhance the non-linear optical properties of ZnO-T, which could be employed in applications like optical switching or frequency conversion [109].

Energy Harvesting and Photovoltaics

ZnO-T, because of its high surface area and light-trapping capabilities, may be utilized in energy harvesting gadgets, including solar cells and photovoltaic devices. The WGM enhances the light absorption efficiency within the tetrapod structure, allowing for improved photoelectric conversion. ZnO-T can improve the performance of thin-film solar cells by enhancing light absorption through WGM, allowing for more efficient conversion of Sunlight into usable electrical energy [47]. The WGM

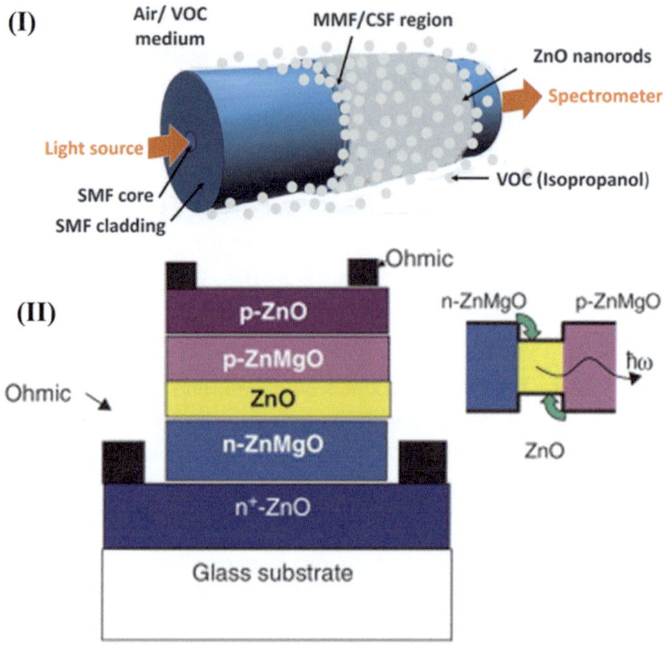

Fig. 22 **I**) Pictorial representation of nanostructured ZnO optical filters in optical communication fibre and, **II**) Light Emitting Diodes (LEDs) [111, 112]

phenomenon in ZnO-T is a promising optical effect that can be harnessed for a multitude of applications, including sensing, biosensing, light emission, laser generation, and energy harvesting. The unique 3D nanostructure of ZnO-T, combined with their high S/V ratio and capability to support light confinement, makes them an ideal candidate for supporting WGM resonances. As studies continue to explore the viability of ZnO-T in nanophotonics and optoelectronics, the ability to tailor their geometry and optical properties opens up exciting possibilities for advanced devices in biomedical sensing, communication technologies, and environmental monitoring.

6 Applications of ZnO Tetrapods

ZnO-T, with their unique three-dimensional (3D) structure comprising a central core and four arms radiating symmetrically, exhibit remarkable aspects that make them perfect for a wide range of applications across various fields, like optoelectronics, sensors, energy harvesting, photocatalysis, and biomedical applications. These properties, including high surface area, piezoelectric behavior, wide bandgap, and excellent optical performance, enhance their functionality compared to other nanostructures of ZnO. In this section, few of the key applications of ZnO-T in various technological and scientific domains are outlined.

6.1 Optoelectronics

ZnO-T having the ability to emit UV light makes it highly potential for several optoelectronic devices. The unique morphology of ZnO-T further boosts its optical characteristics, making it ideal a candidate for several optoelectronic applications [11].

6.1.1 Light-Emitting Diodes and Lasers

ZnO is a viable material for UV LEDs and UV lasers, primarily due to its wide bandgap and efficient excitonic emission. ZnO-T, with their high and varying surface area and enhanced light emission characteristics, can be employed to boost the efficiency and performance of such gadgets. The 3D structure of the tetrapods can help increase the internal quantum efficiency and make them more effective in UV light emission. ZnO-T are particularly suited for UV light sources, as they can emit light in the UV-A and UV-B regions, which is crucial for applications such as sterilization, water purification, and UV sensing (Fig. 22II) [107]. ZnO-T can be used in the fabrication of lasers, particularly in the UV range. Their unique structure allows for enhanced optical gain and emission in specific wavelengths, making them useful in various scientific and industrial applications [113].

6.1.2 Photodetectors

The high surface area and efficient electron transport aspects of ZnO-T also make them excellent candidates for use in photodetectors, especially for UV detection. Their wide bandgap enables them to absorb UV light efficiently, and their 3D nanostructure can improve charge collection efficiency, making them suitable for applications in environmental monitoring, security, and communication systems (Fig. 23).

Fig. 23 Schematics of ZnO-T-based photodetector devices fabricated by flame transport synthesis (FTS) for multifunctional applications with excellent response [114]

6.2 Sensors and Gas Detection

ZnO is well-known for its high sensitivity to gases, making it one of the most frequently used host material for gas sensing. The high S/V ratio of ZnO-T allows for better interaction with the gas species, boosting the sensitivity and response time of gas sensing devices.

6.2.1 Gas Sensors

ZnO-T can be used for the detection of various gases, including nitrogen dioxide (NO_2), ammonia (NH_3), carbon dioxide (CO_2), and hydrogen (H_2). When exposed to these gases, the electrical conductivity of ZnO changes, which can be measured to detect the existence and concentration of the gas (Fig. 24I).

- *NO_2 Detection*: ZnO-T are particularly effective in detecting nitrogen dioxide, a toxic gas commonly found in urban environments and associated with air pollution. Their high surface area and enhanced reactivity enable them to interact more effectively with NO_2, leading to quicker and more sensitive detection [117]. The sponge-based ZnO gas detect can specifically sense NO_2 gas with a limit of detection (LOD) of 8.56 ppb.
- *Ammonia and Hydrogen Sensors*: ZnO-T can also be employed as sensors for ammonia, which is crucial for agricultural applications, and hydrogen, which is important for fuel cell technologies [117]. Pure ZnO subjected to various gas presented good response of 132 for ammonia at 150 °C. Silver doped (6 wt%) ZnO samples were subjected to several gases and a response of 29.5 at 100 ppm was observed towards ammonia [118]. The result imply the highly selective nature of the films towards ammonia compared with other gases.

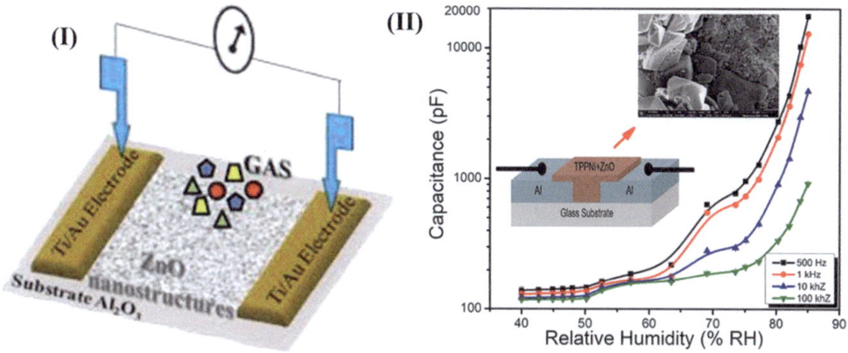

Fig. 24 Pictorial representation of utilizing ZnO nanostructures in, (**I**) gas sensors over an Al_2O_3 substrate [115] and in (**II**) humidity sensors [116]

6.2.2 Humidity Sensors

Due to their high surface area and fast response times, ZnO-T are effective in detecting changes in humidity. The water molecules adsorbed on the surface of the tetrapods can change the electrical properties of the material (Fig. 24 II), making it useful for humidity sensors in industrial processes, environmental monitoring, and smart devices [51, 117].

6.3 Energy Harvesting and Nanogenerators

Piezoelectric aspect of ZnO generate electrical charge under applied mechanical stress. ZnO-T, with their 3D structure, enhance this piezoelectric effect, making them excellent candidates for use in nanogenerators and energy harvesting applications [19].

6.3.1 Piezoelectric Nanogenerators

ZnO-T can be employed in nanogenerators that convert mechanical moments like vibrations or pressure changes, into electrical energy. The piezoelectric properties of ZnO-T are enhanced by their 3D structure, which increases their mechanical strength and surface area, enabling them to generate higher output from smaller mechanical deformations.

- *Vibration Energy Harvesting*: ZnO-T can be used to harvest energy from environmental vibrations, like those found in industrial machinery, transportation systems, or even human motion. This makes them suitable for self-powered sensors and wearable electronics [90].

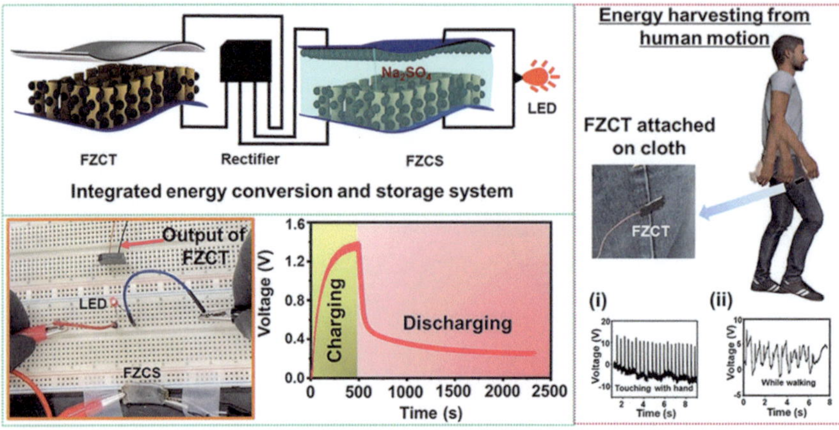

Fig. 25 Pictorial representation of utilization of nanostructured ZnO integrated energy harvesting and storage devices [121]

- *Energy Harvesting from Body Movements*: As part of wearable devices, ZnO-T can be integrated into flexible substrates to harvest energy from body movements (like walking, jumping, jogging, etc.) offering a sustainable solution for powering small sized electronics, gadgets like fitness trackers, health monitors, mobile phones, ear pods, or other portable devices (Fig. 25) [119].

6.4 Photocatalysis and Environmental Remediation

ZnO, especially in nanostructured form like tetrapods, has shown promise as a photocatalyst for environmental remediation, including water splitting, pollutant degradation. The enhanced surface area and catalytic aspects of ZnO-T can significantly improve the photocatalytic efficiency [120].

6.4.1 Photocatalytic Water Splitting

ZnO-T can be employed as photocatalysts for hydrogen generation through water splitting under UV light. This is an important method for producing clean hydrogen, which can serve as a renewable energy source. The high surface area of ZnO-T increases their interaction with water molecules, promoting the photocatalytic reactions necessary for splitting water into H_2 and O_2 gases [122].

6.4.2 Photodegradation of Pollutants

ZnO-T are also effective in the photodegradation of organic pollutants, like as dyes, pesticides, and industrial chemicals, under UV light. The photocatalytic ability of ZnO-T enables them to break down harmful chemicals into less toxic substances, making them valuable in water purification and wastewater treatment applications.

Environmental Clean-up: ZnO-T is used to eliminate contaminants from water/air in environmental clean-up processes, offering a green and cost-effective means for pollution management [16]. ZnO can be employed in water filtration systems, either as a coating for filtration materials or in advanced oxidation processes (AOPs) to purify contaminated water. ZnO-based photocatalysts are particularly useful in treating industrial effluents containing organic pollutants, dyes, and heavy metals. ZnO is used to purify seawater or brackish water by degrading contaminants and microorganisms, making it potable and suitable for agriculture. Portable water purification devices, especially in areas with limited access to clean water, may use ZnO-based materials for on-the-go purification. ZnO can be incorporated into air filters, coatings, or ventilation systems to break down indoor pollutants. ZnO-based photocatalysts are used to treat industrial emissions or to clean air in urban areas, often in the form of coatings on surfaces like walls, facades, or street furniture. ZnO has been used in air purifiers to remove smoke, odours, and volatile compounds in both residential and industrial settings.

6.4.3 CO_2 Reduction

ZnO-T have been explored for their potential use in carbon dioxide reduction. By harnessing solar energy, ZnO-T can convert CO_2 into useful chemicals, such as methane or methanol, providing a potential method for addressing climate change through the reduction of green-house gases [123] (Fig. 26).

6.5 Biomedical Applications

The unique characteristics of ZnO-T, including their biocompatibility, high surface area, and anti-bacterial characteristics, reder them attractive for several biomedical applications.

6.5.1 Drug Delivery and Targeted Therapy

ZnO-T can be sensitized with several biomolecules or drugs to deliver therapeutic agents directly targeting the cells or tissues as depicted in Fig. 27. Their high surface area permits for the attachment of a large number of drug molecules, which can

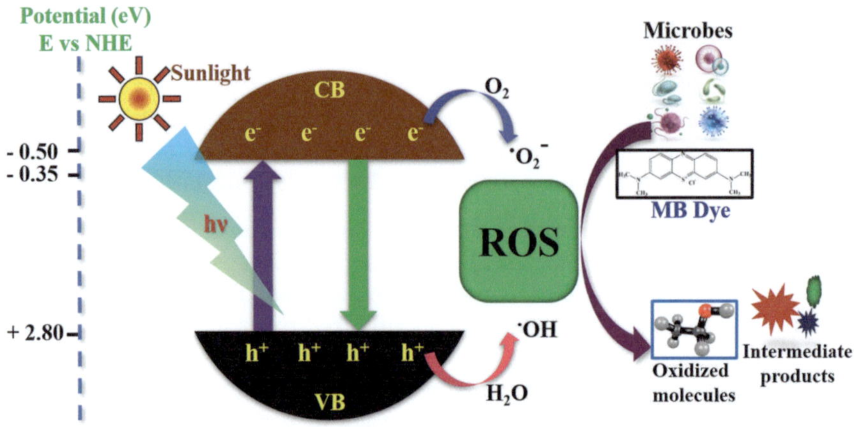

Fig. 26 Plausible general photocatalytic degradation mechanism of using ZnO NPs

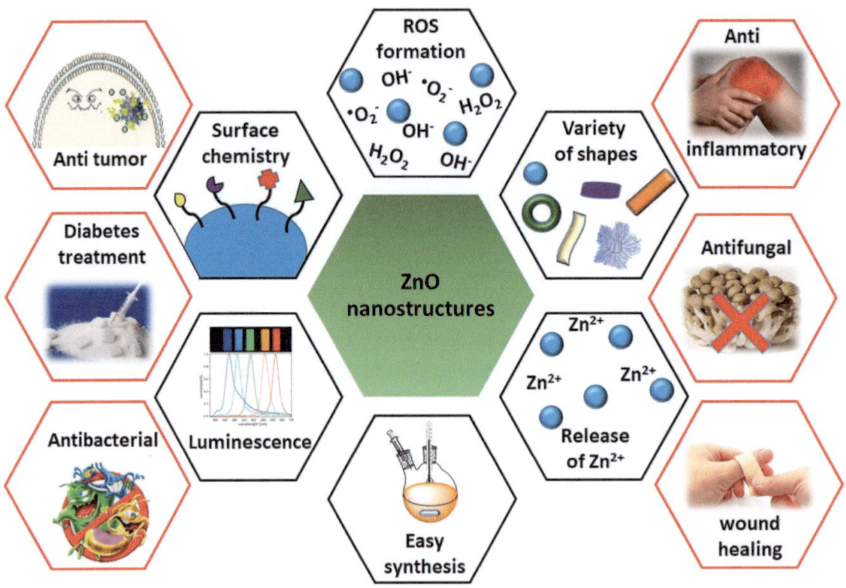

Fig. 27 Pictorial representation of biological applications of various ZnO nanostructures [125]

be released in a controlled manner, improving the efficacy of the treatment and minimizing the detrimental side effects [102, 124].

6.5.2 Antibacterial Coatings

ZnO has inherent anti-bacterial properties, which are enhanced in the tetrapod form due to the increased surface area and reactivity. ZnO-T can be used as antibacterial coatings for medical devices, implants, wound dressings, and surgical instruments to prevent infections and promote faster healing [102]. Upon illumination of UV light, ZnO can liberate electron–hole pairs that lead to the formation of highly reactive species, including hydroxyl radicals (OH•), superoxide anions (O_2^-), and hydrogen peroxide (H_2O_2). These ROS are toxic to bacteria as they can damage bacterial cell walls, membranes, and DNA, leading to death of bacterial cell. ZnO nanostructures can interact with the bacterial cell membrane. Due to their nano-sized dimensions and high surface area, ZnO particles can penetrate the bacterial cell membrane, causing structural damage. The $Zn2^+$ ions released from ZnO can also interact with the bacteria, disrupting cellular functions and causing leakage of intracellular components. ZnO can slowly release zinc ions ($Zn2^+$) into the environment, which can interfere with the bacteria's metabolism. These ions are toxic to bacteria and can disrupt various cellular processes, including protein synthesis and enzyme function. ZnO NPs can pierce the bacterial cell membrane, forming pores that lead to the loss of cellular contents, which ultimately results in the bacteria's death. ZnO's photocatalytic activity under UV light can cause oxidative stress in bacterial cells, further amplifying the antibacterial effect. When exposed to light, ZnO generates ROS that can damage bacterial DNA, proteins, and lipids, contributing to cell death.

6.5.3 Biosensors

ZnO-T can be employed as biosensors for detecting biological molecules, like glucose, proteins, or nucleic acids, in medical diagnostics (Fig. 28). The high surface area and functionalization potential of ZnO-T permits to be tailored for specific biomolecular interactions, making them ideal for sensitive and selective bio-sensing and this will be dealt in detail in chapter-4 [30].

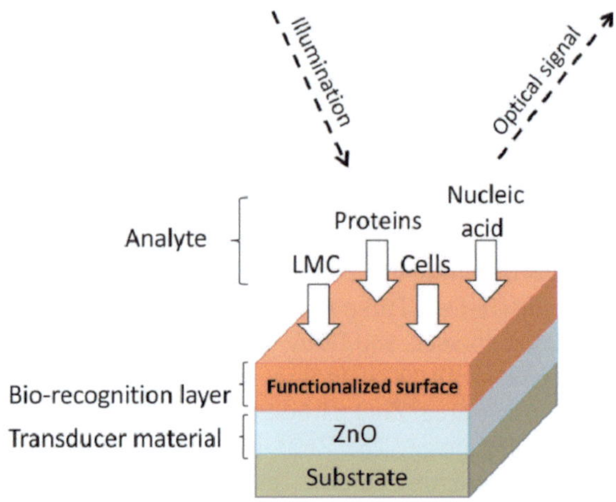

Fig. 28 Schematics of ZnO based biosensors functionalised for detecting proteins, nucleic acid etc. [126]

7 Conclusion

ZnO-T, with their distinctive three-dimensional structure, enhanced optical characteristics, high surface area, and piezoelectric behavior, have a wide range of applications across multiple fields. These applications include optoelectronics (such as UV LEDs and photodetectors), gas sensing, energy harvesting, photocatalysis, and biomedical utility (like drug delivery and antibacterial coatings). The growing research in the synthesis, functionalization, and integration of ZnO-T holds great promise for their widespread use in emerging technologies, particularly in clean energy, environmental remediation, and healthcare. These ZnO-T can be used for environmental monitoring, health monitoring, or in smart textiles. ZnO-based TENGs are integrated into clothing, shoes, or body-worn devices to harvest energy from body movements. They can capture energy from vibrations, mechanical motion, or ambient mechanical sources, which can then be stored for powering small electronic devices. ZnO is a highly promising material for triboelectric nanogenerators, offering enhanced mechanical and electrical properties that make these devices more efficient and versatile in various applications, particularly in wearable and flexible electronics.

As fabrication techniques improve and our understanding of their properties expands, ZnO-T are envisaged to play an increasingly key role in the development of next-generation devices and systems. ZnO-T, with their unique 3D structure consisting of a central body and four extending arms, represent a fascinating class of nanomaterials that offer a wide range of potential use in many fields, including sensing, optoelectronics, photovoltaics, and nanocomposites. The tetrapod morphology offers a large surface area compared to conventional ZnO NPs, which enhances performance in applications like catalysis, sensing, and energy storage. The branched structure facilitates the movement of charge carriers, making ZnO-T suitable for applications in photovoltaics and sensors. The tetrapod structure improves the light harvesting efficiency in photocatalytic and optoelectronic applications, making

ZnO-T effective in solar cells, UV detectors, and photocatalysis. ZnO-T have excellent thermal and chemical stability, which makes them durable and reliable in harsh environments. The large surface area allows for easy functionalization with various molecules, enhancing the performance of ZnO-T in biosensors and drug delivery systems.

The fabrication methods for ZnO-T, including vapor-phase deposition techniques, solution-based processes, and template-assisted approaches, have made it feasible to tune their size, shape, and crystallinity with precision. Techniques like CVD, hydrothermal synthesis, solvothermal methods, and electrochemical deposition offer flexibility in producing high-quality tetrapods, each method contributing specific advantages such as scalability, cost-effectiveness, and enhanced control over nanostructural features.

ZnO-T have presented great promise in a multitude of applications due to its optical and electronic characteristics. For instance, they exhibit strong photoluminescence emission, light trapping capabilities, and high surface area makes them exemplary host for biosensors, gas sensors, and light-emitting devices. Their ability to support WGM resonances further enhances their potential in optical sensing and photonics, where light confinement and enhanced signal sensitivity are crucial. Additionally, their piezoelectric properties enable their integration into energy harvesting devices and micro/nano sensors. Furthermore, the mechanical robustness of ZnO-T, coupled with their excellent thermal stability, makes them apt for a range of high-performance materials in electronics and nanocomposites. They have also demonstrated significant improvements in solar cells and photovoltaic devices, where their capability to boost light absorption and electron transport can help boost energy conversion efficiency. Despite the significant advancements, challenges remain in scaling up the fabrication processes, controlling defect densities, and ensuring reproducibility in large-scale production. Additionally, optimizing the integration of ZnO-T into practical devices and applications will require further research into their interactions with different materials and environments.

In conclusion, ZnO-T offer remarkable versatility and potential across multiple scientific and engineering domains. As fabrication methods continue to improve, these nanostructures are enlarged to play an increasingly key role in advancing nanotechnology, optoelectronics, energy transformation, and sensing technologies. With ongoing innovation and optimization, ZnO-T are poised to become a key material in the development of next-generation devices in both academia and industry.

The following chapters will focus the use of ZnO-T in specific application like photodetectors, photocatalysis and bio-related applications. Apart from these, the theoretical modeling of ZnO-T are very limited which is also addressed by providing a set of theories describing ZnO-T for possible self-assembly towards interdisciplinary applications.

References

1. Mishra, Y.K., Adelung, R.: ZnO tetrapod materials for functional applications. Mater. Today **21**, 631–651 (2018). https://doi.org/10.1016/j.mattod.2017.11.003
2. Schmidt-Mende, L., MacManus-Driscoll, J.L.: ZnO—nanostructures, defects, and devices. Mater. Today **10**, 40–48 (2007). https://doi.org/10.1016/s1369-7021(07)70078-0
3. Wang, Z., Bockstaller, M.R., Matyjaszewski, K.: Synthesis and applications of ZnO/polymer nanohybrids. ACS Materials Lett. **3**, 599–621 (2021). https://doi.org/10.1021/acsmaterialslett.1c00145
4. Sharma, D.K., Shukla, S., Sharma, K.K., Kumar, V.: A review on ZnO: fundamental properties and applications. Mater. Today Proc. **49**, 3028–3035 (2022). https://doi.org/10.1016/j.matpr.2020.10.238
5. Ortiz-Casas, B., Galdámez-Martínez, A., Gutiérrez-Flores, J., Ibañez, A.B., Panda, P.K., et al.: Bio-acceptable 0D and 1D ZnO nanostructures for cancer diagnostics and treatment. Mater. Today **50**, 533–569 (2021). https://doi.org/10.1016/j.mattod.2021.07.025
6. Klingshirn, C.: ZnO: material, physics and applications. Chem. Phys. Chem. **8**, 782–803 (2007). https://doi.org/10.1002/cphc.200700002
7. Raha, S., Ahmaruzzaman, M.: ZnO nanostructured materials and their potential applications: progress, challenges and perspectives. Nanoscale Adv. **4**, 1868–1925 (2022). https://doi.org/10.1039/d1na00880c
8. Mishra, N., Wu, W.-Y., Srinivasan, B.M., Hariharaputran, R., Zhang, Y.-W., Chan, Y.: Continuous shape tuning of nanotetrapods: toward shape-mediated self-assembly. Chem. Mater. **28**, 1187–1195 (2016). https://doi.org/10.1021/acs.chemmater.5b04803
9. Aspoukeh, P.K., Barzinjy, A.A., Hamad, S.M.: Synthesis, properties and uses of ZnO nanorods: a mini review. Int. Nano. Lett. **12**, 153–168 (2021). https://doi.org/10.1007/s40089-021-00349-7
10. Zhou, H., Wang, H., Yue, C., He, L., Li, H., Zhang, H., Yang, S., Ma, T.: Photocatalytic degradation by TiO_2-conjugated/coordination polymer heterojunction: preparation, mechanisms, and prospects. Appl. Catal. B **344**, 123605 (2024). https://doi.org/10.1016/j.apcatb.2023.123605
11. Sha, R., Basak, A., Maity, P.C., Badhulika, S.: ZnO nano-structured based devices for chemical and optical sensing applications. Sens. Actuators Rep. **4**, 100098 (2022). https://doi.org/10.1016/j.snr.2022.100098
12. Shah, N.A., Gul, M., Abbas, M., Amin, M.: Synthesis of metal oxide semiconductor nanostructures for gas sensors. Gas Sens. (2020). https://doi.org/10.5772/intechopen.86815
13. Ching, K.-L., Li, G., Ho, Y.-L., Kwok, H.-S.: The role of polarity and surface energy in the growth mechanism of ZnO from nanorods to nanotubes. Cryst. Eng. Comm **18**, 779–786 (2016). https://doi.org/10.1039/c5ce02164b
14. Guan, S., Hao, L., Murayama, M., Xie, X., Komuro, S., Zhao, X.: Influence of anneal temperature in air on surface morphology and photoluminescence of ZnO thin films. IOP Conf. Ser. Mater. Sci. Eng. **522**, 012004 (2019). https://doi.org/10.1088/1757-899x/522/1/012004
15. Amakali T., Daniel, L.S., Uahengo, V., Dzade, N.Y., de Leeuw, N.H.: Structural and optical properties of ZnO thin films prepared by molecular precursor and sol–gel methods. Crystals **10**, 132 (2020). https://doi.org/10.3390/cryst10020132
16. Ansari, A.A., Lv, R., Gai, S., Parchur, A.K., Solanki, P.R., et al.: ZnO nanostructures—future frontiers in photocatalysis, solar cells, sensing, supercapacitor, fingerprint technologies, toxicity, and clinical diagnostics. Coord. Chem. Rev. **515**, 215942 (2024). https://doi.org/10.1016/j.ccr.2024.215942
17. Silva, T.E.M., Moreira, A.J., Nobrega, E.T.D., Alencar, R.G., Rabello, P.T., et al.: Hierarchical structure of 3D ZnO experimentally designed to achieve high performance in the sertraline photocatalysis in natural waters. Chem. Eng. J. **475**, 146235 (2023). https://doi.org/10.1016/j.cej.2023.146235

18. Muslimov, A.E., Orudzhev, F.F., Grigoriev, Yu.V., Lavrikov, A.S., Kanevsky, V.M.: Luminescence and photocatalytic properties of ZnO tetrapods. J. Surf. Investig. **17**, 1490–1495 (2023). https://doi.org/10.1134/s1027451023060381
19. Phogat, P., Shreya, R.J., Singh, S.: Harnessing ZnO morphologies in energy application and sustainable development. Phys. Scr. **99**, 102004 (2024). https://doi.org/10.1088/1402-4896/ad7990
20. Abou Zeid, S., Leprince-Wang, Y.: Advancements in ZnO-based photocatalysts for water treatment: a comprehensive review. Crystals **14**, 611 (2024). https://doi.org/10.3390/cryst14070611
21. Lin, J., Luo, Z., Liu, J., Li, P.: Photocatalytic degradation of methylene blue in aqueous solution by using ZnO-SnO_2 nanocomposites. Mater. Sci. Semicond. Process. **87**, 24–31 (2018). https://doi.org/10.1016/j.mssp.2018.07.003
22. Reddy, N.P., Santhosh, R., Thogiti, S., Muniramaiah, R., Joseph, D.P., Murali, B.: Nanostructured ternary perovskite oxides as photoconversion efficiency enhancers for DSSC. J. Mater. Chem. C **10**, 1403–1413 (2022). https://doi.org/10.1039/d1tc04584a
23. Matei, E., Şăulean, A.A., Râpă, M., Constandache, A., Predescu, A.M., Coman, G., Berbecaru, A.C., Predescu, C.: ZnO nanostructured matrix as nexus catalysts for the removal of emerging pollutants. Environ. Sci. Pollut. Res. **30**, 114779–114821 (2023). https://doi.org/10.1007/s11356-023-30713-3
24. Bagga, S., Akhtar, J., Mishra, S.: Synthesis and applications of ZnO nanowire: a review. AIP Conf. Proc. **1989**, 020004 (2018). https://doi.org/10.1063/1.5047680
25. Ali, H.M., Ibrahim, S.M., Abo Zeid, E.F., Al-Hossainy, A.F., El-Aal, M.A.: A comparative study of Cu-anchored 0D and 1D ZnO nanostructures for the reduction of organic pollutants in water. RSC Adv. **12**, 16496–16509 (2022). https://doi.org/10.1039/d2ra02515a
26. Baillard, A., Appert, E., Rapenne, L., Jacob, V., Dieulesaint, A., et al.: Significance of the different exposed surfaces of ZnO single crystals and nanowires on the photocatalytic activity and processes. Small Struct. **5**, 2300540 (2024). https://doi.org/10.1002/sstr.202300540
27. Sansenya, T., Masri, N., Chankhanittha, T., Senasu, T., Piriyanon, J., Mukdasai, S., Nanan, S.: Hydrothermal synthesis of ZnO photocatalyst for detoxification of anionic azo dyes and antibiotic. J. Phys. Chem. Solids **160**, 110353 (2022). https://doi.org/10.1016/j.jpcs.2021.110353
28. Su, P., Tu, M., Ameloot, R., Li, W.: Vapor-phase processing of metal-organic frameworks. Acc. Chem. Res. **55**, 186–196 (2021). https://doi.org/10.1021/acs.accounts.1c00600
29. Wang, C., Yan, B., Zheng, J., Feng, L., Chen, Z., Zhang, Q., Liao, T., Chen, J., Jiang, S., Du, C., He, S.: Recent progress in template-assisted synthesis of porous carbons for supercapacitors. Adv. Powder Mater. **1**, 100018 (2022). https://doi.org/10.1016/j.apmate.2021.11.005
30. Krishna, M.S., Singh, S., Batool, M., Fahmy, H.M., Seku, K., Shalan, A.E., Lanceros-Mendez, S., Zafar, M.N.: A review on 2D-ZnO nanostructure based biosensors: from materials to devices. Mater. Adv. **4**, 320–354 (2023). https://doi.org/10.1039/d2ma00878e
31. Wang, Y., Chhowalla, M.: Making clean electrical contacts on 2D transition metal dichalcogenides. Nat. Rev. Phys. **4**, 101–112 (2021). https://doi.org/10.1038/s42254-021-00389-0
32. Wen, F., Zhu, C., Li, L., Zhou, B., Zhang, L., Han, C., Li, W., Yue, Z., Wu, W., Wang, G., Zhang, S.: Enhanced energy storage performance of polymer nanocomposites using hybrid 2D ZnO@MoS2 semiconductive nano-fillers. Chem. Eng. J. **430**, 132676 (2022). https://doi.org/10.1016/j.cej.2021.132676
33. Wang, J., Yang, H., Yang, P.: Photoelectric properties of 2D ZnO, graphene, silicene materials and their heterostructures. Compos. B Eng. **233**, 109645 (2022). https://doi.org/10.1016/j.compositesb.2022.109645
34. Chen, Y., Fan, Z., Zhang, Z., Niu, W., Li, C., Yang, N., Chen, B., Zhang, H.: Two-dimensional metal nanomaterials: synthesis, properties, and applications. Chem. Rev. **118**, 6409–6455 (2018). https://doi.org/10.1021/acs.chemrev.7b00727
35. Feng, C., Wu, Z., Huang, K., Ye, J., Zhang, H.: Surface modification of 2D photocatalysts for solar energy conversion. Adv. Mater. **34** (2022). https://doi.org/10.1002/adma.202200180

36. Li, H., Liu, J., Wang, C., Yang, H., Xue, X.: Oxygen vacancies-enriched and porous hierarchical structures of ZnO microspheres with improved photocatalytic performance. Vacuum **199**, 110891 (2022). https://doi.org/10.1016/j.vacuum.2022.110891
37. Peymanfar, R., Ahmadi, A., Selseleh-Zakerin, E., Ghaffari, A., Mojtahedi, M.M., Sharifi, A.: Electromagnetic and optical characteristics of wrinkled Ni nanostructure coated on carbon microspheres. Chem. Eng. J. **405**, 126985 (2021). https://doi.org/10.1016/j.cej.2020.126985
38. Schneider, R., Facure, M.H.M., Chagas, P.A.M., Andre, R.S., dos Santos, D.M., Correa, D.S.: Tailoring the surface properties of micro/nanofibers using 0D, 1D, 2D, and 3D nanostructures: a review on post-modification methods. Adv. Mater. Interfaces **8** (2021). https://doi.org/10.1002/admi.202100430
39. Chong, W.J., Shen, S., Li, Y., Trinchi, A., Pejak, D., Kyratzis, I.L., Sola, A., Wen, C.: Additive manufacturing of antibacterial PLA-ZnO nanocomposites: benefits, limitations and open challenges. J. Mater. Sci. & Technol. **111**, 120–151 (2022). https://doi.org/10.1016/j.jmst.2021.09.039
40. Wang, Y., Fan, H., Liu, X., Cao, J., Liu, H., Li, X., Yang, L., Wei, M.: 3D ZnO hollow spheres-dispersed $CsPbBr_3$ quantum dots S-scheme heterojunctions for high-efficient Co_2 photoreduction. J. Alloy. Compd. **945**, 169197 (2023). https://doi.org/10.1016/j.jallcom.2023.169197
41. Zhou, X.-Q., Hayat, Z., Zhang, D.-D., Li, M.-Y., Hu, S., Wu, Q., Cao, Y.-F., Yuan, Y.: Zinc oxide nanoparticles: synthesis characterization, modification, and applications in food and agriculture. Processes **11**, 1193 (2023). https://doi.org/10.3390/pr11041193
42. Chakraborty, I., Pandey, A.: A review article on application of ZnO-based nanocomposite materials in environmental remediation. Mater. Today Proc. (2023). https://doi.org/10.1016/j.matpr.2023.02.041
43. Modi, G.: Zinc oxide tetrapod: a morphology with multifunctional applications. Adv. Nat. Sci Nanosci. Nanotechnol. **6**, 033002 (2015). https://doi.org/10.1088/2043-6262/6/3/033002
44. Abdallah, B., Kakhia, M., Obaide, A.: Morphological and structural studies of ZnO nanotube films using thermal evaporation technique. Plasmonics **16**, 1549–1556 (2021). https://doi.org/10.1007/s11468-021-01420-x
45. Hussain, S., Hassan, B., Kanwal, N., Pervaiz, S., Razzaq, M., Khan, I.: Structural and optical properties of Cu-Zno films deposited by thermal evaporation. J. Ovonic Res. **15**, 411–423 (2019)
46. Schwan, A.M., Chwatal, S., Hendler, C., Kopp, D., Lackner, J.M., Kaindl, R., Tscherner, M., Zirkl, M., Angerer, P., Friessnegger, B., Augl, S., Heim, D., Hinterer, A., Stummer, M., Waldhauser, W.: Morphology-controlled atmospheric pressure plasma synthesis of zinc oxide nanoparticles for piezoelectric sensors. Appl. Nanosci. **13**, 6421–6432 (2023). https://doi.org/10.1007/s13204-023-02936-w
47. Yan, L., Uddin, A., Wang, H.: ZnO tetrapods: synthesis and applications in solar cells. Nanomater. Nanotechnol. **5**, 19 (2015). https://doi.org/10.5772/60939
48. Srivastava, A.K., Tawale, J.S., Verma, R., Agarwal, D., Sharma, C., et al.: Morphological evolution driven semiconducting nanostructures for emerging solar, biological and nanogenerator applications. Mater. Adv. **3**, 8030–8062 (2022). https://doi.org/10.1039/d2ma00683a
49. Orudzhev, F., Muslimov, A., Selimov, D., Gulakhmedov, R.R., Lavrikov, A., et al.: Oxygen vacancies and surface wettability: key factors in activating and enhancing the solar photocatalytic activity of ZnO tetrapods. IJMS **24**, 16338 (2023). https://doi.org/10.3390/ijms242216338
50. Omran, M., Fabritius, T., Heikkinen, E.-P., Vuolio, T., Yu, Y., et al.: Microwave catalyzed carbothermic reduction of zinc oxide and zinc ferrite: effect of microwave energy on the reaction activation energy. RSC Adv. **10**, 23959–23968 (2020). https://doi.org/10.1039/d0ra04574h
51. Lei, M., Gao, M., Yang, X., Zou, Y., Alghamdi, A., Ren, Y., Deng, Y.: Size-controlled Au nanoparticles incorporating mesoporous ZnO for sensitive ethanol sensing. ACS Appl. Mater. Interfaces **13**, 51933–51944 (2021). https://doi.org/10.1021/acsami.1c07322

52. Qin, W., Yuan, Z., Gao, H., Zhang, R., Meng, F.: Perovskite-structured $LaCoO_3$ modified ZnO gas sensor and investigation on its gas sensing mechanism by first principle. Sens. Actuators B Chem. **341**, 130015 (2021). https://doi.org/10.1016/j.snb.2021.130015
53. Batterjee, M.G., Nabi, A., Kamli, M.R., Alzahrani, K.A., Danish, E.Y., Malik, M.A.: Green hydrothermal synthesis of zinc oxide nanoparticles for UV-light-induced photocatalytic degradation of ciprofloxacin antibiotic in an aqueous environment. Catalysts **12**, 1347 (2022). https://doi.org/10.3390/catal12111347
54. Rackauskas, S., Klimova, O., Jiang, H., Nikitenko, A., Chernenko, K.A., Shandakov, S.D., Kauppinen, E.I., Tolochko, O.V., Nasibulin, A.G.: A novel method for continuous synthesis of ZnO tetrapods. J. Phys. Chem. C **119**, 16366–16373 (2015). https://doi.org/10.1021/acs.jpcc.5b03702
55. Heo, S.-G., Jo, S.-I., Jeong, G.-H.: Revealing the enhanced photocatalytic properties of ZnO tetrapods produced by atmospheric-pressure microwave plasma jet system. Curr. Appl. Phys. **46**, 46–54 (2023). https://doi.org/10.1016/j.cap.2022.12.004
56. Dien, N.D.: Preparation of various morphologies of ZnO nanostructure through wet chemical methods. Adv. Mater. Sci. **4** (2019). https://doi.org/10.15761/ams.1000147
57. Doicin, I.E., Preda, M.D., Neacsu, I.A., Ene, V.L., Birca, A.C., Vasile, B.S., Andronescu, E.: Tailoring zinc oxide nanoparticles via microwave-assisted hydrothermal synthesis for enhanced antibacterial properties. Appl. Sci. **14**, 7854 (2024). https://doi.org/10.3390/app14177854
58. Shimpi, N.G., Jain, S., Karmakar, N., Shah, A., Kothari, D.C., Mishra, S.: Synthesis of ZnO nanopencils using wet chemical method and its investigation as LPG sensor. Appl. Surf. Sci. **390**, 17–24 (2016). https://doi.org/10.1016/j.apsusc.2016.08.050
59. Zhang, X., Han, Y., Liu, W., Pan, N., Li, D., Chai, J.: A novel synthesis of hexagonal cylinder-like ZnO with an excellent performance by a surfactant-free microemulsion-hydrothermal method. J. Ind. Eng. Chem. **97**, 326–336 (2021). https://doi.org/10.1016/j.jiec.2021.02.019
60. Madhubala, V., Kalaivani, T.: Phyto and hydrothermal synthesis of Fe_3O_4@ZnO core-shell nanoparticles using *Azadirachta indica* and its cytotoxicity studies. Appl. Surf. Sci. **449**, 584–590 (2018). https://doi.org/10.1016/j.apsusc.2017.12.105
61. Smeets, V., Styskalik, A., Debecker, D.P.: Non-hydrolytic sol–gel as a versatile route for the preparation of hybrid heterogeneous catalysts. J. Sol-Gel Sci. Technol. **97**, 505–522 (2021). https://doi.org/10.1007/s10971-021-05486-1
62. Bai, M., Chen, M., Li, X., Wang, Q.: One-step CVD growth of ZnO nanorod/SnO_2 film heterojunction for NO_2 gas sensor. Sens. Actuators B Chem. **373**, 132738 (2022). https://doi.org/10.1016/j.snb.2022.132738
63. Rodwihok, C., Rodwihok, C., Gardchareon, A., Phadungdhitidhada, S., Wongratanaphisan, D., Choopun, S.: Synthesis and characterization of vertically aligned ZnO nanowires by chemical vapor deposition. Thai J. Phys. **10**, 25001 (2014)
64. Zakria, H.S., Othman, M.H.D., Kamaludin, R., Sheikh Abdul Kadir, S.H., Kurniawan, T.A., Jilani, A.: Immobilization techniques of a photocatalyst into and onto a polymer membrane for photocatalytic activity. RSC Adv. **11**, 6985–7014 (2021). https://doi.org/10.1039/d0ra10964a
65. Ejsmont, A., Goscianska, J.: Hydrothermal synthesis of ZnO superstructures with controlled morphology via temperature and pH optimization. Materials **16**, 1641 (2023). https://doi.org/10.3390/ma16041641
66. Li, N., Yang, Y., Shi, Z., Lan, Z., Arramel, A., Zhang, P., Ong, W.-J., Jiang, J., Lu, J.: Shedding light on the energy applications of emerging 2D hybrid organic-inorganic halide perovskites. iScience **25**, 103753 (2022). https://doi.org/10.1016/j.isci.2022.103753
67. Gopinath, C.S., Nalajala, N.: A scalable and thin film approach for solar hydrogen generation: a review on enhanced photocatalytic water splitting. J. Mater. Chem. A **9**, 1353–1371 (2021). https://doi.org/10.1039/d0ta09619a
68. Dubey, R.S., Saravanan, S.: Electrochemical growth and characterisation of ZnO nanostructures for dye-sensitised solar cells. Micro. Nano. Lett. **15**, 840–844 (2020). https://doi.org/10.1049/mnl.2020.0251

69. Rather, A.H., Khan, R.S., Wani, T.U., Beigh, M.A., Sheikh, F.A.: Overview on immobilization of enzymes on synthetic polymeric nanofibers fabricated by electrospinning. Biotech Bioeng. **119**, 9–33 (2021). https://doi.org/10.1002/bit.27963
70. Boukhoubza, I., Matei, E., Jorio, A., Enculescu, M., Enculescu, I.: Electrochemical deposition of ZnO nanowires on CVD-graphene/copper substrates. Nanomaterials **12**, 2858 (2022). https://doi.org/10.3390/nano12162858
71. Wen, L., Xu, R., Mi, Y., Lei, Y.: Multiple nanostructures based on anodized aluminium oxide templates. Nature Nanotech. **12**, 244–250 (2016). https://doi.org/10.1038/nnano.2016.257
72. Manzano, C.V., Bürki, G., Pethö, L., Michler, J., Philippe, L.: Determining the diffusion mechanism for high aspect ratio ZnO nanowires electrodeposited into anodic aluminum oxide. J. Mater. Chem. C **5**, 1706–1713 (2017). https://doi.org/10.1039/c6tc04921d
73. Ma, L.A., Guo, T.L.: Morphology control and improved field emission properties of ZnO tetrapod films deposited by electrophoretic deposition. Ceram. Int. **39**, 6923–6929 (2013). https://doi.org/10.1016/j.ceramint.2013.02.027
74. Kitano, M., Hamabe, T., Maeda, S., Okabe, T.: Growth of large tetrapod-like ZnO crystals. J. Cryst. Growth **102**, 965–973 (1990). https://doi.org/10.1016/0022-0248(90)90867-k
75. Li, W.-J., Shi, E.-W., Zhong, W.-Z., Yin, Z.-W.: Growth mechanism and growth habit of oxide crystals. J. Cryst. Growth **203**, 186–196 (1999). https://doi.org/10.1016/s0022-0248(99)00076-7
76. Lupan, O., Chow, L., Chai, G.: A single ZnO tetrapod-based sensor. Sens. Actuators B Chem. **141**, 511–517 (2009). https://doi.org/10.1016/j.snb.2009.07.011
77. Wang, B.-B., Xie, J.-J., Yuan, Q., Zhao, Y.-P.: Growth mechanism and joint structure of ZnO tetrapods. J. Phys. D Appl. Phys. **41**, 102005 (2008). https://doi.org/10.1088/0022-3727/41/10/102005
78. Lin, H.-F., Liao, S.-C., Hu, C.-T.: A new approach to synthesize ZnO tetrapod-like nanoparticles with DC thermal plasma technique. J. Cryst. Growth **311**, 1378–1384 (2009). https://doi.org/10.1016/j.jcrysgro.2008.12.033
79. Zanotti, L., Calestani, D., Villani, M., Zha, M., Zappettini, A., Paorici, C.: Vapour-phase growth, purification and large-area deposition of ZnO tetrapod nanostructures. Cryst. Res. Technol. **45**, 667–671 (2010). https://doi.org/10.1002/crat.201000081
80. Zhang, X.-D., Luo, R.: The Laplacian eigenvalues of mixed graphs. Linear Algebra Appl. **362**, 109–119 (2003). https://doi.org/10.1016/s0024-3795(02)00509-8
81. Mishra, Y.K., Kaps, S., Schuchardt, A., Paulowicz, I., Jin, X., et al.: Fabrication of macroscopically flexible and highly porous 3D semiconductor networks from interpenetrating nanostructures by a simple flame transport approach. Part. Part. Syst. Charact. **30**, 775–783 (2013). https://doi.org/10.1002/ppsc.201300197
82. Ilickas, M., Mardosaite, R., Cesano, F., Cravanzola, S., Barolo, C., Scarano, D., Viscardi, G., Rackauskas, S.: ZnO tetrapod morphology influence on UV sensing properties. Nanotechnology **35**, 015502 (2023). https://doi.org/10.1088/1361-6528/acfcbf
83. Krasnova, V.V., Muslimov, A.E., Lavrikov, A.S., Zadorozhnaya, L.A., Orudzhev, F.F., et al.: Characterization and photocatalytic properties of ZnO tetrapods synthesized by high-temperature pyrolysis. Crystallogr. Rep. **69**, 439–445 (2024). https://doi.org/10.1134/s1063774524600212
84. Ciobanu, V., Ursaki, V.V., Reimers, A., Mihai, G., Zalamai, V.V., et al.: Controlling hydrophobic/hydrophilic properties of ZnO microtetrapods structures by means of thermal treatment. IFMBE Proc. 284–292 (2023). https://doi.org/10.1007/978-3-031-42775-6_32
85. Balasubramanian, K., Kumar Pandey, A., Abolhassani, R., Rubahn, H.-G., Rahman, S., Kumar Mishra, Y.: Tetrapods based engineering of organic phase change material for thermal energy storage. Chem. Eng. J. **462**, 141984 (2023). https://doi.org/10.1016/j.cej.2023.141984
86. Poschmann, M.P.M., Siebert, L., Lupan, C., Lupan, O., Schütt, F., et al.: Surface conversion of ZnO tetrapods produces pinhole-free ZIF-8 layers for selective and sensitive H2 sensing even in pure methane. ACS Appl. Mater. Interfaces **15**, 38674–38681 (2023). https://doi.org/10.1021/acsami.3c06317

87. Röhr, J.A., Sá, J., Konezny, S.J.: The role of adsorbates in the green emission and conductivity of zinc oxide. Commun Chem **2**, 52 (2019). https://doi.org/10.1038/s42004-019-0153-0
88. Wu, M., Yu, S., He, L., Yang, L., Zhang, W.: High quality transparent conductive Ag-based barium stannate multilayer flexible thin films. Sci. Rep. **7**, 103 (2017). https://doi.org/10.1038/s41598-017-00178-9
89. Giubileo, F., Faella, E., Kumar, A., De Stefano, S., Viscardi, L., et al.: Zinc oxide tetrapods as novel field emitters with low turn-on voltage. Nano Ex. **5**, 045017 (2024). https://doi.org/10.1088/2632-959x/ad9c9f
90. Zhang, Z., Nan, Y., Mishra, Y.K., Willatzen, M., Wang, Z.L.: Understanding the piezoelectric response of ZnO nanotetrapods: detailed numerical calculations. Appl. Phys. Lett. **123** (2023). https://doi.org/10.1063/5.0154454
91. Nguyen, X.-S., Nguyen, M.-Q., Trinh, X.-T., Joita, A.C., Nistor, S.V.: Correlation of native point defects and photocatalytic activity of annealed ZnO nanoparticle studied by electron spin resonance and photoluminescence emission. Semicond. Sci. Technol. **35**, 095035 (2020). https://doi.org/10.1088/1361-6641/aba168
92. Bayan, S.: Recent strategies for development of ZnO-based efficient UV-photodetectors. Adv. Struct. Mater. 165–179 (2024). https://doi.org/10.1007/978-981-99-7848-9_8
93. Kaur, D., Bharti, A., Sharma, T., Madhu, C.: Dielectric properties of ZnO-based nanocomposites and their potential applications. Int. J. Opt. **2021**, 1–20 (2021). https://doi.org/10.1155/2021/9950202
94. Jin, X., Deng, M., Kaps, S., Zhu, X., Hölken, I., et al.: Study of tetrapodal ZnO-PDMS composites: a comparison of fillers shapes in stiffness and hydrophobicity improvements. PLoS ONE **9**, e106991 (2014). https://doi.org/10.1371/journal.pone.0106991
95. Veys, E., Makower, L., Williamson, M., Saure, L.M., Adelung, R., Schütt, F., Pugno, N.M., Marrow, T.J.: In situ observation of compressive deformation of an interconnected network of zinc oxide tetrapods. Scripta Mater. **224**, 115153 (2023). https://doi.org/10.1016/j.scriptamat.2022.115153
96. Tamulevičienė, A., Mardosaitė, R., Ilickas, M., Abakevičienė, B., Tamulevičius, T., et al.: Highly-hydrophobic, transparent, and durable coatings based on ZnO tetrapods with diamond-like carbon nanocomposite. Surf. Coat. Technol. **470**, 129863 (2023). https://doi.org/10.1016/j.surfcoat.2023.129863
97. Karthikeyan, L., Suma, D.D., Nair, S.G., Ratnam, P., Mathew, D., Devasia, R.: Zinc oxide tetrapod-based thermally conducting epoxy systems for aerospace applications. Trans Indian Natl. Acad. Eng. **6**, 71–77 (2020). https://doi.org/10.1007/s41403-020-00179-7
98. Singh, H.H., Khare, N.: Flexible ZnO-PVDF/PTFE based piezo-tribo hybrid nanogenerator. Nano Energy **51**, 216–222 (2018). https://doi.org/10.1016/j.nanoen.2018.06.055
99. Fakhri, P., Eaianli, N., Bagherzadeh, R., Jaleh, B., Kashfi, M., Fausto, R.: Sandwich-type double-layer piezoelectric nanogenerators based on one-and two-dimensional ZnO nanostructures with improved output performance. Sci. Rep. **13** (2023). https://doi.org/10.1038/s41598-023-43047-4
100. Terracciano, M., Račkauskas, S., Falanga, A.P., Martino, S., Chianese, G., et al.: ZnO tetrapods for label-free optical biosensing: physicochemical characterization and functionalization strategies. IJMS **24**, 4449 (2023). https://doi.org/10.3390/ijms24054449
101. Marouf, S., Beniaiche, A., Guessas, H., Azizi, A.: Morphological, structural and optical properties of ZnO thin films deposited by dip coating method. Mat. Res. **20**, 88–95 (2016). https://doi.org/10.1590/1980-5373-mr-2015-0751
102. Haque, M.J., Bellah, M.M., Hassan, M.R., Rahman, S.: Synthesis of ZnO nanoparticles by two different methods & comparison of their structural, antibacterial, photocatalytic and optical properties. Nano Ex. **1**, 010007 (2020). https://doi.org/10.1088/2632-959x/ab7a43
103. Jayavelu, Y., Maharana, G., Rajender, G., Muniramaiah, R., Divyadharshini, S., et al.: Defect-mediated time-efficient photocatalytic degradation of methylene blue and ciprofloxacin using tungsten-incorporated ternary perovskite BaSnO3 nanoparticles. Chemosphere **351**, 141128 (2024). https://doi.org/10.1016/j.chemosphere.2024.141128

104. Villani, M., Rossi, F., Calestani, D., Salviati, G., Fabbri, F.: Evaluating the plasmon-exciton interaction in ZnO tetrapods coupled with gold nanostructures by nanoscale cathodoluminescence. Nano Ex. **2**, 014004 (2021). https://doi.org/10.1088/2632-959x/abe277
105. Padmanaban, D.B., Maguire, P., Mariotti, D.: Non-equilibrium defect chemistry in oxygen-rich zinc oxide nano-tetrapods synthesized using atmospheric pressure microplasma. J. Mater. Chem. A **12**, 9212–9231 (2024). https://doi.org/10.1039/d3ta06821h
106. Naghani, M.E., Neghabi, M., Zadsar, M., Ahangar, H.A.: Synthesis and characterization of linear/nonlinear optical properties of graphene oxide and reduced graphene oxide-based zinc oxide nanocomposite. Sci. Rep. **13** (2023). https://doi.org/10.1038/s41598-023-28307-7
107. Jo, W.-K., Tayade, R.J.: New generation energy-efficient light source for photocatalysis: LEDs for environmental applications. Ind. Eng. Chem. Res. **53**, 2073–2084 (2014). https://doi.org/10.1021/ie404176g
108. Tarasov, A.P., Lavrikov, A.S., Zadorozhnaya, L.A., Kanevsky, V.M.: Low-threshold whispering-gallery mode lasing in large-diameter ZnO microrods. Jetp Lett. **115**, 502–508 (2022). https://doi.org/10.1134/s0021364022100514
109. Zadorozhnaya, L.A., Tarasov, A.P., Kanevsky, V.M.: The growth mechanism luminescence, and lasing of polyhedral ZnO microcrystals with whispering-gallery modes. Photonics **10**, 1328 (2023). https://doi.org/10.3390/photonics10121328
110. Jali, M.H., Rahim, H.R.A., Md Ashadi, M.J., Thokchom, S., Harun, S.W.: Applied microfiber evanescent wave on ZnO nanorods coated glass surface towards temperature sensing. Sens. Actuators A Phys. **277**, 103–111 (2018). https://doi.org/10.1016/j.sna.2018.05.003
111. Chu, S., Zhao, J., Zuo, Z., Kong, J., Li, L., Liu, J.: Enhanced output power using MgZnO/ZnO/MgZnO double heterostructure in ZnO homojunction light emitting diode. J. Appl. Phys. **109** (2011). https://doi.org/10.1063/1.3598136
112. Deep, R., Yoshida, T., Fujita, Y.: Full-color electroluminescence from ZnO nanoparticles-based homojunction diodes. Phys. Rapid Res. Ltrs **18** (2024). https://doi.org/10.1002/pssr.202400149
113. Olivo-Alanís, D., García-González, A., Mueses, M.A., García-Reyes, R.B.: Generalized kinetic model for the photocatalytic degradation processes: validation for dye wastewater treatment in a visible-LED tubular reactor. Appl. Catal. B **317**, 121804 (2022). https://doi.org/10.1016/j.apcatb.2022.121804
114. Mishra, Y.K., Modi, G., Cretu, V., Postica, V., Lupan, O., et al.: Direct growth of freestanding ZnO tetrapod networks for multifunctional applications in photocatalysis, UV photodetection, and gas sensing. ACS Appl. Mater. Interfaces **7**, 14303–14316 (2015). https://doi.org/10.1021/acsami.5b02816
115. Dilonardo, E., Penza, M., Alvisi, M., Di Franco, C., Palmisano, F., Torsi, L., Cioffi, N.: Evaluation of gas-sensing properties of ZnO nanostructures electrochemically doped with Au nanophases. Beilstein J. Nanotechnol. **7**, 22–31 (2016). https://doi.org/10.3762/bjnano.7.3
116. Pearton, S.J., Ren, F.: Advances in ZnO-based materials for light emitting diodes. Curr. Opin. Chem. Eng. **3**, 51–55 (2014). https://doi.org/10.1016/j.coche.2013.11.002
117. Hu, J., Ma, H., Zhou, Y., Ma, L., Zhao, S., Shi, S., Li, J., Chang, Y.: Gas-sensing properties and mechanisms of 3D networks composed of ZnO tetrapod micro-nano structures at room temperature. Materials **17**, 203 (2023). https://doi.org/10.3390/ma17010203
118. Seekaew, Y., Pon-On, W., Wongchoosuk, C.: Ultrahigh selective room-temperature ammonia gas sensor based on tin-titanium dioxide/reduced graphene/carbon nanotube nanocomposites by the solvothermal method. ACS Omega **4**, 16916–16924 (2019). https://doi.org/10.1021/acsomega.9b02185
119. Hasanzadeh, M., Ghahhari, M.R., Bidoki, S.M.: Enhanced piezoelectric performance of PVDF-based electrospun nanofibers by utilizing in situ synthesized graphene-ZnO nanocomposites. J. Mater. Sci. Mater. Electron. **32**, 15789–15800 (2021). https://doi.org/10.1007/s10854-021-06132-w
120. Maghsodian, Z., Sanati, A.M., Mashifana, T., Sillanpää, M., Feng, S., Nhat, T., Ramavandi, B.: Occurrence and distribution of antibiotics in the water, sediment, and biota of freshwater and marine environments: a review. Antibiotics **11**, 1461 (2022). https://doi.org/10.3390/antibiotics11111461

121. Jayababu, N., Kim, D.: ZnO nanorods@conductive carbon black nanocomposite based flexible integrated system for energy conversion and storage through triboelectric nanogenerator and supercapacitor. Nano Energy **82**, 105726 (2021). https://doi.org/10.1016/j.nanoen.2020.105726
122. Kamaruzaman, N.A., Wan Mohamed Zin, W.M.K., Kamarudin, K.H., Saleh, N.M., Yusoff, F.: Recent advances in transition metals- based materials as electrocatalysts for water splitting. Int. J. Electrochem. Sci. **18**, 100187 (2023). https://doi.org/10.1016/j.ijoes.2023.100187
123. Sayed, M., Xu, F., Kuang, P., Low, J., Wang, S., Zhang, L., Yu, J.: Sustained CO_2-photoreduction activity and high selectivity over Mn, C-codoped ZnO core-triple shell hollow spheres. Nat. Commun. **12** (2021). https://doi.org/10.1038/s41467-021-25007-6
124. Anjum, S., Hashim, M., Malik, S.A., Khan, M., Lorenzo, J.M., Abbasi, B.H., Hano, C.: Recent advances in zinc oxide nanoparticles (ZnO NPs) for cancer diagnosis. Target Drug Deliv. Treat. Cancers **13**, 4570 (2021). https://doi.org/10.3390/cancers13184570
125. Martínez-Carmona, M., Gun'ko, Y., Vallet-Regí, M.: ZnO nanostructures for drug delivery and theranostic applications. Nanomaterials **8**, 268 (2018). https://doi.org/10.3390/nano8040268
126. Tereshchenko, A., Bechelany, M., Viter, R., Khranovskyy, V., Smyntyna, V., et al.: Optical biosensors based on ZnO nanostructures: advantages and perspectives. A Rev. Sens. Actuators B Chem. **229**, 664–677 (2016). https://doi.org/10.1016/j.snb.2016.01.099

Dr. D. Paul Joseph presently serving as an Associate Professor, Dept. of Physics, NIT-Warangal completed M. Sc (Loyola College, Chennai), B. Ed, M. Phil, and Ph. D from the Univ. of Madras, India. He was a post-doctoral fellow at National Taiwan Univ., Taipei (2010–2013). His research areas are alternative transparent conducting thin film electrodes, Proximity effect between Superconductor/Half-metallic Ferromagnet, alternative components for dye-sensitized solar cells, nanomaterials for batteries, swift heavy-ion-beam irradiation effects and photocatalytic activity for pollutant degradation. He has 07 contributed book chapters, 80 articles in journals of repute. Five of his X-Ray diffraction data has been included in the ICDD database by invitation. As a PI/Co-PI he has completed several funded research projects (SERB/ANRF, TEQIP-III, NPIU, etc.) and a SPARC project is ongoing. He has several academic laurels to his credit; a notable one is 'Young Scientist Award' (2017) by 'Academy of Sciences-Chennai'. He has co-organized 01 International conference, 09 workshops/FDPs, out of which 02 are National level programs each of one-month duration. He has guided 03 Ph. D scholars and 28 students for their M. Sc (Tech) and M. Tech projects. He has designed several indigenous equipment (automated spray pyrolysis unit, standalone dip-coating unit, high temperature VSM rod adaptor import substitute, Bending-Twisting unit to study flexible films) and real time YBCO and bismuth levitation kits as a teaching aid for students.

Prof. Dr. Yogendra Kumar Mishra is Professor (Full) at Mads Clausen Institute, NanoSYD, University of Southern Denmark (SDU), Denmark. Prior joining to SDU, he worked as Privatdozent/Group Leader at Kiel University, Germany. He did Habilitation (Dr. habil.) in Materials Science from Kiel University in 2015 and Ph. D. in Physics in 2008 from Jawaharlal Nehru University (Inter University Accelerator Centre), New Delhi, India. In Kiel, he introduced a new flame-based process for metal oxide tetrapod nanostructuring and their 3D networks which showed many applications in engineering and biomedical fields. At NanoSYD, he is heading 'Smart Materials' group with the focus to develop new multifunctional 3D nanomaterials for green and sustainable technologies. He has published more than 400 publications which are cited more than 25000 with h-index of over 83. He has served as an Editorial Board Member/Associate Editor for several high impact factor scientific magazines, such as, Advanced Functional Materials, Materials Today Electronics, Critical Reviews in Biotechnology, Energy Materials, Nanotechnologies, Nano Energy, etc., He is Humboldtian, honored with FRSC- Fellow of Royal Society of Chemistry (UK) and has received BHJ Young Innovator award (Denmark). He has attracted significant grants from funding agencies worldwide. He has supervised over 10 Ph. D. students, over 40 M. Sc. students, and over 10 B. Sc. students. He has been teaching Solid State Physics, Semiconductor Optoelectronics, Nanoelectronics, etc. at M. Sc. and B. Sc. levels. He is sitting in the advisory board members at funding agencies, institutes, policy makers, and also conferences. He has delivered over 200 speeches (including several plenary) in the conferences worldwide. He is strong advocator for translational nanomaterials research to bridge the gap from Lab to daily-life technologies. From last decade, he has put a lot of emphasis on AMIL-Advanced Materials for Industrial Leadership to address the future Sustainable Development Goals (SDGs).

A Comprehensive Study on Tetrapod-Structured ZnO for Efficient Photodetector Applications

Joel K. Joseph, Thyda Lavanya, Koppula Naresh, S. Suneetha, and Kuppusamy Thangaraju

Abstract Zinc oxide (ZnO) is a highly versatile semiconducting metal oxide recognized for its remarkable properties including high photosensitivity, tunable specific surface area, nontoxicity, piezo- and pyro-electric effects, wide bandgap (~3.4 eV), large exciton binding energy, and ~85% optical transparency in the visible spectrum. These attributes have made ZnO a focal point in developing the ultraviolet (UV) photodetectors which are increasingly sought after for applications in industrial, environmental, and biological fields. Among the various nano- and microstructures, tetrapod-structures of ZnO have emerged as promising materials for UV photodetection due to their high surface area-to-volume ratio, enhanced charge transport properties, and efficient electron-hole separation capabilities. ZnO tetrapods are typically synthesized using methods such as gas-phase oxidation and vapor transport techniques. Studies indicate that ZnO tetrapod based sensors can exhibit responsivity up to three orders of magnitude higher than that of commercial ZnO nanoparticles. The ZnO tetrapod-based UV photodetectors hold significant promise for diverse applications including environmental monitoring of UV radiation levels, UV dosimetry in healthcare, and optical communication systems. Future research directions will likely focus on enhancing quantum efficiency, stability, and the integration of ZnO tetrapods into flexible and wearable photodetection devices. The size of ZnO tetrapods is critical parameter for achieving quantum confinement effect. For ZnO tetrapods, effective quantum confinement is typically observed within a size range of 10 nm to 50 nm. Tetrapods with leg lengths around 100 nm may not exhibit significant quantum confinement effect, thus, minimizing dimensions while maintaining structural integrity is essential. This overview highlights recent advancements in the synthesis, characterization, and applications of ZnO tetrapods specifically for UV detection, emphasizing their advantages over conventional ZnO nanoparticles and other materials.

J. K. Joseph · T. Lavanya · K. Naresh · S. Suneetha · K. Thangaraju (✉)
Organic Optoelectronics Device Laboratory, DepartmentofPhysics, National Institute of Technology, Warangal, India
e-mail: ktr@nitw.ac.in

K. Thangaraju et al. (eds.), *ZnO Tetrapods*, Engineering Materials,
https://doi.org/10.1007/978-981-95-1614-8_2

1 Introduction

Optoelectronics, an interdisciplinary domain encompasses the study, design, and fabrication of devices capable of converting electrical energy into light and vice versa, fundamentally leveraging semiconductor materials characterized by their unique electronic properties. The optoelectronic devices include an array of components such as light-emitting diodes (LEDs), photodiodes, and solar cells, exploiting the photovoltaic effect—an interaction between electromagnetic radiation and matter that facilitates the generation of charge carriers upon absorption of photons [1]. The burgeoning field of optoelectronics is pivotal in numerous applications spanning telecommunications, medical instrumentation, and military technologies, owing to its ability to enhance data transmission rates through optical fibers and improve the efficiency of energy conversion processes [2]. Semiconductor photodetectors are the key component of optoelectronic technology, enabling the precise and efficient conversion of optical signals into electrical currents. The advanced photodetector devices are based on the photovoltaic effect which occurs when light photons interact with semiconductor materials such as silicon or compound semiconductors such as gallium arsenide, resulting in the formation of electron–hole pairs and the flow of photocurrent [1, 3]. Semiconductor photodetectors have a wide range of design and operational characteristics which are crucial in today's technology landscapes, pushing advancements in optical communication, environmental monitoring, and scientific research [4]. The primary types of effects observed in photodetectors include the photovoltaic effect, photoconductive, and thermal effect [5, 6]. Photovoltaic effect is the fundamental concept behind many photodetectors notably photodiodes and solar cells. When photons impact a semiconductor material, they can sufficiently energize electrons, resulting in the formation of electron–hole pairs. The efficiency of this process is defined by factors such as quantum efficiency which measures the ratio of produced charge carriers to incident photons. In contrast to the photovoltaic effect, the photoconductive effect involves a change in the electrical conductivity of a material upon exposure to light [7, 8]. This effect is exploited in devices like photoconductors where the absorption of light increases the number of charge carriers, thereby enhancing conductivity. The response time and sensitivity of these devices can be influenced by factors such as temperature and applied bias. Thermal detectors operate on the principle that incident light raises the temperature of a material, leading to a measurable change in resistance or voltage. Although generally slower and less sensitive than their photoelectric counterparts, thermal detectors are effective for infrared detection applications. They convert absorbed light into heat, which subsequently alters the electrical properties of the detector material.

Semiconductor photodetectors are classified on their operational principles, structures, and specific applications into photodiodes, Phototransistors, Metal–Semiconductor-Metal (MSM) Photodetectors, Thermal Detectors, and Quantum Dot (QD) Photodetectors. (1) Photodiodes are the most common type of photodetectors, functioning primarily through the photovoltaic effect. They consist of a p–n junction that generates electron–hole pairs when exposed to light. They are further divided into (i) Standard Photodiodes: Basic p–n junction devices that convert light into electrical current, (ii) p-i-n Photodiodes: which incorporate an intrinsic (undoped) layer between the p-type and n-type regions, enhancing performance by extending the depletion region and improving response times, (iii) Schottky Photodiodes: Utilizing a metal–semiconductor junction instead of a p–n junction, these devices offer high-speed response and are suitable for high-frequency applications, and (iv) Avalanche Photodiodes (APDs): Operated under reverse bias close to breakdown voltage, APDs utilize avalanche multiplication to achieve high sensitivity and gain, making them ideal for low-light applications. (2) Phototransistors combine the functionalities of a photodiode and a transistor, allowing them to amplify the photocurrent generated by incident light. They possess three terminals—emitter, base, and collector—and are particularly useful in applications requiring both light detection and signal amplification. (3) Metal–Semiconductor-Metal (MSM) photodetectors consist of a semiconductor layer sandwiched between two metal electrodes, forming interdigitated contacts. This design allows for rapid response times and high bandwidth capabilities, making them suitable for applications in high-speed optical communications. (4) Thermal detectors measure changes in temperature resulting from absorbed light although not semiconductor-based in the traditional sense. They include devices such as bolometers and pyroelectric detectors which convert absorbed radiation into heat, leading to measurable changes in resistance or voltage. These detectors are particularly effective for infrared detection but generally exhibit slower response times compared to semiconductor photodetectors. (5) Quantum dot photodetectors utilize quantum dots—nanoscale semiconductor particles—to achieve sensitivity across a broad wavelength range, including visible and infrared light. Their unique properties allow for efficient light absorption and conversion, making them promising candidates for next-generation photodetection technologies [7, 9].

2 UV Photodetectors

Ultraviolet (UV) photodetectors are essential devices in the realm of optoelectronics, specifically designed to detect UV radiation for a range of applications including environmental monitoring, industrial safety, and biomedical diagnostics (Fig. 1) [10, 11]. These advanced sensors leverage the unique characteristics of wide band gap semiconductor materials such as gallium nitride (GaN), zinc oxide (ZnO), and silicon carbide (SiC) which enable them to function effectively within the UV spectrum while maintaining high sensitivity and rapid response times [12, 13]. One of the key advantages of UV photodetectors is their ability to operate in a "visible-blind" mode, allowing them to accurately measure UV radiation without interference from visible light. Recent innovations in nanostructured materials and hybrid device designs have significantly improved the performance and versatility of these detectors [14–18]. As advancements in technology continue to evolve, UV photodetectors are increasingly recognized for their critical role in enhancing safety, health monitoring, and environmental sustainability.

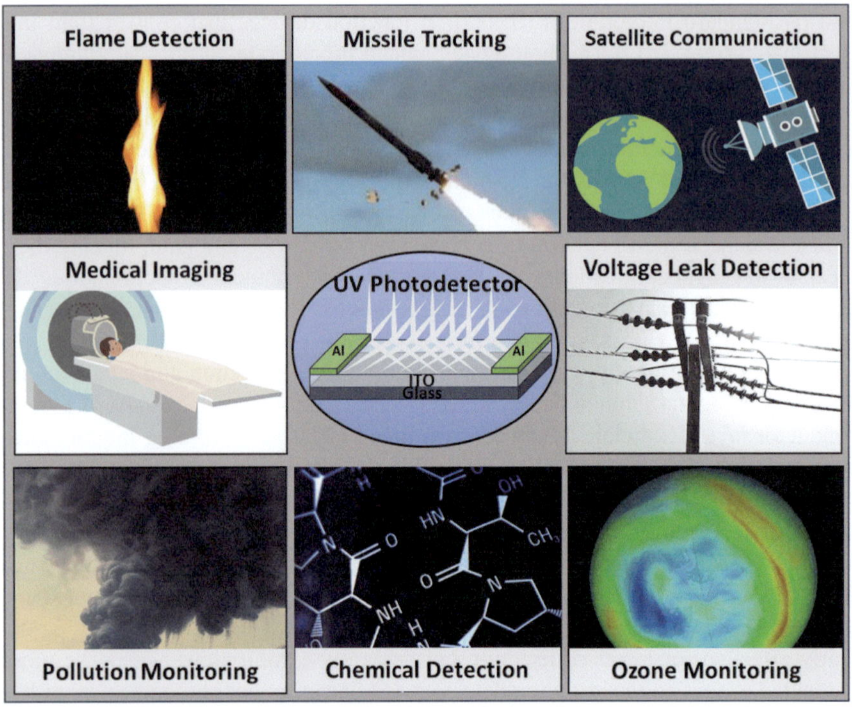

Fig. 1 Various applications of UV photodetectors in different areas of daily life

2.1 History of UV Photodetectors

Ultraviolet photodetectors (UVPDs) are essential in a wide array of applications including engine control systems, solar UV monitoring, astronomy, lithography alignment, secure space communications, and missile detection [19–22]. In recent years, there has been growing interest in UVPDs driven by new requirements from both civil and military sectors for advanced UV instrumentation that can function effectively in extreme temperatures and challenging environments. The UV segment of the electromagnetic spectrum spans the wavelength range from 10 to 400 nm and is commonly categorized into three bands: UVA (400–320 nm), UVB (320–280 nm), and UVC (280–100 nm) [23].

Historically, silicon (Si) has been the primary material used for fabricating the traditional UV photodetectors due to its well-established processing techniques and cost-effectiveness [24]. To address the challenges of Si based photodiodes, researchers have turned their attention to wide bandgap semiconductors such as ZnSe, ZnS, ZnTe, CdSe, CdS, ZnO, TiO_2, SnO_2, GaN, SiC, Ga_2O_3, and diamond for photodetector fabrication [25–29]. These materials exhibit intrinsic solar-blind characteristics and do not require additional filters or refrigeration. The shift toward wide band gap (WBG) semiconductor-based photodetectors marks a significant improvement in achieving high sensitivity and high performance for demanding applications [30, 31]. Semiconductors with band gaps > 2 eV such as SiC and GaN, are gaining recognition for their transformative potential across various applications particularly in power electronics and optoelectronics [30, 32]. WBGs boast superior electrical and thermal properties, allowing them to operate at higher temperatures, voltages, and frequencies compared to traditional silicon-based semiconductors. In addition to their applications in power electronics, WBG semiconductors play a crucial role in optoelectronic devices like UV photodetectors and light-emitting diodes (LEDs) [33, 34]. Their robustness in harsh environments makes them suitable for use in aerospace, military, and industrial applications. The operating principle of UV PDs primarily relies on: when UV light is absorbed by a wide band gap semiconductor with photon energy equal to or higher than band gap energy (Eg), electron–hole (e–h +) pairs are generated. The resulting photoexcited electrons and holes are then collected as photocurrent in an external circuit [35, 36]. It consists of three main steps: first, the absorption of UV energy generates e–h + pairs; second, these charge carriers are separated through a transport mechanism; and third, the electrons and holes are collected as photocurrent. The separation of these photogenerated carriers can be achieved either by applying an external bias voltage or current in conventional designs or by creating an internal built-in electric field in self-powered configurations.

2.2 Characteristics of Photodetector

Dark current (I_{dark}) is the residual electric current that persists in a photodetector under zero illumination conditions. This undesired signal arises from thermal excitations, defects, or intrinsic carrier generation, which can degrade device performance by increasing noise levels. This phenomenon arises from thermally generated charge carriers within the semiconductor material. High levels of dark current can lead to increased noise and a reduced signal-to-noise ratio (SNR), adversely affecting the overall performance of the detector. Several mechanisms contribute to dark current in UV photodetectors [37]. Dark current may be a major issue for materials with a much narrower band gap materials which may compel the operation at much lower temperatures. As a result, certain mid-infrared cameras for instance require a stirling cooler in order to function at temperatures of 100 K or below [38]. One primary source is thermal generation at elevated temperatures. Thermal energy can excite electrons across the band gap, resulting in dark current. Additionally, surface leakage currents may arise from imperfections at the surface or interface of the semiconductor material, particularly in devices with high surface area-to-volume ratios such as nanostructured materials. Managing dark current is crucial for enhancing the sensitivity and reliability of UV photodetectors [39, 40] (Fig. 2). Furthermore, the device's recombination is gauged by the dark current. Large dark currents are found in diodes with high recombination. To minimize dark current, a common strategy involves incorporating a thin charge-blocking layer with a high LUMO level and low mobility of electrons along with reduction of extrinsic defects and trap states in the photoactive layer helping in suppression of unwanted charge injection in the absence of light [41, 42]. On the other hand, the high current produced by the e–h pairs generated under irradiation is known as the photocurrent (I_{ph}). The importance of trap-assisted charge production was emphasized by Chauhan et al. who also showed how maximizing trap concentrations may lower dark current and increase detectivity [43]. The study highlights the interaction of charge-transfer states and traps as crucial elements in the production of dark currents. Wei Tsai et al. demonstrated that incorporating cross-linkable naphthalenediimide (c-NDI) as an interlayer in shortwave infrared (SWIR) organic photodetectors (OPDs) leads to notable performance enhancements. Specifically, c-NDI effectively suppresses I_{dark} and boosts EQE, especially in ultra-narrow band gap systems with pronounced hole-blocking ability. Moreover, devices with c-NDI interlayers exhibit superior long-term stability compared to those using conventional zinc oxide interlayers, emphasizing c-NDI's potential as a robust and adaptable interfacial material for next-generation SWIR photodetectors [44].

Sensitivity refers to a photodetector's capability to detect faint UV signals and is quantified as the ratio of photocurrent (I_{ph}) to the dark current (I_{dark}). Huang et al. reported an exceptionally high sensitivity of 96,000 μV/W in 4H-SiC-based photodetectors when exposed to 365 nm UV light. This enhanced performance was achieved through the transverse Dember effect, which facilitates efficient charge carrier separation [45].

$$S = \frac{\mathrm{Iph}}{\mathrm{Idark}} \tag{1}$$

Fig. 2 Various characteristics of UV photodetectors

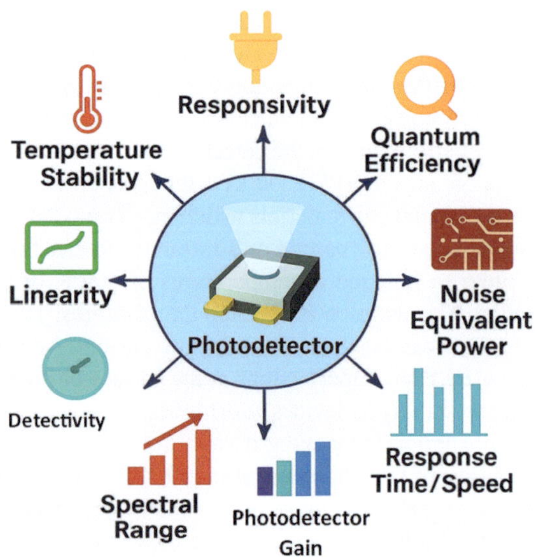

The responsivity (R) which is measured in A W^{-1} or VW^{-1} units, is calculated by dividing the output photocurrent or photovoltaic capacity by the input light power on the photodetector's active region. Han et al. developed a high-performance UV photodetector using amorphous Ga_2O_3, operating under a 25 V bias and illuminated with 240 nm ultraviolet light [46]. At a pressure of 0.5 Pa, the device achieved an impressive photoresponsivity of 436.3 A/W, approaching the peak values typically associated with single-crystal β-Ga_2O_3 fabricated at elevated temperatures. Furthermore, when tested at 1.2 Pa, the detector exhibited an exceptionally high I_{uv}/I_{dark} ratio of 1.07×10^5 and a notably low dark current of 4.9 nA, both measured at the same bias voltage. These performance metrics are on par with more complex MOSFET-based detectors built from crystalline β-Ga_2O_3, highlighting the potential of the amorphous form for efficient and simplified UV detection systems.

$$R = \frac{Iph}{Pin} \tag{2}$$

A higher responsivity indicates that the photodetector can generate a larger photocurrent from a given amount of incident light which is especially important for applications requiring high sensitivity. The choice of semiconductor material significantly affects responsivity such as ZnO and GaN, typically demonstrate higher responsivity due to their effective photon absorption capabilities [47–49]. Additionally, the design and structure of the photodetector including its thickness and layer configuration—play a crucial role in determining how much light is absorbed and how efficiently charge carriers are generated [50, 51]. Furthermore, responsivity varies with wavelength because different semiconductor materials exhibit unique intrinsic properties that affect their ability to absorb light at specific wavelengths.

The response time is a critical parameter that determines a photodetector's ability to track rapid changes in light intensity. This performance metric consists of two key components: rise time and fall time. Rise time refers to the time required for a photodetector's output signal to increase from 10 to 90% of its maximum value upon light exposure. Conversely, fall time represents the duration needed for the signal to decrease from 90 to 10% of peak intensity after the light source is turned off. These parameters characterize the detector's speed in responding to sudden changes in illumination, with shorter times indicating faster response capabilities. The MgZnO thin film/p-Si photodetector fabricated by Kumar et al. demonstrated rapid response characteristics when exposed to 325 nm UV light. Operating at a + 2 V bias, the device achieved notably fast rise and fall times of 0.26 s and 0.25 s, respectively under optimized growth conditions at 500 °C [52]. Transient response photodetectors are often appropriate for applications requiring precise time precision and for the detection of rapidly changing light. For applications like safety systems or high-speed communication technologies that need to detect UV signals quickly, a quicker response time is crucial. The response time can be influenced by several factors, including charge carrier mobility and device capacitance. Higher mobility facilitates faster movement of charge carriers through the semiconductor material, leading to quicker response times. Additionally, the capacitance associated with the device structure affects how rapidly it can charge and discharge when exposed to variations in light intensity; lower capacitance generally results in faster response times.

One important performance metric for comparing photodetectors of various materials and forms is detectivity. Detectivity measures a photodetector's sensitivity to weak light signals by accounting for both its responsivity and noise levels. It evaluates how well the device can detect faint light against its intrinsic noise. Since the noise current increases with the square root of the detector's active area, detectivity becomes independent of device size when properly normalized. 30–40 nm Ag_2S nanocrystals were created via a simple room-temperature synthesis. At normal temperature, the manufactured photodetector demonstrated a high detectivity of 2.7 $\times 10^{10}$ when illuminated with a 750 nm wavelength [53].

$$D = R / \sqrt{2qJ_{dark}} \tag{3}$$

where q is the charge of an electron (C), and J_{dark} is the dark current density (A cm^{-2}).

Noise equivalent power (NEP) is the signal power that produces a signal-to-noise ratio of one in a one hertz output band width. It is used to quantify the sensitivity of a photodetector. It shows the minimum light power that a photodetector can pick up from the ubiquitous sounds such as flicker noise from electrical equipment and shot noises from both dark current transmission and background photon irradiation. The detectivity of an optical detector rises as NEP falls. The ZnO/Ni nanoparticle-based photodetector fabricated on a glass substrate demonstrates remarkable performance characteristics. At an applied bias of 1 V, the device achieves an exceptional responsivity of 121 A W^{-1}, indicating its high sensitivity to light signals. The detector exhibits response times of 5.52 s (rise time) and 12 s (fall time), reflecting its dynamic behaviour under illumination. Furthermore, the system shows excellent noise performance with a noise-equivalent power (NEP) of 1.08×10^{-9} W, confirming its ability to detect weak optical signals [54]. As a result, infrared photodetection places a greater emphasis on the noise problem than UV–Vis detection. In order to illustrate the potential detectivity of infrared photodetectors, the dark current restricted detectivity is frequently evaluated in relation to the dominant noise from dark current. The equation is written as.

$$NEP = \frac{Jn}{R\lambda} \tag{4}$$

The photodetector's conversion efficiency is shown by its external quantum efficiency (EQE) which is the ratio of the number of charge carriers it collects to the number of incident photons. A photodetector's EQE can be represented as a spectrum response using the formula.

$$EQE = R_\lambda hc/q\lambda \tag{5}$$

where c is the speed of light, h is Planck's constant, and λ is the incident light's wavelength. It has been demonstrated that photomultiplication type organic photodetectors may attain high quantum efficiencies, primarily in the visible spectrum. The devices rely on improved hole injection which allows the broadband device to achieve a maximum external quantum efficiency of around 2000% at − 10 V. The absorption zone of the charge-transfer state exhibits the photomultiplicative effect [55]. Another crucial photodetector parameter is gain (G) which it is calculated using the following formula and it indicates how many carriers flow via an external circuit when a specific quantity of incident light illuminates. The photoconductive gain (G) is fundamentally governed by the ratio of the charge carrier lifetime (τ_h) to their transit time (τ_t) through the active layer, expressed as $G = \tau_h/\tau_t$. The transit time, calculated as $\tau_t = d/(\mu V)$, depends on the active layer thickness (d), carrier mobility (μ), and applied bias voltage (V), leading to the complete gain equation $G = (\tau_h \mu V)/d$.

3 Inorganic Zinc Oxide

3.1 Properties of ZnO

ZnO possesses direct band gap (Approx. 3.37 eV) at room temperature which can be tuned between 3.0 eV and 5.0 eV through alloying or doping with elements like magnesium (Mg) or aluminium (Al) [56]. This tunability allows for the optimization of ZnO for specific applications in UV light-emitting diodes (LEDs), lasers, and photodetectors. The electron affinity of ZnO is reported to be around 4.5 eV which is beneficial for forming Schottky barriers with various metals. One of the standout features of ZnO is its high electron mobility which can exceed 100 cm^2/Vs at room temperature and reach values greater than 10^6 cm^2/Vs at low temperatures (2 K) in high-quality epitaxial films [11, 57]. This high mobility is crucial for applications in high-speed electronics and optoelectronic devices as it enables rapid charge transport and improved device performance. ZnO film is highly transparent in the visible region of the electromagnetic spectrum with transmittance values exceeding 90% in the 400–800 nm range. ZnO thin films demonstrated high transparency ($\geq$ 90%) and low sheet resistance ($\approx$10 Ω/sq) when used as transparent conducting oxides (TCOs) [58]. Bulk ZnO exhibits thermal conductivity values around 50 WmK which is lower than that of GaN but still significant for many applications. In nanostructured forms such as ZnO nanowires, thermal conductivity can vary widely due to increased phonon scattering at grain boundaries. Desai et al. found the Young's modulus to be about 21 GPa and the fracture strain to vary from 5 to 15% for ZnO nanorods [59]. Properties of ZnO are listed in Table 1.

Among the many beneficial qualities that set ZnO apart from other metal oxides are its piezoelectric capabilities, biocompatibility, and biosafe qualities. Furthermore, doped ZnO has ferromagnetic characteristics which make it a flexible option for a range of uses. The better radiation resistance and exciton binding energy (60 meV)

Table 1 Properties of zinc oxide

Property	Value/description	Conditions (if applicable)
Chemical formula	ZnO	–
Molar mass	81.38 g/mol	–
Crystal structure	Wurtzite (hexagonal), zinc blende (cubic), rocksalt (high-pressure phase)	Ambient: wurtzite
Lattice parameters	a = 3.25 Å, c = 5.20 Å (wurtzite)	Room temperature
Density	5.606 g/cm^3 (theoretical)	Wurtzite phase
Melting point	~ 1975 °C (decomposes)	At 1 atm
Band gap	3.37 eV (direct)	Room temperature
Exciton binding energy	~ 60 meV	–
Refractive index	~ 2.0 (near-IR to visible)	600 nm wavelength
Dielectric constant	ε_s (static) $\approx$ 8.5–9.0, $\varepsilon\infty$ (high freq.) $\approx$ 4.0	–
Thermal conductivity	40–60 W/(m K) (bulk)	Room temperature
Thermal expansion	α (linear): ~ 4.31 × 10^{-6} K^{-1} (a-axis), ~ 2.49 × 10^{-6} K^{-1} (c-axis)	300 K
Young's modulus	~ 119–140 GPa	Varies with orientation
Hardness (Mohs)	~ 4.5	–
Electrical resistivity	10^{-3}–10^{5} Ω cm (undoped), < 10^{-3} Ω cm (doped, e.g., Al:ZnO)	Depends on doping
Electron mobility	200–300 cm^2/(V s) (bulk), lower in thin films (~10–50 cm^2/(V s))	–
Hole mobility	~ 5–50 cm^2/(V s) (lower than electrons)	–
Solubility in water	~ 1.6 × 10^{-6} g/100 mL (insoluble)	–
Piezoelectric constants	$d_{33} \approx$ 12.4 pm/V, $d_{31} \approx -$ 5.0 pm/V	–
Optical transparency	> 80% (visible range, for thin films)	400–800 nm
Phonon energy (LO)	~ 72 meV (E_2 mode)	–
Specific heat capacity	~ 0.495 J/(g K)	25 °C

of ZnO than that of GaN is one of its major advantages. For detectors made to operate in challenging environmental conditions, this feature is very advantageous. ZnO's affordability as a basic material and the viability of wet-chemical processing techniques that streamline manufacturing and lower production costs further increase its allure. Applications needing effective light interaction benefit greatly from this greater exciton binding energy which makes optical emission and detection more efficient. Furthermore, ZnO can be produced at lower temperatures which enables the creation of better epitaxial films and nanostructures (Fig. 3). The space group $C^4{}_{6v}$'s ($P6_3mc$) hexagonal wurtzite configuration is what defines the crystal structure of zinc oxide (Fig. 4). It has several uses in a variety of industries, including as radiation detection, electronics, and optoelectronics, thanks in part to its structural stability [60, 61].

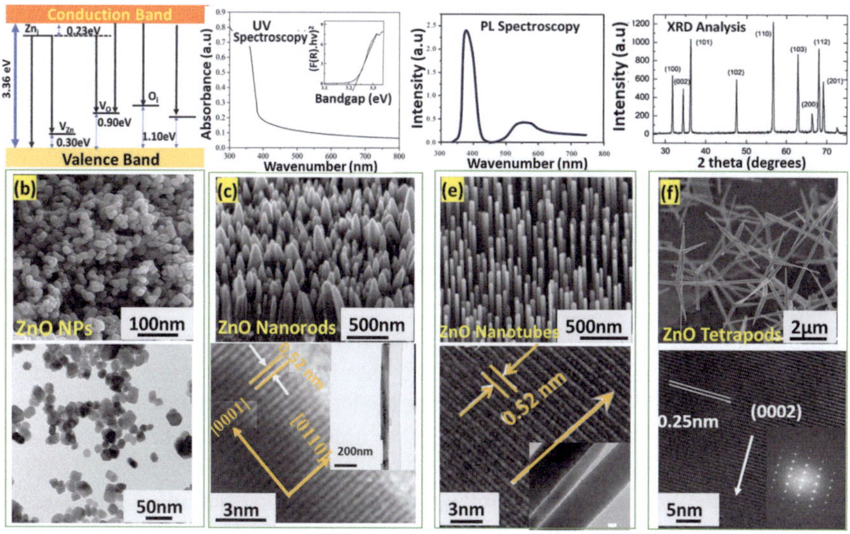

Fig. 3 **a** Band structure of ZnO along with the UV Visible Spectrum, PL emission under 365 nm light and XRD spectrum, **b** ZnO nanoparticles, **c** ZnO nanorods, **d** ZnO nanotubes, **e** ZnO tetrapods and their corresponding TEM images (down)

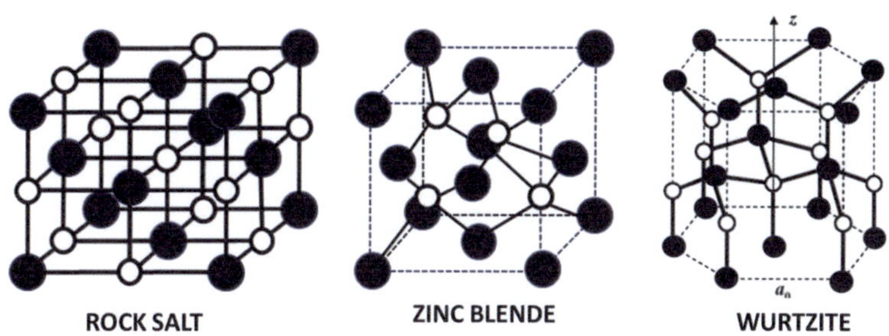

Fig. 4 Different structures exhibited by ZnO

The structure exhibits nine optical phonon modes—three longitudinal optical (LO) and six transverse optical (TO)—along with three acoustic modes: one longitudinal acoustic (LA) and two transverse accoustic (TA). At the Brillouin zone center (Γ point), the first-order optical phonon modes decompose into the irreducible representation:

$$\Gamma = A_1 + 2B_1 + E_1 + 2E_2 \tag{6}$$

Among these, the B_1 modes are Raman silent, while A_1, E_1, and E_2 are Raman active. Due to Coulomb interactions, the polar A_1 and E_1 modes split into LO and TO components. A prominent LO peak near 570 cm^{-1} arises from combined $A_1(LO)$

and E_1(LO) contributions, linked to intrinsic defects like oxygen vacancies (V_O) and zinc interstitials (Zn_i). Second-order Raman modes appear at 154 cm^{-1} [TA(M)], 205 cm^{-1} [2E_2(low)], and 340 cm^{-1} [E_2(high)–E_2(low)]. An additional peak at 786 cm^{-1} may stem from coupling between the E_2(LO) mode and the silent B_1(high) mode induced by defects. A sharp E_2(high) phonon mode dominates at 437 cm^{-1} in the spectra. The hexagonal wurtzite phase of ZnO is characterized by this peak. In contrast to those reported for bulk ZnO, the observed E2 (high) peaks are notably blue shifted which is explained by differing growth directions. This shift, however, is negligible, suggesting that the produced ZnO-T nanostructures have a stress relaxing effect.

In general, the Raman spectra show that the as-synthesised ZnO nanostructures are highly crystalline with a hexagonal wurtzite phase when an intense E2(high) mode at 437 cm^{-1} and a suppressed E1(LO) mode at 580 cm^{-1} are present. In conclusion, ZnO displays unique phonon modes that are indicative of its inherent flaws and structural features. For the purpose of developing ZnO-based devices for a range of applications, the examination of Raman spectra offers important information on the material's characteristics such as crystallinity, defect types, and phonon interactions [62, 63].

Using admittance spectroscopy, Oh et al. detected three electron trap centers near the conduction band edge in ZnO layers grown under different Zn/O flux ratios. These traps, labeled ET1 (33 meV), ET2 (0.14 eV), and ET3 (65 meV), were attributed to intrinsic defects. ET1 and ET2 were associated with singly and doubly ionized zinc interstitials (Zn_i), respectively. Their estimated densities were 3×10^{15} cm^{-3} (ET1), 2×10^{16} cm^{-3} (ET2), and 3×10^{15} cm^{-3} (ET3). The high concentration of Zn interstitials likely contributed to the increased photoresponse time observed in these layers [64] (Fig. 4).

A ternary combination called $Mg_xZn_{1-x}O$ is created by alloying ZnO with MgO, increasing its energy band gap from 3.3 to > 5 eV. $Be_xZn_{1-x}O$ was suggested by Ryu et al. as a viable option for deep UV band gap modulation. The band gap of $Be_xZn_{1-x}O$ can be adjusted from 3.37 to > 5 eV without phase segregation since BeO shares the same wurtzite structure as ZnO [65]. $Zn_{1-x-y}Be_xMg_yO$, a quaternary alloy, continually adjusts the band gap between 3.7 and 4.9 eV [66]. The engineering of ZnO opens up a wide range of application fields, including biodetection in situ combustion monitoring and detection of solar blind and UV-sensitive spaces.

3.2 Mechanism and Photoresponse of ZnO

The photoresponse of ZnO exhibits two distinct components: a fast response and a slow response. The fast response is attributed to reversible solid-state processes such as intrinsic interband transitions or excitonic effects, while the slow response is primarily governed by surface-related phenomena including oxygen adsorption and desorption as well as bulk defect-related recombination processes. In polycrystalline ZnO films, the slow photoconductivity component often dominates the overall photoresponse. Under dark conditions, oxygen molecules adsorb onto the surface at sites such as oxygen deficiency areas, forming negatively charged ions by capturing free electrons from the n-type ZnO [67].

$$O_2(\text{gas}) + e^- = O_2 \tag{7}$$

This process leads to the creation of a depletion region and band bending near the surface which subsequently decreases the film's conductivity. When exposed to ultraviolet (UV) light, photogenerated holes migrate towards the surface due to the electric field present in the depletion region. This movement neutralizes the negatively charged oxygen ions through a photodesorption reaction:

$$h + O_2^- = O_2(\text{gas}) \tag{8}$$

As a result, when free carriers generated by optical absorption are spatially separated, the conductivity of the film increases. The fast photoresponse in ZnO is linked to intrinsic interband photoconductivity where the photons with energy exceeding the band gap excite electrons from the valence band to the conduction band, generating free electrons and holes as shown in Fig. 5. This process can be represented by.

$$hv = h + e^- \tag{9}$$

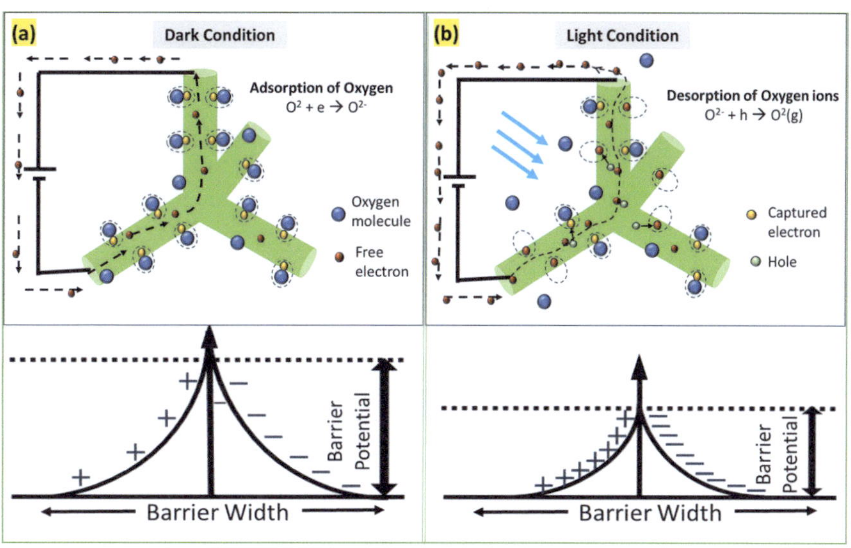

Fig. 5 The working mechanism and the barrier energy of ZnO tetrapod-based photodetector under the applied bias, **a** Dark and, **b** UV illumination conditions

The spatial separation of free carriers in ZnO films leads to an increase in film conductivity. Transient photoconductivity measurements have shown that hole capture in polycrystalline ZnO is a nonthermal process, meaning it is independent of temperature and instead influenced by the ambient oxygen partial pressure. The relaxation of photoconductivity is typically ascribed to electron tunneling into oxygen-related surface states. Moreover, epitaxial ZnO films exhibit slow photoconductivity, which is commonly linked to the presence of surface trap states that prolong carrier recombination dynamics. Defects within the ZnO lattice such as oxygen vacancies (V_O) and zinc interstitials (Zn_i) significantly enhance the performance of photodetectors by creating localized energy states within the bandgap. These states facilitate charge carrier trapping, increasing recombination likelihood and sensitivity to UV and visible light. Tuning these defects through various synthesis methods—like hydrothermal growth or metal doping—can optimize photodetector characteristics. The phenomenon of persistent photoconductivity (PPC) in ZnO is well-documented; it refers to a condition where photocurrent persists even after light exposure ceases. In PPC, photogenerated holes are typically captured by hole traps, which reduces electron–hole recombination rates and extends nonequilibrium electron lifetimes. Consequently, this leads to significant photoconductive gain. However, PPC can adversely affect UV photodetector performance by lowering the UV/visible rejection ratio and slowing photoresponse speed. Interestingly, negative photoconductivity has been observed in assemblies of ZnO nanowires (NWs) such as composites with anodic alumina membranes (AAM) or carbon nanotubes (CNT). In these cases, electrical conductivity decreases upon light exposure—contrary to standard behavior expected from photoconductive materials—which typically show increased conductivity under illumination. This complex interplay of mechanisms highlights both opportunities and challenges in utilizing ZnO for advanced optoelectronic applications [67, 68].

When light excites electrons into the conduction band of ZnO, these electrons can recombine with holes more quickly than they can contribute to electrical conductivity, leading to a decrease in overall conductivity. This effect is particularly evident in materials with numerous defect states that can trap charge carriers which reduces the number of free carriers available for conduction. Under the illuminated conditions, defects such as oxygen vacancies (V_O) and zinc vacancies (V_{Zn}) can capture the excited electrons, preventing them from participating in the conduction process. The implications of these behaviors are significant for potential applications in integrated nanoscale optoelectronics including optical logic circuits and high-resolution image sensors. By understanding and manipulating these mechanisms, researchers can enhance the performance and functionality of ZnO-based devices in various optoelectronic applications.

4 Tetrapod-Structures of ZnO

ZnO is particularly effective in detecting UV radiation, making it suitable for applications in environmental monitoring, flame detection, and biomedical fields. Because of its special qualities—such as its high efficiency, low cost, non-toxicity, stability, and capacity to function at high temperatures. It also exhibits environmental compatibility which makes it appropriate for many uses.

The three main crystal forms of zinc oxide are rock salt, zinc blende, and wurtzite (Fig. 4). The wurtzite phase which has a hexagonal structure and is a member of the P63mc space group, is the most thermodynamically stable state of ZnO at room temperature [69, 70]. For wurtzite ZnO, the lattice parameters are a = 0.3296 nm and c is 0.5206 nm. This structure is non-centrosymmetric because to the hexagonal close-packed (hcp) arrangement of zinc (Zn) and oxygen (O) atoms which alternate along the c-axis. Each Zn atom is encircled by four O atoms in a tetrahedral shape. Because of its exceptional radiation hardness and chemical stability, ZnO can function well in challenging conditions. Numerous synthesis techniques including sol–gel and hydrothermal procedures which make it easier to produce nanostructures with higher surface area and sensitivity, further promote its cost-effectiveness [71, 72]. One-dimensional (1D) ZnO nanostructures like nanowires and nanorods have been found to have higher photosensitivity. These structures can be altered by doping them with metals like copper (Cu) or manganese (Mn) or by adorning them with gold nanoparticles or cadmium sulfide (CdS). These changes increase sensitivity to visible light by introducing mid-gap electronic levels that produce charge carriers when exposed to visible light. Polytypism the phenomenon where multiple stable crystal phases coexist facilitates the growth of branched nanostructures in which nanorods of one crystal phase extend from a core or "trunk" of another phase within the same semiconductor material [73, 74]. This characteristic is particularly evident in II–VI compound semiconductors such as cadmium selenide (CdSe), cadmium telluride (CdTe), zinc sulfide (ZnS), and ZnO. The simplest form of these structures is the tetrapod which consists of four rod-like arms connected at tetrahedral angles to a central core. A significant benefit of tetrapods is their ability to orient spontaneously with one arm perpendicular to the substrate, enhancing electron extraction in photovoltaic applications compared to one-dimensional nanorods or bulk materials. The unique geometric arrangement of tetrapod nanostructures contributes to their superior performance in various applications. Each arm of the tetrapod is a defect-free single crystal aligned along the c-axis ([0001]). To exploit the advantageous one-dimensional properties of these structures, it is crucial to maintain the diameter of the ZnO-T "legs" below 20 nm for effective quantum confinement. Studies indicate that ZnO-T with legs measuring 17 nm exhibit more than double the sensitivity to humidity compared to those with larger diameters, such as 100 nm [75]. Furthermore, reducing the diameter of these legs results in a blue shift in photoluminescence, reflecting changes in their electronic properties. Research conducted by Ronning et al. has investigated the central core of these tetrapods, suggesting that a structural phase transition occurs from cubic to hexagonal during growth [76]. The formation of the four ZnO legs may begin from a multiply twinned seed crystal shaped like a tetrahedron, containing four equivalent hexagonal ZnO grains aligned with the tetrahedron's apices. Photoluminescence measurements reveal that ZnO-T's exhibit strong luminescence, characterized by two primary spectral regions within the range of 1.8–3.43 eV at 12 K. The first region corresponds to near-band edge emissions peaking at approximately 3.363 eV, associated with donor-bound excitons and several longitudinal optical phonon replicas linked to an unresolved ionized donor around 3.37 eV. The second region features deep level emissions often referred to as green luminescence, peaking at about 2.48 eV and exhibiting intensity significantly lower than that of near-band-edge emissions. The photoluminescence spectrum distinctly shows a strong ultraviolet (UV) emission band centered at 394 nm and a weaker green emission band at 518 nm. The UV emission arises from band-edge transitions or exciton transitions typical of ZnO, while the observed shift of this emission stands separate from bulk ZnO's band gap (around 380 nm) because of quantum confinement effects found in the tetrapod structures while the green emission presents structural defects including oxygen vacancies and ZnO interstitials [77, 78]. X-ray diffraction (XRD) analysis confirms that ZnO T's are well-aligned with typical wurtzite-type crystal structures, showing lattice constants consistent with standard data for wurtzite ZnO crystals (JCPDS card No. 79–2205) as reported by N.T. Loan et al. with peaks at 32 (100), 34 (002), 36 (101), 48 (102), 56 (110), 63 (103), 66 (200), 68 (112) and 69 (201). High-resolution transmission electron microscopy (HRTEM) further demonstrates that each arm grows preferentially along the [0001] direction and that individual ZnO nanorods are single crystals [79]. The optical band gap (3.2 eV) of the samples were found to be lower than that of bulk ZnO materials (3.37 eV).

The distinctive architecture of tetrapods consisting of 4 rod-shaped structures joined at tetrahedral angles to a central core, enhances directional charge transport to the collection substrate. In polymer-nanocrystal solar cells, tetrapod-shaped ZnO nanocrystals, typically around 100 nm in size, offer advantages over larger vapor-deposited structures by enabling the integration of thin polymer layers. The formation of these tetrapods is believed to initiate with the nucleation of a zinc blende core, followed by the growth of wurtzite-phase arms. Among various structural models, the octa-twin model by Takeuchi et al. is widely accepted [80]. This model proposes that the nucleus forms from eight tetrahedral crystals, each with three [$1\bar{1}22$] pyramid facets and one [0001] basal plane, assembling into a single octahedral core. The antisymmetric arrangement of adjacent tetrahedrons results in alternating [0001] surfaces, where Zn-terminated planes, being highly reactive, promote directional growth. The combustion-based synthesis of ZnO-T involves heating zinc particles, melting, zinc vaporization, and an exothermic reaction with oxygen, culminating in steady-state combustion and tetrapod growth [80, 81].

Among various ZnO nanostructures, randomly assembled nanoparticles and nanorods often suffer from numerous grain boundaries that impede electron transport. Additionally, nanowire arrays present challenges in control and cost-effectiveness when attempting to enhance the required roughness factor. Tetrapod structures present a significant advantage in solar cell applications because they are composed of four one-dimensional nanorod arms, which naturally assemble into a network with minimal grain boundaries. This architecture enhances electron transport efficiency toward the charge collection electrode. Unlike randomly aligned nanorods that lie flat and underutilize their 1-D transport potential, tetrapods orient their arms in tetrahedral directions, promoting vertical electron movement perpendicular to the substrate. Additionally, these nanotetrapods are compatible with low-cost, scalable printing methods for photoelectrode fabrication, making them a practical and efficient alternative for next-generation solar energy devices. Their design not only addresses limitations found in other nanostructures but also aligns well with modern manufacturing techniques aimed at producing efficient and cost-effective energy conversion devices.

4.1 Synthesis of ZnO Tetrapods

Several growth theories have been proposed on the synthesis of ZnO-T. According to Yu et al. each arm of the ZnO-T grows from a tetrahedral zinc blende core, supporting the idea that the tetrapod structure forms through the nucleation of this core followed by the growth of wurtzite arms [55]. This mechanism postulates that the ZnO nucleus is formed from eight tetrahedral crystals, each with three [1122] pyramid facets and one [0001] basal plane facet. The synthesis parameters such as growth time and temperature can be adjusted to manipulate the arm dimensions and enhance the multifunctional properties of the tetrapods [80].

ZnO-T were discovered in smoke from zinc-smelting plants and first studied in chemical vapor deposition systems. The synthesis process mostly takes place from the solid state even though wet chemical etching from liquid phase is also an important process. The complexity of the process often requires hazardous chemicals, high pressure and precise control over multiple parameters such as pH, temperature, and concentration to achieve the desired morphology and purity. Additionally, the wet chemical methods can involve lengthy processing times due to precipitation, filtration, drying, and calcination which may result in agglomerated powder samples. The solid-state processes such as thermal evaporation, catalytic oxidation and microwave plasma synthesis however suffer from limitations like fabricating nanostructures with any desired size, shape, and quantity even though widely used [80, 82–85] The synthesis methods involving combustion of Zn particles in air as proposed by S. Rackauskas et al. and Flame Transport synthesis (FTS) by Mishra et. al were able to overcome these limitations and the tetrapods of high crystallinity with an average leg diameter of 15 ± 5 nm and a length of 200 ± 100 nm has been synthesized [75, 81, 86].

The vapour transport synthesis undergoes the evaporation of Zn metal in powder or foil form, oxidation, nucleation and growth, at high temperatures (> 600 °C) [87, 88]. Zinc powder is directly placed into a reactor for thermal evaporation and oxidation with no added catalyst. The oxygen can be supplied along with an inert gas like argon and depending upon the ratio of the gases, the product dimensions and conversion ratio varies. In atmosphere-controlled oxidation techniques, water vapour or an oxidising agent is reacting with the Zn sample creating hydrogen gas favouring the production of sub oxide of zinc. The tetrapods developed in this manner will have longer arm lengths as compared to the thermal evaporation. Microwave synthesis of ZnO-T involves oxidizing zinc in a tube furnace at 4 Torr and temperatures up to 600 °C followed by microwave treatment at 400 W and 2.45 kHz reaching temperatures of 1000 °C. The reaction takes place for 30 min under the vacuum with carrier gases such as argon and oxygen, resulting in tetrapod arms measuring 0.5 mm in length and 50 nm in diameter with a yield of 75% and a broad size distribution [89]. FTS is a novel method for producing ZnO-T, leveraging the combustion of a sacrificial polymer to facilitate the transformation of zinc microparticles into nanostructures. In this process, zinc microparticles are mixed with polyvinyl butyral (PVB) powder and heated in a muffle furnace at approximately 900 °C where metallic Zn microparticles are directly transformed into ZnO-T within the flame at high temperatures through a solid–vapor–solid (SVS) growth mechanism. The combustion of PVB generates a flame that transports the zinc particles upward where they are vaporized into individual zinc atoms. These unstable atomic species then nucleate and grow into tetrapod structures in the presence of ambient oxygen. The FTS method allows for high yields upto 99.9% and produces ZnO-T with uniform morphologies, characterized by arm diameters ranging from 200 nm to 1 μm and lengths between 5 μm and 30 μm [90]. The process of catalytic oxidation deals with incorporating elements such as Pb, Cd, Sn, Al, Ag, and Sb into pure zinc materials. ZnO-T were formed after oxidation with modified nucleation and growth processes of whiskers [91, 92].

Dai et al. used a thermal evaporation and oxidation process in air to create ZnO nanotetrapods using 99% pure Zn powder. The heating source was an induction heater with a 60 kW maximum output power and a 60 kHz steady frequency. Carbon paste was used to create the devices through the use of screen-printing technique [93]. Lastly, ZnO-T are made using carbon paste that has been printed by electrostatic spraying and baked for 30 min at 110 °C under atmospheric pressure. Another study used atmospheric pressure plasma from a zinc metal sacrificial wire as the starting material and He gas as the carrier gas to demonstrate the synthesis of rich ZnO NPs and nanoscale, inherent oxygen vacancy-free ZnO nanotetrapods. The synthesis circumstances have a direct impact on the creation of nanotetrapods. Minor modifications in synthesis conditions—specifically gas flow rates and plasma power—significantly influence the morphology and defect structure of ZnO nanostructures. These adjustments yield distinct crystalline forms, particularly the hexagonal wurtzite phase growing along the (0002) plane, which forms the polar surfaces of nanotetrapod arms. Structural analysis reveals a predominance of spherical nanoparticles (average diameter: 13.8 nm $\pm$ 4.2 nm), with a minority exhibiting rod-like shapes. The nanotetrapod arms display average dimensions of 65.6 nm $\pm$ 16.9 nm in length and 11.5 nm $\pm$ 2.3 nm in diameter, resulting in aspect ratios varying between 2 and 26. These tetrapodal structures can be synthesized through zinc powder vaporization, with collection occurring either from reactor walls or via filtration of the exhaust stream. The diameter of ZnO-T's "legs" must be regulated and maintained below 20 nm in order to meet the requirements for quantum confinement and use the intriguing 1D features. Photoluminescence shifts to the blue when ZnO-T's diameter decreases. The combination of Zn and C is heated to 900 °C in a crucible that is positioned in the center of the quartz tube furnace. 500 sccm of Ar carrier gas with 0.5–5% O_2 concentration passes through the quartz tube. The quartz tube's downstream section has a number of baffles that help with vapor mixing. The growing nanocrystals are harvested by mounting a bare Si substrate outside the baffles. The temperature in this area of the furnace is 600 °C [94, 95]. The size and shape of the tetrapod nanocrystals (width 15 $\pm$ 5 nm and length 200 $\pm$ 100 nm) may be controlled by altering growth conditions such the O_2 concentration. By altering the growth circumstances, we have created homogeneous tetrapods with arm lengths ranging from 200 nm to 10 μm and cross-sectional widths between 50 and 500 nm. It is known that during synthesis, ZnO-T nucleation and growth take place in the vapor phase. Therefore, in order to achieve ZnO-T, it is crucial to maintain a high Zn vapor pressure by mixing with O_2 in order to achieve preferred development at the majority of reactive sites. The synthesis of ZnO-T by Rackauskas et al. was conducted under thermodynamic control, influenced by the low zinc vapor pressure and the mixing velocity within the system [75]. To address the challenge of size control during vapor transport, they developed a vertical flow reactor specifically designed for the continuous mass production of ZnO-T via Zn powder combustion in air. A custom vibratory powder feeder was employed to introduce Zn powder at a controlled rate of 0.42 g/min, enabling rapid evaporation. The zinc vapor then reacted with ambient oxygen, leading to the formation of fine ZnO-T through supersaturated vapor condensation. The system also utilized a 2.0 L/min vacuum airflow drawn from the reactor top, ensuring efficient mixing and oxidation within the heated zone. Tetrapod "legs" with an average diameter of 15 $\pm$ 5 nm, a length of 200 $\pm$ 100 nm and a high yield at 750 − 850 °C was obtained by this way [75, 96]. The synthesis methods of ZnO-T are illustrated in Fig. 6

5 ZnO Based UV Photodetectors

ZnO based photodetectors have emerged as a significant area of research due to their unique properties and versatility in UV detection applications. ZnO a wide bandgap semiconductor with a bandgap of approximately 3.37 eV has favoured for its excellent radiation hardness, high chemical stability, and low cost. Based on their device designs and operating principles, ZnO-based UV photodetectors may be generally divided into many groups. Among these, metal–semiconductor junctions are used by Schottky barrier UV photodetectors to provide high sensitivity and quick reaction times. With their interdigitated electrode architectures, MSM photodetectors provide effective charge transfer and low dark currents. For efficient charge separation, homojunction-based UV photodetectors use intrinsic or extrinsically doped ZnO layers to create a p–n junction. On the other hand, ZnO is combined with other semiconductors in p-n heterojunction UV photodetectors to design based on the band alignment for increased carrier transport and photoresponse. Furthermore, ZnO's

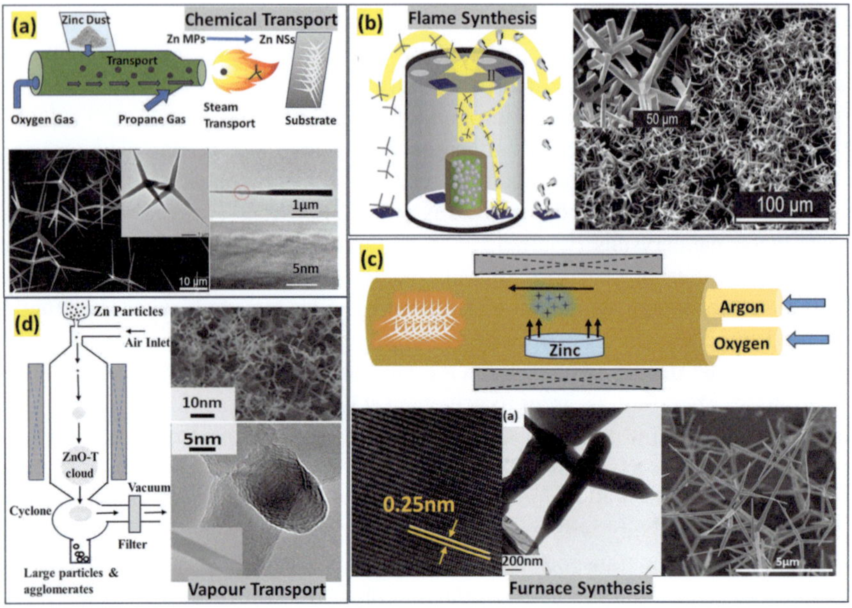

Fig. 6 Synthesis methods of ZnO tetrapods, **a** Chemical transport, **b** Flame synthesis, **c** Furnace synthesis, and, **d** Vapour transport method. SEM and TEM images of the obtained ZnO-T [75, 81]

high mobility and organic materials' tunability are used by organic/inorganic hybrid photodetectors to provide versatile, affordable, and extremely sensitive UV detection. These classifications highlight the versatility of ZnO in photodetection applications, making it a key material for next-generation UV sensing technologies. Additionally, advancements in doping techniques, such as incorporating magnesium into ZnO to modify its bandgap, further expand the range of achievable cut-off wavelengths and improve overall detector performance. This diverse classification underscores the adaptability of ZnO in meeting the demands of modern optoelectronic applications.

5.1 Schottky Barrier Photodetectors

One of the most basic unipolar photovoltaic devices is the Schottky barrier (SB) photodetector. The intrinsic electric field generated at the Schottky barrier established at the semiconductor surface collects the nonequilibrium minority carriers generated within the semiconductor bulk under optical irradiation. High speed, low noise performance, and simple fabrication procedures are only a few benefits of Schottky photodiodes [97]. Fabricius et al. used gold (Au) (a top thin sputtered layer of polycrystalline ZnO) to create UV-sensitive Schottky barrier photodiodes [98]. However, because of the large number of recombination sites in the polycrystalline ZnO layer, these photodiodes had poor quantum efficiencies of just 1% and sluggish operating speeds with rise and decay durations of 20 ms and 30 ms, respectively [116]. Chemical interactions at the metal–semiconductor interface and metal migration into the semiconductor have a major impact on the creation of Schottky connections. Compared to III–V nitrides, this effect is more noticeable for II–VI oxides. In the case of ZnO, Schottky connections for UV photodetectors are made from silver (Ag), a less reactive metal. Schottky diodes were studied using Ag–ZnO–Al circular structures, while Schottky-type metal–semiconductor-metal (MSM) photodetectors were studied using Ag–ZnO–Ag structures [99]. For comparative purposes, Al–ZnO–Al structures were also fabricated as MSM photoconductive detectors. Electron beam evaporation was employed for metal deposition with both Al and Ag layers having a thickness of 2000 Å. Photoluminescence (PL) analysis revealed excitonic near-band-edge (NBE) emission from ZnO, peaking at 3.21 eV (386.3 nm) for bare tetrapods [100]. When coupled with gold nanostructures of varying morphologies, a blue shift and quenching of the NBE emission were observed. The excitonic emission from ZnO is significantly influenced by its proximity to gold nanostructures, resulting in a decrease in integrated intensity, broadening, and a blue shift of the UV NBE emission alongside funneling effects from localized surface plasmon resonance (LSPR) emission. This interaction arises from the formation of a Schottky barrier between gold nanostructures and ZnO [100]. In a Schottky diode, the general I–V characteristics are represented by:

$$J = J_0 \exp(ZkTqV)^{-1} \tag{10}$$

where Jo is the saturation current density $J_0 = AT^2\exp(kTFB)$, Z is the ideality factor, k is Boltzmann's constant, T is the absolute temperature, A is the effective Richardson coefficient, and FB is the barrier height [100].

5.2 *Metal–Semiconductor-Metal Photodetectors*

The architecture of metal–semiconductor-metal (MSM) photodetectors (PDs) plays a crucial role in enhancing the efficiency of their photoconduction mechanisms (Fig. 7). Typically, these devices consist of two metal electrode contacts positioned at either end of a semiconductor, forming an ohmic contact that facilitates the generation and separation of electron–hole pairs upon light absorption [101]. This configuration allows for effective transfer of electrons to one electrode and holes to the other, significantly improving photocurrent generation which is closely related to the applied potential across the device. The photoconductive effect in MSM photodetectors is characterized by their simple planar structure, often featuring interdigitated electrodes. This design not only simplifies integration with other optoelectronic components but also minimizes transit times for photogenerated charge carriers, resulting in high sensitivity and rapid response times [102]. Performance is influenced by several factors including charge carrier mobility and lifetime. Recent research has focused on optimizing MSM photodetectors through various doping techniques. Cheng et al.

reported a ZnO nanowire-based MSM UV–VIS-NIR photodetector that incorporates two back-to-back Schottky junctions [103]. The introduction of nickel (Ni) doping in ZnO nanorods significantly enhanced sensitivity under UV illumination by reducing oxygen vacancies, thereby increasing the concentration of free electrons in the conduction band. Similarly, iron (Fe) doping was explored in ZnO nanorods with silver (Ag) serving as Schottky electrodes, further boosting responsivity due to enhanced electron availability [104]. Recent advancements in MSM photodetector technology underscore their potential for high sensitivity and rapid response times across various applications. For instance, studies have shown that MSM photodetectors can achieve responsivities exceeding 10^4 A/W under specific conditions, demonstrating their effectiveness in detecting weak optical signals while maintaining low noise levels. These developments pave the way for integrating MSM photodetectors into the advanced optoelectronic systems, enhancing their applicability in fields such as telecommunications and environmental monitoring. Homojunction diodes are formed by combining p-type and n-type regions of the same semiconductor material, where doping variations create the junction. Under forward bias, electrons from the heavily doped n-side and holes from the lightly doped p-side diffuse across the junction, allowing current flow. In reverse bias, the depletion region widens as majority carriers move away, blocking current. In contrast, heterojunction photodetectors use different semiconductor materials with distinct bandgaps to enhance performance. For example, ZnO-based heterojunctions improve responsivity and reduce dark current by leveraging favorable band alignment, which promotes efficient charge separation and minimizes recombination. The built-in potential at the heterojunction interface enhances carrier collection, leading to higher quantum efficiency. While homojunctions rely on internal electric fields for charge separation, heterojunctions benefit from additional interfacial effects, making them more efficient for optoelectronic applications. This fundamental difference highlights how material choice and junction design influence device performance in semiconductor technology.

5.3 Homojunction Photodetectors

When the work function of a metal is higher than that of ZnO nanowires (NWs) or nanorods (NRs), electrons transfer from the ZnO to the metal until their Fermi levels align, achieving thermal equilibrium. This electron transfer creates a space charge region near the ZnO surface, where ionized donor atoms remain due to electron depletion, while electrons accumulate at the metal interface. The resulting charge distribution generates a stable internal electric field with a fixed potential difference and a well-defined depletion width. As a consequence, the energy bands of the ZnO nanostructures bend upward, forming a Schottky barrier that restricts electron flow [105, 106]. In Schottky-based ultraviolet (UV) photodetectors, this built-in electric field and potential barrier facilitate the separation of photogenerated electron–hole pairs, enhancing device performance (Fig. 7) [107].

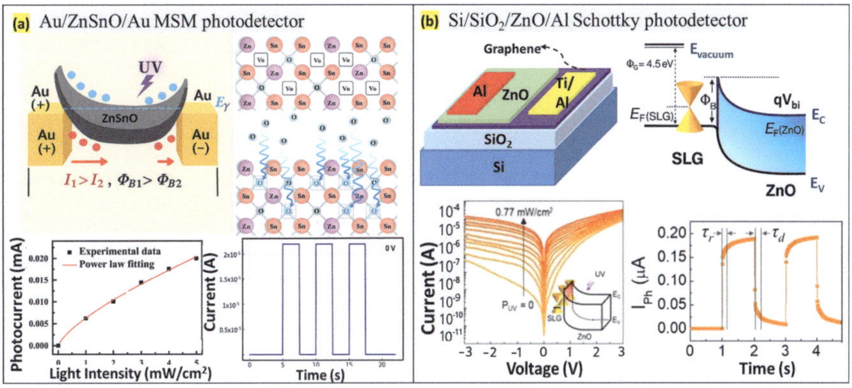

Fig. 7 **a** Au/ZnSnO/Au MSM photodetector [102] and **b** $Si/SiO_2/ZnO/Al$ Schottky photodetector and their photodetectivity measurements [107]

The absorption of incident UV light in a p–n junction's depletion region depends on the semiconductor's bandgap energy. While silicon homojunctions dominate commercial photovoltaic devices, their relatively narrow bandgap restricts their UV sensitivity. This limitation has driven research into wide-bandgap semiconductors, such as oxides and nitrides, for constructing homojunction-based self-powered UV photodetectors. Recent advancements in ZnO-based homojunctions demonstrate their potential for efficient UV detection, leveraging their intrinsic bandgap properties and improved charge separation mechanisms [108–110]. These materials are favored for their high carrier mobility, exceptional stability, and well-established synthesis processes. Self-powered UV PDs utilizing homojunctions benefit from a perfect lattice match that minimizes structural defects at the p-n interface and ensures a consistent built-in electric field. Moreover, these devices demonstrate resilience against chemical and thermal degradation under extreme operating conditions. Among various wide bandgap semiconductors, ZnO has emerged as a widely studied material for fabricating p-n homojunctions in self-powered UV PDs [111, 112]. With a bandgap of 3.37 eV, ZnO exhibits excellent photosensitivity in the UV spectral range while allowing most visible light to pass through. A vertical p-ZnO:Cu/n-ZnO NW device was made by Hsu et al. with I_{UV} almost 6 times than that of I_{dark} [113] (Fig. 8).

One-dimensional (1D) ZnO nanostructures, such as nanowires (NWs) and nanorods (NRs), have attracted considerable attention in optoelectronics due to their unique optical and electrical characteristics. The enhanced electron–hole interactions in these nanostructures arise from a pronounced electronic density of states near the van Hove singularity, which significantly improves their optoelectronic performance. In bulk ZnO, the exciton Bohr radius is confined to a narrow range of 1.8–2.3 nm, meaning that quantum confinement effects become prominent in ZnO NWs when their diameters approach this scale [114]. Interestingly, when the NW diameter exceeds the exciton Bohr radius yet remains below the optical wavelength,

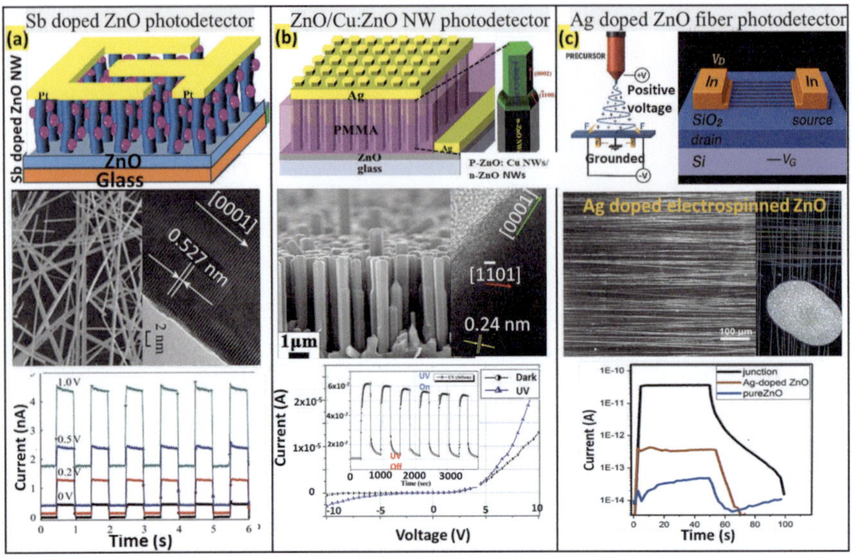

Fig. 8 (**a**) Sb doped ZnO NW photodetector [125], (**b**) ZnO/Cu:ZnO NW photodetector [113] and (**c**) Ag doped electrospinned ZnO photodetector [126] and their photodetectivity measurements, along with the SEM and TEM morphological studies

a giant oscillator strength effect emerges, dramatically boosting their UV detection capabilities [115, 116]. Furthermore, the high S/V ratio of ZnO NWs introduces a substantial density of surface trap states, primarily associated with oxygen vacancies. These defects play a dual role: they extend the lifetime of photogenerated carriers by trapping them, while also influencing effective carrier mobility through scattering mechanisms [117, 118].

However, achieving p-type doping in ZnO remains challenging, complicating the fabrication of p-n homojunctions. Shan Chong-Xin et al. synthesized p-type ZnO through lithium-nitrogen co-doping methods, laying a solid foundation for developing p-n homojunctions in self-powered UV photodetectors [119]. In these devices, electrons captured by silver (Ag) nanoparticles lead to increased hole collection at the electrode and reduced rise time. Conversely, when illumination ceases, an expanded depletion region inhibits free carrier flow, resulting in rapid declines in current and decay time. Furthermore, increasing Ag doping concentration enhances Ag nanoparticle formation, creating more internal Schottky barriers that improve device responses. The ongoing exploration into different configurations and approaches continues to enhance UV detection technologies while showcasing the versatility and potential applications of ZnO-based materials in optoelectronic devices [57, 120]. The current density in nanowires (NWs) can be described by the equation,

$$J_{ph} = vqD_{nph} \tag{11}$$

where q represents the elemental charge, D_{nph} is the density of photogenerated carriers, and v is the carrier velocity [121]. Assuming a uniform carrier generation profile within the optical absorption depth α_l, the expression for D_{nph} can be formulated as,

$$D_{nph} = hFV_{ph}t(F) = hF\alpha_l WLt(F) \tag{12}$$

where h is the quantum efficiency, F is the photon absorption rate, V_{ph} is the photo generation volume, α_l is the optical absorption coefficient and t is the carrier lifetime, which is a function of photon absorption rate F.

The continuity equation gives: $\partial D_{nph}/\partial t = g_{ph}$ where g_{ph} is the carrier photogeneration rate [122]. Zhang et al. investigated a metal–semiconductor-metal (MSM) ultraviolet (UV) photodetector constructed from arrays of ZnO nanowires (NWs) mounted on a flexible polyester substrate. Their results indicated that at a bias voltage of 5 V, the dark current measured as 25 nA, while the photocurrent reached to 150 nA, yielding a sensitivity of approximately 5. At a bias of around − 1 V, the photocurrent and sensitivity increase by 82% and 130%, respectively, at a 0.2% tensile strain [123].

Yang et al. developed an innovative surface-ionic-gate (SIG) modulation technique to optimize the performance of ZnO nanowire (NW)-based Schottky barrier photodetectors [124]. By applying an external electric field, they induced the migration of positive ions toward the cathode while oxygen ions (O^{2-}) and electrons moved toward the anode, leading to localized ion formation on the ZnO NW surface. Experiments conducted in different atmospheres confirmed that O^{2-} ions and electrons generated via gas discharge can accumulate on the nanowire surfaces. Prior to SIG modulation, energy band bending occurred at the ZnO NW surface due to Fermi-level misalignment between silver (Ag) and ZnO. After modulation, negative charges from corona discharge migrated to the Ag/ZnO interface, shifting the ZnO Fermi level below the surface state energy. This adjustment significantly improved device performance, yielding a photocurrent of 1.1 A and an on/off ratio of 630 at a 5 V bias. Notably, the recovery time was drastically reduced from 87 s to 0.3 s and a 290-fold enhancement in response speed is observed.

In a complementary approach, Wang et al. demonstrated a self-powered UV photodetector using antimony (Sb)-doped ZnO [125]. Under UV illumination, the device generated an open-circuit voltage (V_{oc}) of 9 mV and operated without external bias. At a minimal reverse bias of − 0.1 V, it exhibited a photocurrent of 1.65 nA and a dark current of 60 pA, highlighting its efficiency as a zero-power-consumption UV sensor.

Similarly, using electrospun ZnO nanofiber arrays, a new self-powered UV photodetector is developed by Ning et.al with aligned pure ZnO nanofibers and Ag-doped p-type ZnO nanofibers [126]. Because the nanofibers are naturally spaced apart, the device is completely transparent on a quartz substrate. Ag doping is shown to help decrease reaction time by forming Ag nanoparticles in ZnO nanofibers and converting ZnO to p-type conductivity which enables the construction of this self-powered photodetector. The improvements in performance are attributed to the creation of an Ag-ZnO Schottky barrier which increases the depletion region within the Ag-doped ZnO nanofibers due to differences in Fermi levels that hinder carrier transport. When exposed to UV light, the width of this depletion region decreases along with carrier recombination, allowing photogenerated electrons to be captured by Ag nanoparticles. This process results in a higher collection of holes at the electrode, thereby reducing rise time. Conversely, when UV illumination is turned off, an expanded depletion region restricts free carrier flow, leading to a rapid decline in current and increased decay time. Furthermore, increasing Ag doping concentration enhances Ag nanoparticle formation, creating more internal Schottky barriers that improve device performance. This UV detector operated with a quick rise/decay duration of 3.90 s/4.71 s and an on–off ratio up to 10^4 at zero bias [126].

5.4 PN Heterojunction Photodetectors

Schottky barrier-based ultraviolet (UV) photodetectors are known for their rapid response speeds. In comparison, pn photodetectors offer advantages such as high speed and low dark current. A systematic review of ZnO nanowires (NWs) and nanorods (NRs)-based vertical pn heterojunction UV photodetectors reveals that when n-type ZnO NWs or NRs contact a p-type semiconductor material, a p-n heterojunction is formed due to the diffusion and drift of electrons and holes. This interaction leads to the creation of a space charge region near the heterointerface under thermal equilibrium conditions, establishing a stable internal electric field.

The separation of photogenerated charge carriers in ZnO-based p-n heterojunction UV photodetectors is primarily governed by the built-in electric field at the interface. ZnO nanostructures (NWs/NRs) can form heterojunctions with various inorganic materials like silicon (Si), GaN, and cesium lead bromide ($CsPbBr_3$), as well as organic materials such as PEDOT: PSS, creating either n–n or n-p configurations [126–128]. Among these, Si has been particularly advantageous for all-inorganic designs due to its established fabrication processes and natural abundance. Recent advances have introduced novel enhancement mechanisms including the pyro-phototronic effect, which synergistically combines pyroelectric and photoelectric phenomena—where light-induced temperature gradients generate polarization charges in ZnO that actively optimize charge carrier separation, transport, and recombination dynamics. Similarly, the piezo-phototronic effect utilizes strain-induced piezoelectric potentials to fine-tune energy band alignment at interfaces. These innovative approaches have demonstrated significant improvements in photodetector performance beyond what can be achieved through conventional field-driven separation alone, offering new dimensions for device optimization through the strategic coupling of multiple physical phenomena. The ability to harness these effects while maintaining compatibility with diverse material systems highlights the versatility of ZnO-based heterostructures for advanced UV detection applications. Despite significant advancements in self-powered UV photodetectors utilizing the p-n homojunction configuration, challenges remain that hinder large-scale applications. Achieving effective doping is essential for forming a p-n homojunction, but this remains a major challenge for some wide bandgap semiconductors due to low dopant solubility, self-compensation by native defects, and deep impurity levels. Particularly, achieving p-type doping in ZnO has proven difficult [129]. Furthermore, the high costs associated with doping processes required to maintain rigorous fabrication standards for p-n homojunctions, have shifted the focus of device engineers toward p-n heterojunction-based self-powered UV photodetectors. Unlike p-n homojunctions, heterojunctions provide greater flexibility in selecting various semiconductor materials, allowing for adjustments in spectral response that enhance the performance of self-powered UV PDs. To date, several device configurations based on p-n heterojunctions have been successfully demonstrated as self-powered UV PDs. Examples include p-GaN/n-ZnO, p-NiO/n-ZnO, p-CuSCN/n-ZnO, p-PANI/n-MgZnO, p-PANI/n-TiO_2, and p-Ga_2O_3/n-ZnO [130–134]. ZnO forms the basis for many optoelectronic devices, but its grain boundaries can significantly degrade performance by introducing charge-trapping sites and recombination centers. Additionally, the cutoff wavelength and signal-to-noise ratio (SNR) of self-powered UV photodetectors are strongly influenced by defects in tunable semiconductor alloys, such as AlGaN, which are often employed to tailor spectral response.

Among alternative wide-bandgap semiconductors, titanium dioxide (TiO_2) stands out due to its high UV absorption coefficient, making it a promising candidate for highly sensitive self-powered UV detection. However, conventional TiO_2-based photodetectors suffer from high dark currents and slow response speeds, primarily due to carrier recombination at surface and defect states [135, 136]. To mitigate these issues, researchers are exploring crystallinity improvements and novel device architectures, such as TiO_2/Au/TiO_2 heterostructures, which leverage built-in electric fields at metal–semiconductor interfaces to enhance charge separation.

Recent advancements in ZnO:Sb/Au and TiO_2/Au/TiO_2 configurations have successfully demonstrated self-powered operation, where the built-in electric field at the interface drives photocurrent generation without external bias as shown in Fig. 9. These developments highlight the importance of interface engineering and defect passivation in optimizing the performance of next-generation UV photodetectors. Future research may focus on hybrid material systems and nanostructured designs to further improve sensitivity, response speed, and dark current suppression [137].

Interface engineering has become a pivotal strategy for enhancing the performance of ZnO nanowire (NW)-based ultraviolet (UV) photodetectors [138, 139]. By carefully selecting an appropriate inserting layer material and embedding it between ZnO NWs and Schottky contact electrodes, researchers can modify the distribution of interface energy bands. This adjustment helps reduce the density of surface states and strengthens the built-in electric field, leading to more efficient separation and extraction of photogenerated electron–hole pairs. Consequently, this results in improved photocurrent and reduced dark current. Zheng et al. investigated the role of interface engineering in improving the photoresponse of ZnO nanowire (NW)/Au-based Schottky UV photodetectors [138]. By introducing a titanium dioxide (TiO_2) interlayer between the ZnO NWs and gold (Au) electrode, they constructed

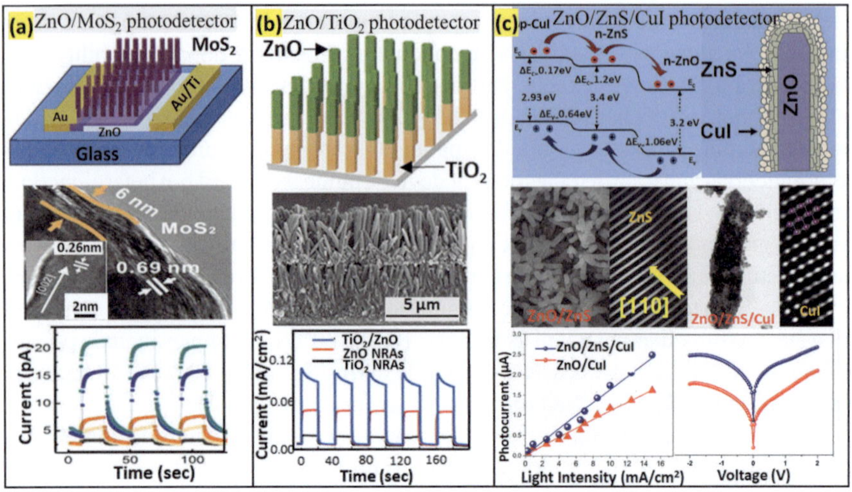

Fig. 9 **a** MoS_2/ZnO NW [145], **b** ZnO/TiO_2 NW [138] and **c** ZnO/ZnS/CuI photodetectors [146] and their SEM and TEM morphological studies, along with the photodetector measurements

an Au-TiO_2-ZnO (ATZ) heterostructure, which significantly enhanced device performance. At a bias voltage of 4 V, the sensitivity of the photodetector without TiO_2 was measured at 749, while the sensitivity for the device with TiO_2 reached 1483. The dark currents were recorded at 37.7 μA and 0.034 μA with the ATZ photodetector achieving a photocurrent value of 51 μA.

Meng et al. reported on a fiber-based ZnO NW network UV photodetector that considered interface and surface gating effects to investigate its optoelectronic performance [140]. Due to their one-dimensional structure, ZnO NWs can become completely depleted, resulting in a significant increase in the distance between the Fermi level and conduction band which dramatically decreases carrier density in the dark. The experimental results showed that this device achieved an on/off ratio of 4.61×10^7 due to surface gating effects. Effective surface and interface gating effects are indicated by the low dark current of 1.14×10^{-14} A at 10 V. When exposed to 1.0 mW cm^{-2} of UV light, the current at 10 V approaches 2.30×10^{-6} A, demonstrating a very high on/off ratio of 1.98×10^8. Even under UV light with an intensity as low as 0.5 nW cm^{-2}, the photodetector exhibits a wide response range to light intensity, ranging from 0.5 nW cm^{-2} to 1.0 mW cm^{-2} [140]. In another study, Yin et al. developed a ZnO nanorods (NRs)/NiO/Si heterojunction-based UV photodetector (Fig. 10). The incorporation of a NiO layer effectively reduced leakage current compared to conventional ZnO/Si heterostructures, thereby enhancing the pyro-phototronic effect [141, 142]. Flemban et al. studied about p-Si/ZnO nanotube heterojunction device [143]. The high photo-induced carrier density combined with the elevated surface-to-volume ratio of ZnO nanostructures and the extremely low density of defect trapping states, contributes to exceptional responsivity levels in the hundreds of amperes per watt range around 101.2 AW^{-1}. Under 365 nm ultraviolet

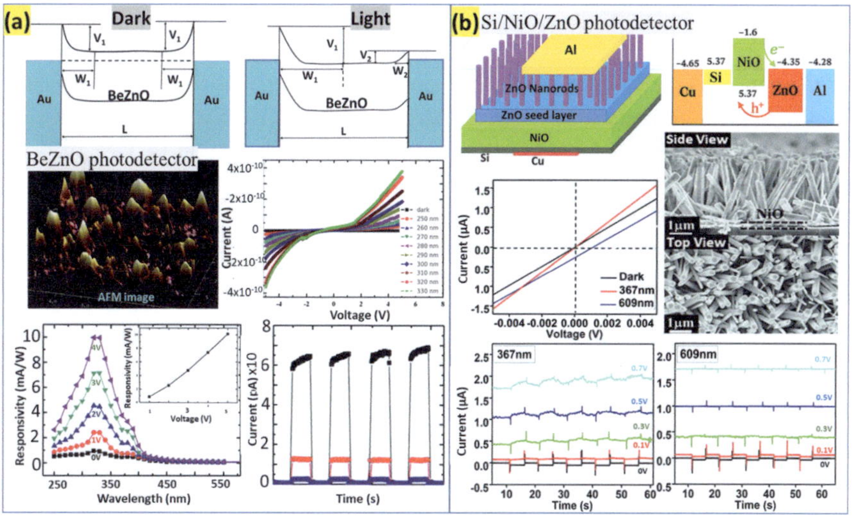

Fig. 10 **a** Be-alloyed ZnO photodetector [147], AFM images of the alloy and photodetector measurements and, **b** Cu/Si/NiO/ZnO/Al photodetector [141], band energy levels and their photodetectivity measurements, along with the SEM images of the films

illumination at an intensity of 1 mW cm^{-2} and a bias voltage of − 2 V, a responsivity of 21.51 AW^{-1} is achieved in the UV region. This performance corresponds to a specific detectivity (D) of 1.26×10^{12} cm $Hz^{1/2}$ W^{-1} and an external quantum efficiency (EQE) of 73.1×10^2%, with an effective area of approximately 0.79 mm^2.

In a similar work of NiO/ZnO heterojunction arrays-based UV photodetectors with designed tunable Fermi level of ZnO, the devices with ZnO NRs grown in 0.5% Al^{3+} ion additions in the precursor solution exhibit high detectivity D (1.737 $\times 10^{12}$ cm $Hz1^{/2}$/W), quick response speed (~2 ms), and good UV spectrum selectivity (FWHM of 9 nm) under weak light exposure without external bias. By reducing carrier recombination and facilitating the separation and transport of photo-generated carriers, the larger built-in electric field created at the interface of ZnO and NiO nanoflakes with 0.5% Al^{3+} ion additions enhances device performance in terms of photocurrent, signal-to-noise ratio, photoresponsivity, and detectivity [144]. A study by Ling et al. reported a p-MoS_2/n-ZnO nanoparticle heterojunction device exhibiting exceptional broadband photoresponse characteristics, achieving over 60 times the photocurrent obtained compared to planar heterojunction counterpart under UV illumination (Fig. 9) [145]. In parallel work, an insertion of ZnS layer in the p-CuI/ZnO nanorod heterojunction device improved the photocurrent response from 2.21×10^{12} to 3.84×10^{12} [146]. The insertion reduced the band offsets and hence contributed to 43.5 mA/W current and 3.82×10^{14} Jones detectivity for an incident light of wavelength 372 nm.

Fang et al. explored alloying ZnO with beryllium oxide (BeO) as shown in Fig. 10 to modulate its bandgap from near-UV to deep-UV regions, allowing BeZnO-based photodetectors to respond at different cut-off wavelengths [147]. Asymmetric Schottky contacts made from wide Ti/Au and narrow Ti/Au were deposited onto alloyed BeZnO films, resulting in an extremely low dark current of 1×10^{-2} pA and a photocurrent approximately 1000 times greater due to defect suppression from incorporating BeO with ZnO. This low dark current contributes to high detectivity levels: 1×10^{10} Jones at 330 nm and 1.6×10^{9} Jones at 240 nm under a reverse bias of 2 V. Ultrafast response speeds were recorded at this device with rise times of just 35 μs and decay times of 880 μs when intermittently exposed to a UV light source at a wavelength of 355 nm. However, challenges remain regarding crystal quality as increasing Mg content in MgZnO alloys can lead to phase segregation that negatively impacts device performance. In contrast, $Mg_xZn_{1-x}O$ offers tunable bandgap properties suitable for UV detection by varying x composition while maintaining environmental friendliness and biocompatibility [147]. Xie et al. demonstrated a self-powered ultraviolet (UV) photodetector based on the ternary alloy magnesium zinc oxide (MgZnO) which exhibited a faster response time compared to beryllium zinc oxide (BeZnO) [148]. The Mott-type VIS-blind UV photodetector with quick photoresponse and high gain, was based on a C-MgZnO/W-MgZnO vertical structure with an active anti-reflection layer. Under 30 V, the highest responsiveness is 431 mA/W. When exposed to UV light, the detector responds quickly with rise time of 0.6 ms and fall time of 1.5 ms. However, incorporating magnesium into zinc oxide can lead to deterioration in crystal quality when Mg concentration exceeds certain levels, resulting in phase segregation that negatively impacts device performance.

5.5 *Organic/Inorganic UV Photodetectors*

Zhu et al. constructed a fiber-shaped UV photodetector using ZnO nanorods (ZnO NRs) in combination with graphene, demonstrating that the built-in electric field at the ZnO NR/graphene interface facilitates the efficient separation and transport of photogenerated electrons and holes [149]. They applied strain engineering by bending the PET substrate which effectively modulated the band structure of the ZnO NR/graphene heterostructure. This strain increased the height of the Schottky barrier, thereby enhancing the built-in electric field near the heterointerface and promoting the separation and transportation of photogenerated carriers, ultimately leading to improved device performance. The photoresponsivity of the device reaches an impressive 1.92 A W^{-1}. Furthermore, the performance can be enhanced through the application of external compressive strain, leveraging the piezoelectric properties of ZnO. Specifically, under a compressive strain of $-$ 0.33%, the responsivity of the photodetector can increase by up to 12.5%.

Fang Yi et al. has successfully demonstrated the self-powered ultraviolet photodetector based on a p–n junction using ZnO and PEDOT:PSS [150]. When the p–n heterojunction was lit by the 325 nm UV light (0.16 mW) at zero bias, the detector displayed an on/off ratio of around 1100 with a rise time of approximately 3.5 s and a decay time of approximately 4.5 s. The photovoltaic effect drove the self-powered properties which had a fill factor of around 25%, an open-circuit voltage of about 0.2 V, and a short-circuit current of about 1.1 nA. The authors indicated that this device not only shows potential for self-powered UV detection but could also serve as an environmentally sustainable ultralow power source for various sensors. Recent research by Thyda et al. explored hybrid organic–inorganic photodetectors incorporating ZnO nanorods with carbon nanostructures [151]. The study compared devices employing carbon quantum dots (CQDs), revealing superior performance in the quantum dot-based configuration Fig. 11. The CQD-modified devices demonstrated significantly enhanced charge separation efficiency, attributed to the quantum dots' ability to effectively control photogenerated carrier transport. This improvement resulted in exceptional detector performance, with the CQD/ZnO NR hybrid achieving a remarkable highest photocurrent density of 504 μA/cm^2 and EQE of 48.4%. These findings position carbon quantum dots as a promising alternative to conventional polymeric charge transport layers, offering advantages in both device performance and potential fabrication simplicity for UV photodetection applications. The study highlights the importance of interfacial engineering in hybrid photodetector systems and suggests new pathways for developing high-performance optoelectronic devices through nanomaterial integration.

Nguyen et al. reported a UV photodetector based on Cu-doped ZnO nanorods (ZnO:Cu NRs)/ PEDOT:PSS with ZnO:Cu NRs with high visible absorption acting as material responsible for electron hole generation under UV light [152]. Recent studies

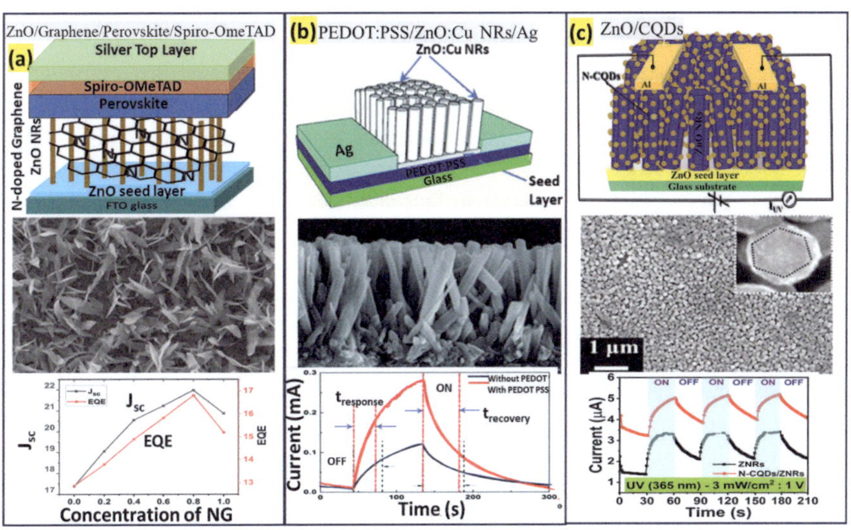

Fig. 11 The schematics of structures, and the device characteristics of various UV photodetectors and their photoresponses, **a** ZnO/nitrogen doped Graphene/Perovskite/Spiro-OmeTAD/Ag [161], **b** PEDOT:PSS/ZnO:Cu NRs/Ag [152], **c** ZnO/CQDs/Al photodetectors [151]

demonstrate that incorporating graphene quantum dots (GQDs) in photodetector designs enhances device performance through two mechanisms: increased Schottky barrier height and improved UV light absorption. This modification effectively minimizes charge carrier recombination while boosting quantum efficiency. Researchers have also fabricated an alternative design using aluminum-doped ZnO nanorods coated with CdS combined with chemically-modified PEDOT:PSS to create efficient Schottky barriers. In parallel developments, bulk heterojunction structures employing carefully matched organic semiconductors have achieved breakthrough performance. These systems leverage optimized energy level alignment to enable charge separation efficiencies exceeding 80% even without applied voltage, representing a significant advancement in photodetector technology.

While this dual-component approach enhances exciton dissociation, it introduces spectral complexity as both donor and acceptor materials contribute to photogeneration across overlapping absorption ranges. Researchers have exploited the tunable bandgaps of organic semiconductors through conjugation length engineering, enabling broadband organic photodetectors (OPDs) with response spectra spanning ultraviolet to near-infrared wavelengths [153, 154]. Notably, specialized visible-blind OPDs have been developed with exceptional UV responsivity, $> 10^{10}$ dynamic range, and substantial photoconductive gain. Parallel advancements in colloidal quantum dot (CQD) systems have revealed multiple exciton generation phenomena—where supra-bandgap photon energy creates multiple electron–hole pairs—offering potential pathways for exceeding traditional efficiency limits in photodetection. These complementary strategies (BHJ optimization, bandgap engineering, and quantum confinement effects) collectively address the fundamental trade-offs between spectral selectivity, responsivity, and quantum efficiency in next-generation photodetectors [155]. This phenomenon has been explored across various material systems and device architectures. Current advancements highlight that the interface engineering plays a crucial role in enhancing the performance of ZnO NW-based UV photodetectors. By strategically inserting layer materials between ZnO NWs and Schottky contact electrodes, researchers can modify energy band distributions at the interface [156, 157]. This adjustment reduces surface state densities and strengthens built-in electric fields, leading to more efficient separation and extraction of photogenerated carriers.

Recent advances in hybrid photodetector designs have demonstrated remarkable performance enhancements through strategic material integration and strain engineering. Li et al. developed an innovative n-Si/n-ZnO NWs/PEDOT:PSS tri-layer photodetector that leverages the piezo-phototronic effect, showing exceptional strain-dependent responsivity [158]. Under 405 nm illumination at 2 V bias, the photocurrent increased dramatically from 21.51 μA (0% strain) to 627.43 μA at 10.78% compressive strain, with peak responsivity reaching 0.218 A/W and detectivity of 2.12×10^{19} Jones as shown in the Fig. 11. This represents a 3000% enhancement for 405 nm and 1800% for 648 nm illumination compared to unstrained operation, attributed to strain-induced band structure modifications that optimize carrier separation. Further improvements were achieved by incorporating $CsPbBr_3$

perovskite into ZnO nanowire/graphene systems, where the resulting p-n heterojunction creates favorable band bending that enhances built-in electric fields. This architecture not only improves interfacial charge transport but also expands the effective carrier collection area, yielding an impressive light-to-dark current ratio of 10^3. The synergistic combination of these approaches—strain modulation through the piezo-phototronic effect and band engineering via heterojunction formation—highlights new pathways for developing ultra-sensitive photodetectors across multiple spectral ranges [158]. In a study conducted by Zhao et al. a novel dual-band photodetector (PD) was developed through the deposition of ZnO, the organic polymer poly-(N-vinylcarbazole) (PVK), and copper oxide (Cu_2O) semiconductor materials as shown in the Fig. 12 [159]. This innovative ZnO/PVK/Cu_2O PD is capable of selectively converting responses between two distinct wavelength bands: 285–492 nm and 492–645 nm, depending on the applied bias voltage. Under zero-bias conditions, the photodetector achieved a peak responsivity of 0.19 A/W at a wavelength of 450 nm. Interestingly, when a reverse bias voltage of − 2.5 V was applied, the peak responsivity shifted to 0.20 A/W at 556 nm. This ability to modulate the response based on bias voltage illustrates the potential of this dual-band photodetector for applications requiring selective light detection across different spectral ranges.

Chen et al. investigated a heterojunction formed by electroplating polyaniline (PANI) onto ZnO (Fig. 12) [160]. This innovative ZnO/PANI hetero-diode achieved an impressive rectification ratio of 7.5×10^5 ($I_{+3\ V}/I_{-3\ V}$), demonstrating exceptional diode characteristics. Additionally, the device exhibited a maximum rejection ratio of 3556 (I_{350nm}/I_{dark}) at a bias voltage of − 3 V. These results indicate the potential of this heterojunction for applications as rectified switches, highlighting its effectiveness in managing current flow and enhancing performance in electronic devices. Under a bias

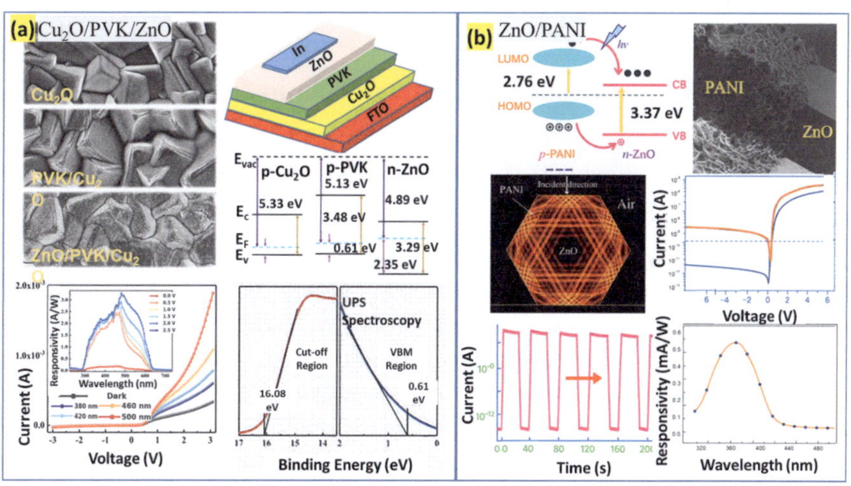

Fig. 12 **a** Cu_2O/PVK/ZnO hybrid photodetector [159] and UPS spectroscopy results, **b** ZnO/PANI photodetector [160] and simulated optical field distribution of PANI/ZnO core shell microwire along with the photodetectivity measurements and SEM images

voltage of 0 V, the self-powered photodetector demonstrates a maximum responsivity of 0.56 mA/W along with a rapid response speed of 0.11 ms for the rise time and 1.45 ms for the decay time. This performance indicates the device's capability to efficiently convert light into electrical signals while responding quickly to changes in illumination. Chandrasekhar et al. developed a nitrogen-doped graphene/ZnO nanorod composite photoanode with optimal amount of graphene contributing to improved device characteristics [161].

The organic component Spiro-MeOTAD is noted for its low toxicity, while nitrogen doping improved ZnO's photoresponse (Fig. 11). Furthermore, Tran et al. studied graphene quantum dot (GQD)-decorated ZnO (ZnO:GQD) which exhibited reduced surface roughness and nanoparticle size, leading to enhanced carrier transport and extended carrier lifetimes [162]. The optimized GQD concentration in ZnO:GQD/poly(N,N′-bis(4-butylphenyl)-N,N′-bisphenylbenzidine)(poly-TPD) photodiodes resulted in improved energy structure balance, increased mobility, and minimized charge recombination. The photodetector demonstrated a strong photoresponse to 365 nm UV light at a low operating voltage. Under an illumination of 48 $\mu W/cm^2$ UV light, the measured parameters were as follows: responsivity (R) of 21.6 A/W at 3.6 V, the photocurrent gain (G) of 85.8 at 4.0 V. Additionally, the response time (T_r) of 1.9 s and the decay time (T_d) of 51.1 s [163]. Li et al. also reported on the piezo-phototronic effect in constructing advanced UV photodetectors using ZnO nanowires (NWs) and nanorods (NRs) [164]. Their study revealed that applying external compressive strains significantly influenced device performance with the photocurrent increasing from 21.51 μA at zero strain to 627.43 μA at 10.78% strain before decreasing at higher strains. The maximum photoresponsivity reached to 0.218 A/W under optimal strain conditions with detectivity peaking at 2.12×10^{19} Jones, attributed to the piezo-phototronic effect enhancing charge carrier dynamics.

6 Optoelectronic Applications of ZnO Nanostructures

The integration of carbon dots (CDs) with ZnO significantly enhances photodetector performance by addressing the inherent limitation of slow oxygen desorption kinetics in pure ZnO devices. When CDs are incorporated at optimal concentrations (e.g., CQD/ZnO ratio of 1.2×10^{-3}), they form a conformal surface layer that serves dual functions: (1) reducing oxygen adsorption sites, thereby accelerating response time by two orders of magnitude (from 329.05 s to 0.77 s decay time), and (2) creating alternative electron transport pathways. High-resolution TEM studies reveal a nanocomposite structure with well-dispersed CDs (1–5 nm) among ZnO grains (10–20 nm), where the quantum dots act as electron barriers that force carriers to traverse lower-resistance routes, extending diffusion distances while paradoxically reducing decay times [10]. This architecture also enables broadband detection, expanding the photoresponse from UV (300–360 nm) to visible wavelengths (400–600 nm), with exceptional spectral selectivity demonstrated by rejection ratios of 6.40

$\times 10^6$ (365 nm/880 nm) and 7.47×10^3 (470 nm/880 nm). The collective improvements—including 30-fold faster rise times (5.94 s vs. 180.93 s) and suppressed dark current—stem from synergistic effects between the ZnO matrix and CDs, illustrating the potential of hybrid designs to overcome fundamental material limitations [165]. ZnO nanostructures have garnered significant attention for their exceptional charge transport characteristics, outperforming conventional polycrystalline films in electronic applications. Single-crystalline ZnO nanowires exhibit remarkably high electron mobility (80 cm^2/V s) compared to thin-film transistors (7 cm^2/V s), as demonstrated by field-effect measurements [166]. Recent advancements in surface engineering have further enhanced these properties—a simple polyimide coating on ZnO nanorods was shown to improve device performance dramatically, increasing transconductance from 140 nS to 1.8 mS and current on/off ratios from 7 to 10^4–10^5 [167]. This surface treatment reduces electron scattering and trapping, yielding record mobilities of 1000–1200 cm^2/V s. However, the development of reliable p-type ZnO remains a fundamental challenge for complete ZnO-based optoelectronic systems. Innovative doping strategies, such as Ga-N co-doping, have achieved p-type conductivity with resistivities as low as 0.5 Ω cm, enabling the fabrication of ZnO homojunction diodes and LEDs from complementary n-type and p-type nanowires [168–170].

Recent studies have demonstrated that plasma treatment can significantly modify the electrical properties of ZnO nanowires (NWs) [171–173]. Oxygen plasma exposure leads to oxygen ion adsorption, inducing upward band bending and reduced conductivity. Conversely, hydrogen plasma treatment promotes hydrogen ion adsorption and formation of O–H bonds with ZnO, resulting in downward band bending and the creation of a highly conductive surface layer.

In a related work, researchers have fabricated Schottky diodes using ZnO nanorods (NRs) grown by chemical bath deposition. Initial devices exhibited high ideality factors and poor rectification. However, surface treatment with H_2O_2 substantially improved the metal–semiconductor contact quality [174]. For silver (Ag) contacts, this optimization yielded a rectification ratio of 10^6 at 2 V and an ideality factor of 2.4, representing significant enhancement in diode performance characteristics. These findings highlight the importance of surface engineering in controlling the electronic properties of ZnO-based devices.

Recent studies on doped ZnO nanostructures reveal significant advancements in photodetection performance. Co-doped ZnO nanobelts exhibit negative photoconductivity under 630 nm illumination, with current decreasing from 1.3 to 0.25 μA in humid environments due to photoinduced water desorption [175]. In contrast, Cu-doped ZnO nanowires demonstrate enhanced multispectral sensitivity via avalanche photomultiplication, enabling efficient UV and visible light detection [176]. Contact engineering further improves device performance; Pt-modified ZnO nanowires achieve ultralow contact resistance (1.1×10^5 Ω cm^2) and high photoconductive gain (~10^8), while optimized FET configurations (V_{ds} = 1 V, V_{gs} = − 1 V) yield exceptional photo-to-dark current ratios (~10^6) [177]. Additionally, alignment techniques such as electrospinning with Cu electrodes enable long-range uniaxial ordering of nanowires. Doping strategies also play a critical role—Al^{3+} incorporation (5%) boosts ZnO nanowire conductivity from 2.55×10^{-5} S/cm to 9.73×10^{-3} S/cm by providing additional conduction band electrons through Zn^{2+} site substitution [178].

6.1 *ZnO Tetrapods for Photodetection*

Photodetectors are integral components in a wide range of applications, including binary switching in imaging systems, light-wave communications, and emerging memory storage solutions as well as optoelectronic circuits. Among nanoscale devices, one-dimensional (1D) metal-oxide nanostructures particularly ZnO nanowires and nanotetrapods have demonstrated exceptional sensitivity to light [179, 180]. This sensitivity is primarily attributed to their high surface-to-volume ratios and Debye lengths that are comparable to their small dimensions [181]. Experimental investigations have consistently shown that the presence of oxygen vacancies significantly influences the electronic properties of these nanostructures, akin to their bulk forms. When an electric field is applied, unpaired electrons migrate toward the anode, enhancing conductivity and resulting in the formation of a low-conductivity depletion layer near the surface of the semiconductor. Under illumination, photogenerated holes move to the surface where they become trapped, leaving behind unpaired electrons that contribute to the photocurrent [182]. In ZnO nanowires, the lifetime of these unpaired electrons is further extended by the desorption of oxygen molecules from the surface when holes neutralize oxygen ions. ZnO with a tetrapod morphology has garnered significant interest for its potential in flexible solar cell applications due to its effective vectorial charge transport and excellent charge collection capabilities, even when processed using low-temperature film fabrication techniques [183, 184]. However, dye-sensitized solar cells (DSSCs) based on ZnO-T have encountered challenges with lower power conversion efficiencies when compared to those utilizing ZnO nanoparticles. This limitation is primarily due to their reduced specific surface areas [183, 185, 186]. To address this issue, researchers have developed composite photoanodes by integrating small nanoparticles into the hollow nanotetrapod network. This innovative approach has led to DSSCs on flexible plastic substrates achieving efficiencies of 4.91%. The incorporation of a ZnO shell on SnO_2 enhances contact between nanoparticles and improves open-circuit voltage (V_{oc}) which can be activated under mild NH_3 treatment conditions (Fig. 13).

The symmetrically branching 1D structure of ZnO nanotetrapods not only facilitates easy network formation on flexible substrates with improved mechanical durability but also positively impacts electron transport, resulting in better charge collection efficiency compared to larger ZnO particles [187]. Chen et al. demonstrated that DSSCs utilizing ZnO nanotetrapod photoanodes with thicknesses exceeding 30 μm can achieve impressive open-circuit voltage (V_{oc}) and fill factor (FF) without the need for calcination. This distinctive combination of thick photoanodes and the absence

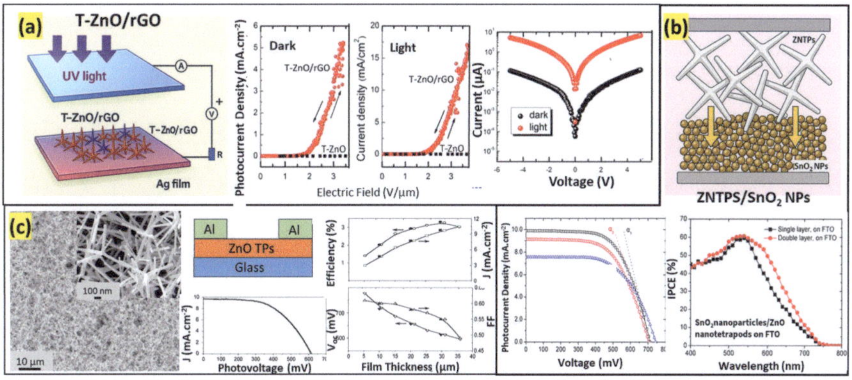

Fig. 13 **a** ZnO-Ts/rGO [189], **b** ZnO-Ts/SnO_2 [187], **c** ZnO-Ts photodetectors [188] and their photodetector characteristics

of calcination differentiates ZnO nanotetrapod photoanodes from most nanoparticle-based alternatives, offering significant advantages for flexible solar cell applications. The short-circuit current density (J_{sc}) reached 10.31 mA/cm^2, leading to an efficiency improvement to 3.27% [188]. SEM images reveal that both pure SnO_2 films and SnO_2 nanoparticles/ZnO nanotetrapod composite films exhibit notable porosity. As the proportion of ZnO nanotetrapods increases, the films become hollower and more optically opaque due to enhanced light scattering effects. Additionally, the branched structure of the nanotetrapods stabilizes the composite film by alleviating locally accumulated stress during the drying process on fluorine-doped tin oxide (FTO) glass substrates. However, charge collection losses in composite films with higher ZnO nanotetrapod content and larger thicknesses are identified as a primary limiting factor affecting incident photon-to-current efficiency (IPCE) and Jsc. To further investigate performance upon incorporating nanoparticles, an ultrathin layer of ZnO was spontaneously deposited onto SnO_2 nanoparticles. This configuration allows the ZnO nanotetrapod network to function as a global electron transport highway while enhancing surface roughness and contributing to local charge transport. In another innovative approach, graphene sheets are incorporated into ZnO/graphene heterojunctions, serving as a buffer layer between ZnO and the bottom electrode in field-emission devices. Hydrothermally grown ZnO nanorod arrays on flexible reduced graphene sheets exhibit a turn-on field of 2.1 V/μm and an emitting current density of 470 mA/cm^2 at 3 V/μm, attributed to reduced barrier resistance at the ZnO/graphene-aluminum heterojunction [189]. Experimental results confirm that electrons can effectively traverse Schottky barriers formed at this heterojunction through a field-emission process. Moreover, analysis of the O 1 s signal from ZnO-T/reduced graphene oxide (rGO) nanocomposites reveals three Gaussian components at binding energies of 530.28 eV, 531.54 eV, and 532.98 eV (Fig. 13). The component at 531.54 eV likely corresponds to an O–C bond due to differing electron affinities among carbon, zinc, and oxygen atoms. These findings indicate that graphene in ZnO-T/rGO nanocomposites engages in strong chemical interactions with ZnO

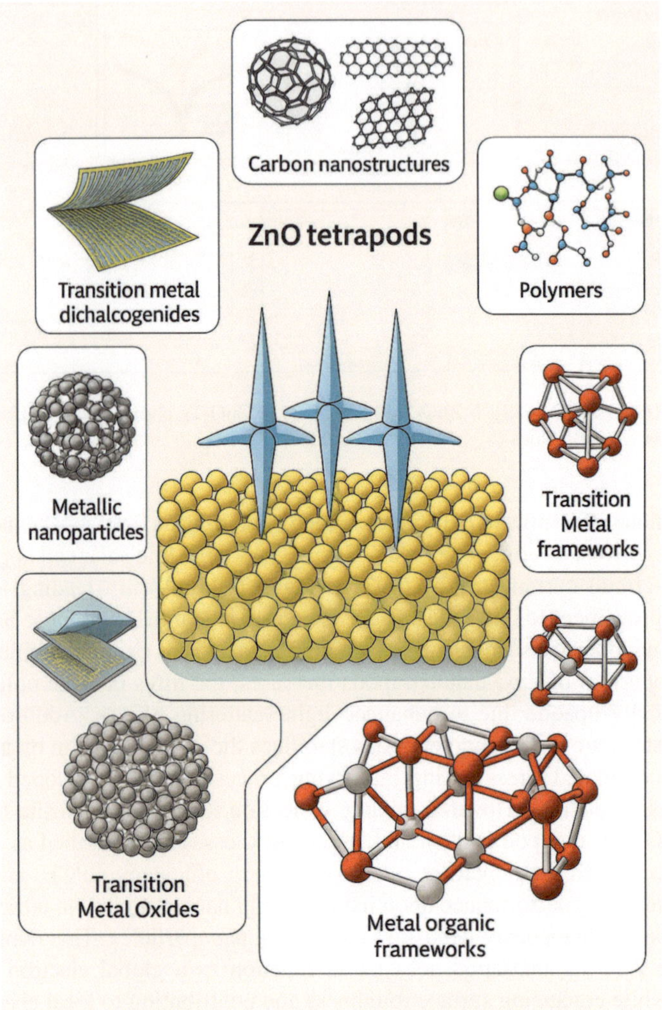

Fig. 14 Various materials/ZnO-T structures for the photodetection applications

rather than merely serving as a physisorption coating [189]. While the corrosion of ZnO nanotetrapods can create new pore channels that disrupt continuity within the nanoparticle network, this effect is less pronounced in SnO_2/ZnO (2:1) composite films due to their lower content of ZnO nanotetrapods. This comprehensive understanding of material interactions and structural advantages underscores the potential for innovative designs in future solar cell technologies.

ZnO-T (mean-arm length $\approx$ 26 μm) were processed into porous ceramic disks (2.4 mm thickness) using polystyrene microspheres as pore-forming agents, creating uniform micron-scale porosity in the ITO/porous ZnO/Ag photodetector architecture

[190]. The device exhibited enhanced thermoelectric performance through controlled temperature gradients generated by a TE module, quantified by the figure of merit $zT = S^2\sigma T/\rho\kappa$ where S, σ, T, ρ, and κ are the Seebeck coefficient, electrical conductivity, the absolute temperature, the electrical resistivity, and the thermal conductivity, respectively. Upon illumination, photothermal heating at the ITO interface established a temperature gradient ($\Delta T \approx -2.4$ K), driving electron diffusion toward the cooler Ag electrode and generating a built-in thermopotential (V_{TE}). This process simultaneously induced band bending—elevating conduction/valence bands at the ZnO/Ag interface while depressing them at the ITO/ZnO interface—which synergistically promoted charge separation. The thermo-phototronic coupling effect significantly boosted device performance, yielding 27.46% higher photocurrent and 21.78% greater sensitivity under 0.408 W cm^{-2} illumination compared to conventional operation. The porous ZnO matrix facilitated this enhancement through its combined properties of high surface area for efficient photon absorption, interconnected pathways for charge transport, and temperature-responsive band structure modulation, demonstrating the potential of morphology-engineered ZnO for advanced optoelectronic applications.

ZnO-T demonstrate significant promise for chemical gas sensing applications through surface chemisorption mechanisms [191]. When gas molecules chemically adsorb onto the ZnO surface, they form covalent bonds that either donate or accept electrons from the semiconductor matrix, thereby modulating its electrical properties. Experimental studies reveal a direct correlation between oxygen concentration and electrical resistance in tetrapodal structures—the resistance exhibits monotonic enhancement with increasing O_2 partial pressure, even at trace concentrations (measurements taken after 10-min stabilization periods). This sensitivity to oxygen chemisorption manifests particularly at room temperature, where sensor responsivity shows an inverse relationship with oxygen pressure. The pressure-dependent behavior suggests that surface charge transfer processes dominate the sensing mechanism, with oxygen acting as an electron acceptor that modifies the depletion layer characteristics This is mainly due to the increasing depletion region of the electrons in ZnO. Oxygen is chemisorbed to ZnO surface forming oxygen negative ions, thus leading to increase in the electrical resistance and surface charge depletion region. The persistence of photoconductivity in ZnO nanostructures exhibits strong environmental dependence. Under inert atmospheres (N_2 or Ar) or reduced pressure conditions, UV-induced photocurrent demonstrates prolonged decay kinetics (Fig. 13). In contrast, oxygen-containing environments induce a biphasic decay profile - an initial rapid decrease followed by extended persistent photoconductivity. This behavior originates from competing surface processes: chemisorbed oxygen species act as electron traps, extracting charge carriers from the conduction band and causing the initial current drop. However, progressive surface charge accumulation creates an upward band bending that eventually inhibits further electron trapping. This self-limiting mechanism explains the characteristic long-tail decay kinetics, where complete recovery to baseline dark conductivity may require several hours. The phenomenon is particularly pronounced in tetrapodal ZnO structures, where high surface-to-volume ratios amplify these surface-mediated effects.

Postica et al. investigated four morphologically distinct ZnO nanostructures for UV sensing applications: (i) sheet-decorated tetrapods (ZnO-I, 3–8 μm leg length, 250–500 nm diameter), (ii) low-sheet nanowires (ZnO-II, 50–300 nm diameter), (iii) large tetrapods (ZnO-III, 10–30 μm arm length, 1–3 μm diameter), and (iv) small tetrapods with complex arm geometry featuring abrupt diameter transitions to ultra-thin wire-like segments (ZnO-IV) [192]. The study revealed an inverse relationship between applied bias voltage and UV response across all samples, attributed to self-heating effects that promote electron–hole recombination. This dependence was most pronounced in ZnO-II and ZnO-IV due to their smaller diameters enhancing Joule heating. Tetrapod-based devices (ZnO-III/IV) exhibited faster response times than nanowire networks (ZnO-II), with ZnO-IV's thin wire-like segments (50–150 nm, comparable to the Debye length) creating numerous potential barriers that significantly modulated depletion regions. The ZnO-IV morphology demonstrated superior performance, achieving responsivities of 5.2×10^{-3}, 16×10^{-3}, and 58×10^{-3} A/W at 5, 10, and 20 V respectively, along with a peak UV response of ~3100 at 5 V. This enhancement stemmed from its high porosity and optimal barrier density facilitating efficient oxygen photodesorption—where photogenerated holes discharge surface-adsorbed oxygen species while electrons accumulate in the conduction channel, reducing barrier heights. In contrast, ZnO-III's larger dimensions limited barrier formation, compromising sensitivity but improving response speed. The nanostructures also showed promise in photovoltaics, with nanotetrapods exhibiting exceptional charge transport (Jsc = 10.31 mA/cm^2, diffusion lengths > 36 μm) in DSSC applications, while SnO_2/ZnO core–shell configurations boosted Voc through conduction band engineering. In a study by Alsulthany et al., the UV detection performance of an Al/ZnO-T/Al device was evaluated by analyzing current–voltage (I–V) characteristics under dark conditions and 365 nm UV illumination (0.22 mW/cm^2 intensity, 5 V bias) [193]. The linear I–V response in both forward and reverse bias directions confirmed ohmic contact formation between the ZnO-T and Al electrodes—a critical factor for improving photodetector efficiency. Ohmic contacts enhance device sensitivity by reducing junction resistance. At 5 V bias, the dark current measured 0.158 A, significantly lower than values reported in previous studies. This improvement is likely due to the tetrapod morphology's high aspect ratio, which reduces the active volume compared to traditional detector designs. Under UV exposure, the device exhibited a photocurrent of 24.2 A, achieving a sensitivity (S) of 15.22×10^3 and a current gain of 153.2 [194, 195].

In another study, Yi et al. explored the heterojunction properties of ZnO-T combined with PEDOT: PSS as an organic layer [150]. In this configuration, electrons from ZnO migrate into PEDOT: PSS until electronic equilibrium is established, creating a depletion region at the interface. When exposed to UV light, photogenerated electron–hole pairs in this depletion zone are separated by the built-in electric field with electrons directed towards the ZnO side, contributing to a photocurrent in reverse bias conditions even at zero bias. The device exhibited significant increases in current under UV irradiation, demonstrating photovoltaic properties with an open-circuit voltage (V_{oc}) of approximately 0.2 V and a short-circuit current (I_{sc}) of about 1.1 nA. The maximum power output measured under a 325 nm UV laser beam

(0.16 mW) was approximately 55 pW at $V_m = 0.1$ V and $I_m = 0.55$ nA, yielding a fill factor (FF) of around 25%. The rise and decay times for I_{sc} were approximately 3.5 s and 4.5 s, respectively, indicating that the device can operate without external power (self-powering capability) when illuminated by UV light. The high hydrophobicity and 3D architecture of ZnO-Ts facilitate the attachment of carbon nanodots (CNOs) onto the ZnO arms through hydrophobic interactions. The differences in surface energies between ZnO-T and CNOs where are related to oxygen vacancies in ZnO-T and surface defects in CNOs, promote long-range attractive interactions that stabilize smaller CNOs on the flat side faces of ZnO-T. Although ZnO-T does not easily suspend in water, the ZnO-T/CNO hybrid composite can be readily dispersed in water and remains stable for extended periods. The rate constant for the ZnO-T/CNO composite is approximately seven times higher (0.01834 min^{-1}) than that of ZnO-T alone (0.00274 min^{-1}), and about thirty-six times higher than that of CNOs (0.0005 min^{-1}). A visible-light-sensitive hybrid composite was successfully fabricated through a straightforward one-step process, allowing easy access to the beneficial features of both components. The complementary properties of ZnO-T and CNOs lead to enhanced photocatalytic activity even under visible light, attributed to improved absorption capabilities from CNO sensitization and effective charge separation facilitated by the unique tetrapod morphology [196, 197]. Newton et al. constructed a single-ZnO nanotetrapod p–n junction diode that demonstrated improved sensitivity to UV light, achieving a remarkable responsivity of 1.7×10^4 A/W. This enhancement in sensitivity is attributed to the photoinduced charge carriers and the widening of conduction channels [198]. The interconnections within the tetrapod arms facilitate continuous paths for carrier migration, resulting in a rapid response time of 32 ms and a decay time of 200 ms.

6.2 *Optimization of ZnO Tetrapod Based Devices*

- To fully exploit ZnO's capabilities in energy and nanoelectronics, researchers have implemented multiple modification strategies. Common approaches involve doping with transition metals or nonmetals, forming heterostructures with complementary semiconductors, functionalizing surfaces with organic dyes, depositing noble metal nanoparticles, and integrating carbon-based nanomaterials. An emerging technique focuses on intrinsic defect engineering, where controlled introduction of vacancies and interstitials tailors ZnO's properties without external dopants [199, 200]. Engineering techniques have been developed to enhance the performance of ZnO ultraviolet photodetectors (PDs), focusing on both material properties and device architecture. Key methods include:
- Introducing Imperfections: Creating specific defects within the crystal lattice can adjust the band gap and electronic configuration, improving charge carrier concentration and mobility. This leads to significant enhancements in electrical, optical, and magnetic properties.

- Nonstoichiometric ZnO: This variant shows considerable promise for use in emerging nanoscale devices due to its modified characteristics.
- Among the various defects in ZnO, oxygen vacancies (Vo) and zinc vacancies (V_{Zn}) are particularly noteworthy. These defects are characterized by low formation energy, making them prevalent in nominally undoped ZnO. Experimental and theoretical studies indicate that these vacancies significantly influence the modification of ZnO properties [76]. The presence of oxygen vacancies enhances the utilization of ultraviolet light and extends the absorption spectrum into the visible range. This improvement is crucial for photocatalytic processes such as hydrogen production from water splitting and the degradation of organic pollutants. Conversely, zinc vacancies contribute to room-temperature ferromagnetism which could be advantageous for developing spintronic devices like magnetoresistive sensors or memory systems [201, 202].
- During thermal treatment, hydrogen atoms react with surface oxygen and zinc atoms in ZnO, generating vacancies while incorporating interstitial and substitutional hydrogen. This mechanism occurs in multiple steps. First, hydrogen adsorbs onto available oxygen and zinc vacancy sites, partially passivating and ionizing existing oxygen vacancies (V_O) and zinc vacancies (V_{Zn}). Computational studies indicate that atomic hydrogen stabilizes when adsorbed on oxygen sites or combined oxygen-zinc sites, whereas adsorption solely on zinc sites is energetically unfavourable. The creation of oxygen vacancies in hydrogenated ZnO is often accompanied by zinc vacancy formation, suggesting these processes are interdependent. Additionally, hydrogenation induces a dimensional contraction in ZnO due to the ejection of oxygen and zinc atoms. The final distribution of these released species—whether they fully desorb or redeposit onto the ZnO surface—depends on the experimental conditions [76, 77, 202].

To mitigate the need for high temperatures and excessive energy consumption during the hydrogenation process using H_2 gas, alternative methods have been developed for creating defective ZnO. Among II—VI semiconductors, ZnO exhibits relatively high stability and is less susceptible to degradation than similar materials. Moderate-energy electron bombardment can induce defects, particularly zinc vacancies, with little surface damage [203, 204]. To recover stoichiometry, treating ZnO with nitrogen gas (N_2) at temperatures under ~ 200 °C is effective. Unlike electron bombardment, ion sputtering can produce oxygen vacancies on ZnO surfaces even at low energies. However, creating detectable bulk vacancies via sputtering demands much higher energies—ranging from hundreds of keV to several MeV. The defects produced under ion irradiation tend to be more stable than those created by electron bombardment and are not easily removed by low-temperature annealing [205, 206]. Doping is essential for the application of optoelectronic devices, particularly due to the shift towards reduced dimensionality structures such as quantum wells. Historically, research on ZnO focused on bulk samples, addressing various topics including growth, doping, transport properties, deep centers, band structure, excitons, polaritons, luminescence, and lasing phenomena. Among these factors, doping plays a crucial role in enhancing the sensory response of ZnO nanostructures. Transition metal ions like cobalt (Co), nickel (Ni), copper (Cu), and iron (Fe) have been prominently utilized as dopants in ZnO nanostructures [207–209]. These dopants not only enhance the material's activity but also modify its resistance. Furthermore, doping reduces the operating temperature of sensors while improving selectivity, stability, and response time to target gases. For instance, films of Cu-doped ZnO nanorod (NR) arrays demonstrate stable and repeatable photocurrent cycles without degenerative effects during detection. Additionally, Cu-doped ZnO NR arrays exhibit high efficiency in ultraviolet (UV) detection [210]. A photodetector based on gallium-doped ZnO films produced via radio frequency sputtering showed a photoresponsivity of approximately 2.6 A/W at a 10 V bias in the UV region, characterized by high responsiveness and rapid rise and decay times of 10 ns and 960 ns, respectively [211]. Moreover, research by Shinde et al. indicated that gallium-doped ZnO thin-film detectors achieved a high responsivity of about 1125 A/W at a 5 V bias with a maximum wavelength of 365 nm [212].

To optimize ZnO for energy and nanoelectronics applications, several advanced techniques are employed. Key strategies include defect engineering (introducing oxygen and zinc vacancies), nonstoichiometric ZnO synthesis, hydrogenation, electron bombardment, and ion sputtering. These approaches modify the crystal structure to enhance electrical, optical, and magnetic properties. Additionally, doping with transition metals like Co, Ni, Cu, Fe, and Ga significantly boosts ZnO's photodetection and sensing capabilities by improving carrier concentration, stability, and response time. Collectively, these methods enable the fine-tuning of ZnO for high-performance optoelectronic and spintronic devices.

7 Conclusion

In this chapter, a comprehensive study is focused on tetrapod-structured ZnO for efficient ultraviolet photodetector applications. The chapter discusses the fundamental mechanisms of photodetectors, different types including Schottky barrier, metal–semiconductor-metal, homojunction, and heterojunction devices, and the critical performance parameters such as responsivity, detectivity, and response time. UV photodetectors are critical in areas like environmental monitoring, flame sensing, and space communications. Traditional silicon-based photodetectors fall short due to their narrow bandgap, which necessitates the use of external filters and leads to suboptimal performance in UV-specific applications. This has led researchers to explore wide bandgap (WBG) materials like GaN, SiC, and notably ZnO, which provide intrinsic solar-blind capabilities and better thermal and radiation tolerance. Of these, ZnO stands out due to its cost-effectiveness, high electron mobility, and ease of synthesis. ZnO is also a highly versatile semiconducting metal oxide with remarkable properties including high photosensitivity, tunable specific surface area, nontoxicity, piezo- and pyroelectric effects, wide bandgap (3.4 eV), large exciton binding energy. It also has the potential to exist in multiple structures and morphologies with ZnO-T being the one under highlight due to their unique characteristics like high surface area-to-volume ratio, enhanced charge transport properties and efficient electron–hole separation capabilities. Various synthesis techniques can be used for fabricating ZnO-T like vapor transport, microwave synthesis, flame transport synthesis (FTS), and thermal evaporation with FTS standing out for its scalability, high yield and ability to produce tetrapods with high crystallinity and well-controlled dimensions. Each leg of the tetrapod is defect-free and aligned along the c-axis, promoting high carrier mobility and reducing recombination losses. These factors make tetrapod structures particularly beneficial for high-gain, low-noise UV photodetection. In comparison to nanorods or nanoparticles, tetrapod-based networks form naturally interconnected 3D matrices with minimal grain boundary resistance. This structure supports vertical charge extraction and improves the interaction with electrode materials. Additionally, the orientation of one arm toward the substrate facilitates efficient anchoring, aiding in device fabrication and mechanical stability. Tetrapod-based devices demonstrate exceptional performance metrics: low dark current, fast rise/fall times (< 1 s in optimized systems), and high UV/visible rejection ratios. These properties make ZnO-T suitable not only for traditional UV sensors but also for next-generation flexible, transparent, and wearable photodetectors. Maintaining the diameter of tetrapod "legs" below 20 nm is important for quantum confinement and enhanced sensitivity. Thinner diameters can lead to higher responsivity and faster response times due to increased potential barriers and modulation of the depletion region. Homojunction devices using p-n ZnO layers face challenges due to the difficulty of stable p-type doping, but when successful, they enable self-powered operation. Heterojunction photodetectors, which combine ZnO with other materials such as NiO or GaN, exploit band alignment and interfacial electric fields to improve charge separation and reduce recombination losses, while organic/inorganic hybrid structure is of wide attention for its ability to combine the adavantages of both organic and inorganic materials.

Introduction of transition metals (e.g., Cu, Ga) and nonmetals can modify ZnO's properties, enhance activity, alter resistance, lower operating temperature, and improve selectivity and response time. Controlling defects like oxygen vacancies (V_O) and zinc vacancies (V_{Zn}) is crucial. Oxygen vacancies enhance UV light utilization and extend absorption into the visible range, beneficial for photocatalysis. Surface sensitization with organic dyes, decoration with noble metals, and hybridization with carbon materials are some other optimization techniques.

References

1. Shaker, L.M., Al-Amiery, A., Isahak, W.N.R.W.: Optoelectronics' quantum leap: Unveiling the breakthroughs driving high-performance devices. Green Technol. Sustain. **2**, 100111 (2024). https://doi.org/10.1016/j.grets.2024.100111
2. Butt, M.A., Janaszek, B., Piramidowicz, R.: Lighting the way forward: the bright future of photonic integrated circuits. Sens. Int. **6**, 100326 (2025). https://doi.org/10.1016/j.sintl.2025.100326

3. Bernardo, C.P.C.V., Lameirinhas, R.A.M., de Melo Cunha, J.P., Torres, J.P.N.: A revision of the semiconductor theory from history to applications. Discov. Appl. Sci. **6**, 316 (2024). https://doi.org/10.1007/s42452-024-06001-1
4. Rogalski, A., Bielecki, Z., Mikołajczyk, J., Wojtas, J.: Ultraviolet photodetectors: from photocathodes to low-dimensional solids. Sensors **23**, 4452 (2023). https://doi.org/10.3390/s23094452
5. Ma, T., Xue, N., Muhammad, A., Fang, G., Yan, J., Chen, R., Sun, J., Sun, X.: Recent progress in photodetectors: from materials to structures and applications. Micromachines **15**, 1249 (2024). https://doi.org/10.3390/mi15101249
6. Molaei-Yeznabad, A., Abedi, K.: Optimal design of graphene-based plasmonic enhanced photodetector using PSO. Sci. Rep. **14**, 15291 (2024). https://doi.org/10.1038/s41598-024-65311-x
7. Vai, A.T., Kuznetsov, V.L., Dilworth, J.R., Edwards, P.P.: UV-induced improvement in ZnO thin film conductivity: a new in situ approach. J. Mater. Chem. C **2**, 9643–9652 (2014). https://doi.org/10.1039/c4tc01749h
8. Srivathsa, M., Kumar, P., Rajendra, B.V.: Ultraviolet photoconductivity and photoluminescence properties of spray pyrolyzed ZnO nanostructure: effect of deposition temperature. Opt. Mater. **131**, 112726 (2022). https://doi.org/10.1016/j.optmat.2022.112726
9. L. Nanver, T. Knežević, Optical detectors. Adv. Semicond. Technol. 211–229, (2022). https://doi.org/10.1002/9781119869610.ch11
10. Thyda, L., Naresh, K., Joseph, J.K., Suneetha, S., Jeyanthi, C.E., Amaladass, P., Selvaraju, C., Thangaraju, K.: Improved ultraviolet photodetector performances using solution-processed nitrogen-doped carbon quantum dots/ZnO hybrid thin films. Thin Solid Films **790**, 140221 (2024). https://doi.org/10.1016/j.tsf.2024.140221
11. Kazmi, J., Abbas, A., James Young, D., Hussain Shah, J., Ahmad, W., Shah, S.S.A., Raza, S.R.A., Mohamed, M.A., Govoeov, A.O.: ZnO nanowire UV photodetectors: at the intersection of flexibility, biocompatibility, and visible blindness. Mater. Today **82**, 139–180 (2025). https://doi.org/10.1016/j.mattod.2024.11.004
12. Alaie, Z., Mohammad Nejad, S., Yousefi, M.H.: Recent advances in ultraviolet photodetectors. Mater. Sci. Semicond. Process. **29**, 16–55 (2015). https://doi.org/10.1016/j.mssp.2014.02.054
13. Jia, L., Zheng, W., Huang, F.: Vacuum-ultraviolet photodetectors. PhotoniX **1**, 22 (2020). https://doi.org/10.1186/s43074-020-00022-w
14. Yuvaraja, S., Khandelwal, V., Tang, X., Li, X.: Wide bandgap semiconductor-based integrated circuits. Chip **2**, 100072 (2023). https://doi.org/10.1016/j.chip.2023.100072
15. Lutz, G., Richter, R.H., Strüder, L.: Novel pixel detectors for X-ray astronomy and other applications. Nucl. Instrum. Methods Phys. Res. Sect. A **461**, 393–404 (2001). https://doi.org/10.1016/s0168-9002(00)01258-4
16. Singh, B.J., Chakraborty, A., Sehgal, R.: A systematic review of industrial wastewater management: evaluating challenges and enablers. J. Environ. Manage. **348**, 119230 (2023). https://doi.org/10.1016/j.jenvman.2023.119230
17. Al-Hardan, N.H., Abdul Hamid, M.A., Jalar, A., Firdaus-Raih, M.: Unleashing the potential of gallium oxide: a paradigm shift in optoelectronic applications for image sensing and neuromorphic computing applications. Mater. Today Phys. **38**, 101279 (2023). https://doi.org/10.1016/j.mtphys.2023.101279
18. Chen, H., Liu, K., Hu, L., Al-Ghamdi, A.A., Fang, X.: New concept ultraviolet photodetectors. Mater. Today **18**, 493–502 (2015). https://doi.org/10.1016/j.mattod.2015.06.001
19. Cai, Q., You, H., Guo, H., Wang, J., Liu, B., Xie, Z., Chen, D., Lu, H., Zheng, Y., Zhang, R.: Progress on AlGaN-based solar-blind ultraviolet photodetectors and focal plane arrays. Light Sci. Appl. **10** (2021). https://doi.org/10.1038/s41377-021-00527-4
20. Wang, X., Ding, K., Huang, L., Li, X., Ye, L., Luo, J., Jiang, J., Li, H., Xiong, Y., Ye, L., Pang, D., Tang, Y., Li, W., Zhang, H., Kong, C.: Enhancing the performance of self-powered deep-ultraviolet photoelectrochemical photodetectors by constructing α-Ga_2O_3@a-Al_2O_3 core-shell nanorod arrays for solar-blind imaging. Appl. Surf. Sci. **648**, 159022 (2024). https://doi.org/10.1016/j.apsusc.2023.159022

21. Tsai, S.-H., Shen, Y.-C., Huang, C.-Y., Horng, R.-H.: Deep-ultraviolet schottky photodetectors with high deep-ultraviolet/visible rejection based on a $ZnGa_2O_4$ thin film. Appl. Surf. Sci. **496**, 143670 (2019). https://doi.org/10.1016/j.apsusc.2019.143670
22. Suchikova, Y., Nazarovets, S., Popov, A.I.: Ga_2O_3 solar-blind photodetectors: from civilian applications to missile detection and research agenda. Opt. Mater. **157**, 116397 (2024). https://doi.org/10.1016/j.optmat.2024.116397
23. Tang, X., Yang, T., Yu, D., Xiong, H., Zhang, S.: Current insights and future perspectives of ultraviolet radiation (UV) exposure: friends and foes to the skin and beyond the skin. Environ. Int. **185**, 108535 (2024). https://doi.org/10.1016/j.envint.2024.108535
24. Liu, J., Cristoloveanu, S., Wan, J.: A review on the recent progress of silicon-on-insulator-based photodetectors. Phys. Status Solidi **218** (2021). https://doi.org/10.1002/pssa.202000751
25. El-Mahalawy, A.M., Abdrabou, M.M., Wassel, A.R., El-Salam, M.A., Amin, F.M.: Plasmonic enhanced ultraviolet photodetection performance of n-TiO_2/p-Si anisotype heterojunction with aluminum patterned array. J. Phys. Chem. Solids **170**, 110943 (2022). https://doi.org/10.1016/j.jpcs.2022.110943
26. Song, Z., Liu, Y., Wang, Q., Yuan, S., Yang, Y., Sun, X., Xin, Y., Liu, M., Xia, Z.: Self-powered photodetectors based on a ZnTe–TeO_2 composite/Si heterojunction with ultra-broadband and high responsivity. J. Mater. Sci. **53**, 7562–7570 (2018). https://doi.org/10.1007/s10853-018-2093-y
27. Xu, J., Rechav, K., Popovitz-Biro, R., Nevo, I., Feldman, Y., Joselevich, E.: High-Gain 200 ns photodetectors from self-aligned CdS–CdSe core-shell nanowalls. Adv. Mater. **30**, 1800413 (2018). https://doi.org/10.1002/adma.201800413
28. Weng, C., Gao, A., Huang, J., Yang, L., He, S.: High-performance self-powered transparent metal–semiconductor-metal ultraviolet photodetector based on Sub-10 nm thick dual-asymmetric interdigitated electrodes. ACS Photonics **11**, 561–569 (2024). https://doi.org/10.1021/acsphotonics.3c01435
29. Adegoke, O., Nyokong, T., Forbes, P.B.C.: Deposition of CdS, CdS/ZnSe and CdS/ZnSe/ZnS shells around CdSeTe alloyed core quantum dots: effects on optical properties. Luminescence **31**, 694–703 (2015). https://doi.org/10.1002/bio.3013
30. Dasi, G., Lavanya, T., Sathiyan, G., Gupta, R.K., Garg, A., Amaladass, P., Thangaraju, K.: Improved hole injection/extraction using PEDOT:PSS interlayer coated onto high temperature annealed ITO electrode for efficient device performances. Superlattices Microstruct. **156**, 106953 (2021). https://doi.org/10.1016/j.spmi.2021.106953
31. Dasi, G., Ramarajan, R., Joseph, D.P., Vijayakumar, S., Shim, J.-J., Arivananthan, M., Jayavel, R., Thangaraju, K.: Enhanced UV emission of solution processed highly transparent Alq_3/ZnO hybrid thin films. Thin Solid Films **710**, 138265 (2020). https://doi.org/10.1016/j.tsf.2020.138265
32. Roccaforte, F., Fiorenza, P., Greco, G., Lo Nigro, R., Giannazzo, F., Iucolano, F., Saggio, M.: Emerging trends in wide band gap semiconductors (SiC and GaN) technology for power devices. Microelectron. Eng. **187–188**, 66–77 (2018). https://doi.org/10.1016/j.mee.2017.11.021
33. Almalawi, D., Lopatin, S., Mitra, S., Flemban, T., Siladie, A.-M., Daudin, B., Roqan, I.S.: Enhanced UV emission of GaN nanowires functionalized by wider band gap solution-processed p-MnO quantum dots. ACS Appl. Mater. Interfaces **12**, 34058–34064 (2020). https://doi.org/10.1021/acsami.0c07029
34. Li, Z., Yan, T., Fang, X.: Low-dimensional wide-bandgap semiconductors for UV photodetectors. Nat. Rev. Mater. **8**, 587–603 (2023). https://doi.org/10.1038/s41578-023-00583-9
35. Jia, M., Wang, F., Tang, L., Xiang, J., Teng, K.S., Lau, S.P.: High-performance deep ultraviolet photodetector based on NiO/β-Ga_2O_3 heterojunction. Nanoscale Res. Lett. **15**, 47 (2020). https://doi.org/10.1186/s11671-020-3271-9
36. Praveen, S., Veeralingam, S., Badhulika, S.: A flexible self-powered UV photodetector and optical UV filter based on β-Bi_2O_3/SnO_2 quantum dots schottky heterojunction. Adv. Materials Inter. **8**, 2100373 (2021). https://doi.org/10.1002/admi.202100373

37. Kublitski, J., Hofacker, A., Boroujeni, B.K., Benduhn, J., Nikolis, V.C., Kaiser, C., Spoltore, D., Kleemann, H., Fischer, A., Ellinger, F., Vandewal, K., Leo, K.: Reverse dark current in organic photodetectors and the major role of traps as source of noise. Nat. Commun. **12**, 551 (2021). https://doi.org/10.1038/s41467-020-20856-z
38. Shi, L., Zhu, Y., Li, G., Ji, T., Wang, W., Zhang, Y., Wu, Y., Hao, Y., Wang, K., Yuan, J., Zou, Y., Ong, B.S., Zhu, F., Cui, Y.: Atomic-level chemical reaction promoting external quantum efficiency of organic photomultiplication photodetector exceeding 108% for weak-light detection. Sci. Bull. **68**, 928–937 (2023). https://doi.org/10.1016/j.scib.2023.04.015
39. Eun, H.J., Lee, H., Shim, Y., Seo, G.U., Lee, A.Y., Park, J.J., Heo, J., Park, S., Kim, J.H.: Strain-durable dark current in near-infrared organic photodetectors for skin-conformal photoplethysmographic sensors. *iScience* **25**, 104194 (2022). https://doi.org/10.1016/j.isci.2022.104194
40. Liang, W., Wang, L., Li, Y., Zhang, F., Chen, X., Wu, D., Tian, Y., Li, X., Shan, C., Shi, Z.: Stable and ultraviolet-enhanced broadband photodetectors based on Si nanowire arrays-$Cs_3Cu_2I_5$ nanocrystals hybrid structures. Mater. Today Phys. **18**, 100398 (2021). https://doi.org/10.1016/j.mtphys.2021.100398
41. Ma, X., Janssen, R.A.J., Gelinck, G.H.: Trap-assisted charge generation and recombination in state-of-the-art organic photodetectors. Adv Mater. Technol. **8**, 2300234 (2023). https://doi.org/10.1002/admt.202300234
42. Hammond, W.T., Mudrick, J.P., Xue, J.: Balancing high gain and bandwidth in multilayer organic photodetectors with tailored carrier blocking layers. J. Appl. Phys. **116**, 214501 (2014). https://doi.org/10.1063/1.4902149
43. Chauhan, D., Perera, A.G.U., Li, L.H., Chen, L., Linfield, E.H.: Dark current and photoresponse characteristics of extended wavelength infrared photodetectors. J. Appl. Phys. **122**, 024501 (2017). https://doi.org/10.1063/1.4992075
44. Tsai, K., Chen, M., Suthar, G., Hsiao, Y., Cheng, L., Liao, C., Chen, F., Chu, C., Chang, Y.: Suppressing the dark current while improving the quantum efficiency in shortwave infrared organic photodetectors through naphthalenediimide-based interlayer. Adv. Opt. Mater. **12**, 2302435 (2024). https://doi.org/10.1002/adom.202302435
45. Huang, Y., Ji, A., Zhang, X., Wang, K., Wang, Y.: Highly sensitive UV detector based on the transverse dember effect of 4H-SiC single crystal. Appl. Phys. Lett. **126**, 023904 (2025). https://doi.org/10.1063/5.0243564
46. Han, S., Huang, X., Fang, M., Zhao, W., Xu, S., Zhu, D., Xu, W., Fang, M., Liu, W., Cao, P., Lu, Y.: High-performance UV detectors based on room-temperature deposited amorphous Ga_2O_3 thin films by RF magnetron sputtering. J. Mater. Chem. C **7**, 11834–11844 (2019). https://doi.org/10.1039/c9tc03613j
47. So, H., Lim, J., Suria, A.J., Senesky, D.G.: Highly antireflective AlGaN/GaN ultraviolet photodetectors using ZnO nanorod arrays on inverted pyramidal surfaces. Appl. Surf. Sci. **409**, 91–96 (2017). https://doi.org/10.1016/j.apsusc.2017.02.139
48. Waghadkar, Y.B., Umarji, G., Kekade, S.S., Rane, S., Chauhan, R., Ashokkumar, M., Gosavi, S.W.: Synthesis and characterization of indium-doped ZnO nanoparticles by coprecipitation method for highly photo-responsive UV light sensors. Sens. Int. **5**, 100271 (2024). https://doi.org/10.1016/j.sintl.2023.100271
49. Zhang, Y., Wu, Z.-F., Gao, P.-F., Fang, D.-Q., Zhang, S.-L.: Enhanced visible light absorption in ZnO/GaN heterostructured nanofilms. J. Alloy. Compd. **704**, 478–483 (2017). https://doi.org/10.1016/j.jallcom.2017.02.096
50. Abbas, K., Ji, P., Ullah, N., Shafique, S., Zhang, Z., Ameer, M.F., Qin, S., Yang, S.: Graphene photodetectors integrated with silicon and perovskite quantum dots. Microsyst. Nanoeng. **10**, 81 (2024). https://doi.org/10.1038/s41378-024-00722-4
51. Hosen, R., Sikder, S., Uddin, Md.S., Haque, Md.M., Mamur, H., Bhuiyan, M.R.A.: Effect of various layers on improving the photovoltaic efficiency of Al/ZnO/CdS/CdTe/Cu_2O/Ni solar cells. J. Alloy. Metall. Syst. **4**, 100041 (2023). https://doi.org/10.1016/j.jalmes.2023.100041
52. Kumar, M., Dhar, J.C.: Enhanced UV photodetector performance using sputtered Mg-doped ZnO thin film. Opt. Mater. **157**, 116059 (2024). https://doi.org/10.1016/j.optmat.2024.116059

53. Roshan, H., Ravanan, F., Sheikhi, M.H., Mirzaei, A.: High-detectivity near-infrared photodetector based on Ag_2S nanocrystals. J. Alloy. Compd. **852**, 156948 (2021). https://doi.org/10.1016/j.jallcom.2020.156948
54. Elkamel, I.B., Hamdaoui, N., Mezni, A., Ajjel, R., Beji, L.: High responsivity and 1/f noise of an ultraviolet photodetector based on Ni doped ZnO nanoparticles. RSC Adv. **8**, 32333–32343 (2018). https://doi.org/10.1039/c8ra05567j
55. Kublitski, J., Fischer, A., Xing, S., Baisinger, L., Bittrich, E., Spotore, D., Benduhn, J., Vandewal, K., Leo, K.: Enhancing sub-bandgap external quantum efficiency by photomultiplication for narrowband organic near-infrared photodetectors. Nat. Commun. **12**, 4259 (2021). https://doi.org/10.1038/s41467-021-24500-2
56. Panda, D., Tseng, T.-Y.: One-dimensional ZnO nanostructures: fabrication, optoelectronic properties, and device applications. J. Mater. Sci. **48**, 6849–6877 (2013). https://doi.org/10.1007/s10853-013-7541-0
57. Wagh, S.S., Kadam, V.S., Jagtap, C.V., Salunkhe, D.B., Patil, R.S., Pathan, H.M., Patole, S.P.: Comparative studies on synthesis, characterization and photocatalytic activity of Ag doped ZnO nanoparticles. ACS Omega **8**, 7779–7790 (2023). https://doi.org/10.1021/acsomega.2c07499
58. Fernández, S., Naranjo, F.B.: Optimization of aluminum-doped zinc oxide films deposited at low temperature by radio-frequency sputtering on flexible substrates for solar cell applications. Sol. Energy Mater. Sol. Cells **94**, 157–163 (2010). https://doi.org/10.1016/j.solmat.2009.08.012
59. Desai, A.V., Haque, M.A.: Mechanical properties of ZnO nanowires. Sens. Actuators A **134**, 169–176 (2007). https://doi.org/10.1016/j.sna.2006.04.046
60. Xu, X., Chen, Y., Zhang, G., Bian, H., Zhao, M., Ma, S.: Optical properties and the bandgap variation in diverse $Zn_{1\text{-}x}Sn_xO$ nanostructures. Superlattices Microstruct. **123**, 349–357 (2018). https://doi.org/10.1016/j.spmi.2018.09.021
61. Elangovan, P., John Peter, A.: Size dependent effective band gap, optical absorption coefficients and refractive index changes in a wide $ZnS/Zn_{1-x}Mg_xS$ strained quantum well. Phys. B Condens. Matter **407**, 2578–2583 (2012). https://doi.org/10.1016/j.physb.2012.03.073
62. Rodrigues, J., Ben Sedrine, N., Correia, M.R., Monteiro, T.: Photoluminescence investigations of ZnO micro/nanostructures. Mater. Today Chem. **16**, 100243 (2020). https://doi.org/10.1016/j.mtchem.2020.100243
63. Sima, M., Mihut, L., Vasile, E., Sima, M., Logofatu, C.: Optical properties of Mn doped ZnO films and wires synthesized by thermal oxidation of ZnMn alloy. Thin Solid Films **590**, 141–147 (2015). https://doi.org/10.1016/j.tsf.2015.07.072
64. Oh, D.C., Suzuki, T., Kim, J.J., Makino, H., Hanada, T., Cho, M.W., Yao, T.: Electron-trap centers in ZnO layers grown by molecular-beam epitaxy. Appl. Phys. Lett. **86**, 032909 (2005). https://doi.org/10.1063/1.1849852
65. Ryu, Y., Lee, T.-S., Lubguban, J.A., White, H.W., Kim, B.-J., Park, Y.-S., Youn, C.-J.: Next generation of oxide photonic devices: ZnO-based ultraviolet light emitting diodes. Appl. Phys. Lett. **88**, 241108 (2006). https://doi.org/10.1063/1.2210452
66. Yang, C., Li, X.M., Gao, X.D., Cao, X., Yang, R., Li, Y.Z.: Effects of the oxygen pressure on the structural and optical properties of ZnBeMgO films prepared by pulsed laser deposition. J. Cryst. Growth **312**, 978–981 (2010). https://doi.org/10.1016/j.jcrysgro.2009.12.068
67. Zhong, J., Lu, Y.: ZnO-Based Ultraviolet Detectors, Zinc Oxide Materials for Electronic and Optoelectronic Device Applications, pp. 285–329. Wiley (2011). https://doi.org/10.1002/9781119991038.ch11
68. Özdoğan, M., Çelebi, C., Utlu, G.: Mechanisms behind slow photoresponse character of pulsed electron deposited ZnO thin films. Mater. Sci. Semicond. Process. **107**, 104863 (2020). https://doi.org/10.1016/j.mssp.2019.104863
69. Abdioglu, B.U.: Structure and morphology modification of ZnO nanocrystals via substitutional cobalt doping for enhanced electrochemical activity. Physica B **684**, 415968 (2024). https://doi.org/10.1016/j.physb.2024.415968

70. Baruah, S., Dutta, J.: PH-dependent growth of zinc oxide nanorods. J. Cryst. Growth **311**, 2549–2554 (2009). https://doi.org/10.1016/j.jcrysgro.2009.01.135
71. Srinet, G., Sharma, S., Guerrero-Sanchez, J., Garcia-Diaz, R., Ponce-Perez, R., Siqueiros, J.M., Herrera, O.R.: Room-temperature ferromagnetism on ZnO nanoparticles doped with Cr: an experimental and theoretical analysis. J. Alloy. Compd. **849**, 156587 (2020). https://doi.org/10.1016/j.jallcom.2020.156587
72. Mansouri, I., Saib, F., Boulahlib, S., Laachachi, A., Özacar, M., Bessekhouad, Y.: Effect of Sr-doping toward the optoelectrical ZnO properties. Inorg. Chem. Commun. **162**, 112236 (2024). https://doi.org/10.1016/j.inoche.2024.112236
73. Aebersold, A.B., Fanni, L., Hessler-Wyser, A., Nicolay, S., Ballif, C., Hébert, C., Alexander, D.T.L.: Zinc blende–wurtzite polytypism in nanocrystalline ZnO films. Acta Mater. **130**, 240–248 (2017). https://doi.org/10.1016/j.actamat.2017.03.021
74. Menad, A., Benmalti, M.E., Zaoui, A., Ferhat, M.: Impact of polytypism on the ground state properties of zinc oxide: a first-principles study. Results Phys. **18**, 103316 (2020). https://doi.org/10.1016/j.rinp.2020.103316
75. Rackauskas, S., Klimova, O., Jiang, H., Nikitenko, A., Chernenko, K.A., Shandakov, S.D., Kauppinen, E.I., Tolochko, O.V., Nasibulin, A.G.: A novel method for continuous synthesis of ZnO tetrapods. J. Phys. Chem. C **119**, 16366–16373 (2015). https://doi.org/10.1021/acs.jpcc.5b03702
76. Cammi, D., Ronning, C.: Persistent photoconductivity in ZnO nanowires in different atmospheres. Adv. Condens. Matter Phys. **2014**, 1–5 (2014). https://doi.org/10.1155/2014/184120
77. Van de Walle, C.G.: Defect analysis and engineering in ZnO. Physica B **308–310**, 899–903 (2001). https://doi.org/10.1016/s0921-4526(01)00830-4
78. Gurylev, V., Perng, T.P.: Defect engineering of ZnO: review on oxygen and zinc vacancies. J. Eur. Ceram. Soc. **41**, 4977–4996 (2021). https://doi.org/10.1016/j.jeurceramsoc.2021.03.031
79. Loan, N.T., Quynh, L.M., Dai, N.X., Long, N.N.: Electrochemical biosensor for glucose detection using zinc oxide nanotetrapods. Int. J. Nanotechnol. **8**, 300 (2011). https://doi.org/10.1504/ijnt.2011.038207
80. Takeuchi, S., Iwanaga, H., Fujii, M.: Octahedral multiple-twin model of tetrapod ZnO crystals. Philos. Mag. A **69**, 1125–1129 (1994). https://doi.org/10.1080/01418619408242243
81. Mishra, Y.K., Adelung, R.: ZnO tetrapod materials for functional applications. Mater. Today **21**, 631–651 (2018). https://doi.org/10.1016/j.mattod.2017.11.003
82. Yu, W.D., Li, X.M., Gao, X.D.: Homogeneous-catalytic synthesis of tetrapodlike ZnO nanocrystals and their photoluminescence properties. Chem. Phys. Lett. **390**, 296–300 (2004). https://doi.org/10.1016/j.cplett.2004.04.032
83. Manna, L., Milliron, D.J., Meisel, A., Scher, E.C., Alivisatos, A.P.: Controlled growth of tetrapod-branched inorganic nanocrystals. Nature Mater. **2**, 382–385 (2003). https://doi.org/10.1038/nmat902
84. Kitano, M., Hamabe, T., Maeda, S., Okabe, T.: Growth of large tetrapod-like ZnO crystals. J. Cryst. Growth **102**, 965–973 (1990). https://doi.org/10.1016/0022-0248(90)90867-k
85. Qi, C., Dam-Johansen, K., Weinell, C.E., Wu, H.: Synthesis of micro-structured zinc particles by thermal evaporation and their application in zinc containing coatings for steel corrosion protection. Prog. Org. Coat. **187**, 108143 (2024). https://doi.org/10.1016/j.porgcoat.2023.108143
86. Gedamu, D., Paulowicz, I., Kaps, S., Lupan, O., Wille, S., Haidarschin, G., Mishra, Y.K., Adelung, R.: Rapid fabrication technique for interpenetrated ZnO nanotetrapod networks for fast UV sensors. Adv. Mater. **26**, 1541–1550 (2013). https://doi.org/10.1002/adma.201304363
87. Zhou, Z., Deng, H., Yi, J., Liu, S.: A new method for preparation of zinc oxide whiskers. Mater. Res. Bull. **34**, 1563–1567 (1999). https://doi.org/10.1016/s0025-5408(99)00183-x
88. Delaunay, J.-J., Kakoiyama, N., Yamada, I.: Fabrication of three-dimensional network of ZnO tetratpods and its response to ethanol. Mater. Chem. Phys. **104**, 141–145 (2007). https://doi.org/10.1016/j.matchemphys.2007.02.099

89. Chamoli, P., Shukla, R.K., Bezbaruah, A.N., Kar, K.K., Raina, K.K.: Microwave-assisted rapid synthesis of honeycomb core-ZnO tetrapods nanocomposites for excellent photocatalytic activity against different organic dyes. Appl. Surf. Sci. **555**, 149663 (2021). https://doi.org/10.1016/j.apsusc.2021.149663
90. Mishra, Y.K., Modi, G., Cretu, V., Postica, V., Lupan, O., Reimer, T., Paulowicz, I., Hrkac, V., Benecke, W., Lorenz Kienle, R.: Adelung, direct growth of freestanding ZnO tetrapod networks for multifunctional applications in photocatalysis, UV photodetection, and gas sensing. ACS Appl. Mater. Interfaces **7**, 14303–14316 (2015). https://doi.org/10.1021/acsami.5b02816
91. Yang, D., Gopal, R.A., Lkhagvaa, T., Choi, D.: Oxidizing agent impacting on growth of ZnO tetrapod nanostructures and its characterization. Environ. Res. **197**, 111032 (2021). https://doi.org/10.1016/j.envres.2021.111032
92. Castillejos, E., Bacsa, R., Guerrero-Ruiz, A., Rodríguez-Ramos, I., Datas, L., Serp, P.: Catalytic activity of gold supported on ZnO tetrapods for the preferential oxidation of carbon monoxide under hydrogen rich conditions. Nanoscale **3**, 929–932 (2011). https://doi.org/10.1039/c0nr00724b
93. Dai, Y., Zhang, Y., Bai, Y.Q., Wang, Z.L.: Bicrystalline zinc oxide nanowires. Chem. Phys. Lett. **375**, 96–101 (2003). https://doi.org/10.1016/s0009-2614(03)00823-6
94. Zhang, Z., Yuan, H., Gao, Y., Wang, J., Liu, D., Shen, J., Liu, L., Zhou, W., Xie, S., Wang, X., Zhu, X., Zhao, Y., Sun, L.: Large-scale synthesis and optical behaviors of ZnO tetrapods. Appl. Phys. Lett. **90**, 153116 (2007). https://doi.org/10.1063/1.2712512
95. Newton, M.C., Warburton, P.A.: ZnO tetrapod nanocrystals. Mater. Today **10**, 50–54 (2007). https://doi.org/10.1016/s1369-7021(07)70079-2
96. Rackauskas, S., Mustonen, K., Järvinen, T., Mattila, M., Klimova, O., Jiang, H., Tolochko, O., Lipsanen, H., Kauppinen, E.I., Nasibulin, A.G.: Synthesis of ZnO tetrapods for flexible and transparent UV sensors. Nanotechnology **23**, 095502 (2012). https://doi.org/10.1088/0957-4484/23/9/095502
97. Janardhanam, V., Jyothi, I., Zumuukhorol, M., Yuk, S.-H., Shim, K.-H., Choi, C.-J.: High gain GaN ultraviolet Schottky photodetector with Al-doped ZnO interlayer. Surf. Interfaces **27**, 101405 (2021). https://doi.org/10.1016/j.surfin.2021.101405
98. Fabricius, H., Skettrup, T., Bisgaard, P.: Ultraviolet detectors in thin sputtered ZnO films. Appl. Opt. **25**, 2764 (1986). https://doi.org/10.1364/ao.25.002764
99. Liang, S., Sheng, H., Liu, Y., Huo, Z., Lu, Y., Shen, H.: ZnO Schottky ultraviolet photodetectors. J. Cryst. Growth **225**, 110–113 (2001). https://doi.org/10.1016/s0022-0248(01)00830-2
100. Litton, C.W., Reynolds, D.C., Collins, T.C.: Zinc Oxide Materials for Electronic And Optoelectronic Device Applications. Wiley 29–60 (2011). https://doi.org/10.1002/9781119991038
101. Karagoz, E., Altaf, C.T., Yaman, E., Yildirim, I.D., Erdem, E., Fidan, M., Sankir, M., Sankir, N.D.: Flexible metal/semiconductor/metal type photodetectors based on manganese doped ZnO nanorods. J. Alloy. Compd. **959**, 170474 (2023). https://doi.org/10.1016/j.jallcom.2023.170474
102. Huang, C.Y., Chen, K.C., Chang, C.J.: Realization of a self-powered ZnSnO MSM UV photodetector that uses surface state controlled photovoltaic effect. Ceram. Int. **47**, 1785–1791 (2021). https://doi.org/10.1016/j.ceramint.2020.09.004
103. Hsu, C.-L., Wang, Y.-C., Chang, S.-P., Chang, S.-J.: Ultraviolet/visible photodetectors based on p–n NiO/ZnO nanowires decorated with Pd nanoparticles. ACS Appl. Nano Mater. **2**, 6343–6351 (2019). https://doi.org/10.1021/acsanm.9b01333
104. Katiyar, A., Kumar, N., Shukla, R.K., Srivastava, A.: Synergistic effect of Fe and Ag co-doping on the persistent photoconductivity of vertical ZnO nanorods. Ceram. Int. **48**, 23002–23015 (2022). https://doi.org/10.1016/j.ceramint.2022.04.277
105. Nishimura, T.: Understanding and controlling band alignment at the metal/germanium interface for future electric devices. Electronics **11**, 2419 (2022). https://doi.org/10.3390/electronics11152419

106. Tseng, Y.-W., Hung, F.-Y., Lui, T.-S., Chen, Y.-T., Xiao, R.-S., Chen, K.-J.: Electrical crystallization mechanism and interface characteristics of nanowire ZnO/Al structures fabricated by the solution method. J. Nanomater. **2012**, 208362 (2012). https://doi.org/10.1155/2012/208362
107. Lee, Y., Kim, D.Y., Lee, S.: Low-power graphene/ZnO schottky UV photodiodes with enhanced lateral schottky barrier homogeneity. Nanomaterials **9**, 799 (2019). https://doi.org/10.3390/nano9050799
108. Fang, W., Li, Q., Li, J., Li, Y., Zhang, Q., Chen, R., Wang, M., Yun, F., Wang, T.: Deep ultraviolet photodetector: materials and devices. Crystals **13**, 915 (2023). https://doi.org/10.3390/cryst13060915
109. Su, L., Yang, W., Cai, J., Chen, H., Fang, X.: Self-powered ultraviolet photodetectors driven by built-in electric field. Small **13**, 1701687 (2017). https://doi.org/10.1002/smll.201701687
110. Mishra, M., Gundimeda, A., Garg, T., Dash, A., Das, S., Gupta, G.: ZnO/GaN heterojunction based self-powered photodetectors: influence of interfacial states on UV sensing. Appl. Surf. Sci. **478**, 1081–1089 (2019). https://doi.org/10.1016/j.apsusc.2019.01.192
111. Wan, P., Jiang, M., Xu, T., Liu, Y., Kan, C.: High-mobility induced high-performance self-powered ultraviolet photodetector based on single ZnO microwire/PEDOT: PSS heterojunction via slight ga-doping. J. Mater. Sci. Technol. **93**, 33–40 (2021). https://doi.org/10.1016/j.jmst.2021.03.047
112. Boonyarattanakalin, K., Sinornate, W., Mimura, H., Mekprasart, W., Pecharapa, W.: Ultraviolet response of ZnO nanorod/Sb-doped ZnO thin film homojunction structure prepared by all-chemically wet process. Radiat. Phys. Chem. **227**, 112240 (2025). https://doi.org/10.1016/j.radphyschem.2024.112240
113. Hsu, C.L., Gao, Y.D., Chen, Y.S., Hsue, T.J.: Vertical p-type Cu-doped ZnO/n-type ZnO homojunction nanowire-based ultraviolet photodetector by the furnace system with hotwire assistance. ACS Appl. Mater. Interfaces **6**, 4277–4285 (2014). https://doi.org/10.1016/j.ipha.2024.08.004
114. Aguilar, N., Rozas, S., Escamilla, E., Rumbo, C., Martel, S., Barros, R., Marcos, P.A., Bol, A., Aparicio, S.: Theoretical multiscale study on the properties, aqueous solution behavior and biological impact of zinc oxide nanoparticles. Surf. Interfaces **46**, 103965 (2024). https://doi.org/10.1016/j.surfin.2024.103965
115. Yang, P., Yan, H., Mao, S., Russo, R., Johnson, J., Saykally, R., Morris, N., Pham, J., He, R., Choi, H.-J.: Controlled growth of ZnO nanowires and their optical properties. Adv. Funct. Mater. **12**, 323 (2002). https://doi.org/10.1002/1616-3028(20020517)12:5%3c323::aid-adfm323%3e3.0.co;2-g
116. Özgür, Ü., Alivov, Y.I., Liu, C., Teke, A., Reshchikov, M.A., Doğan, S., Avrutin, V., Cho, S.-J., Morkoç, H.: A comprehensive review of ZnO materials and devices. J. Appl. Phys. **98**, 041301 (2005). https://doi.org/10.1063/1.1992666
117. Jiang, H., Liu, S., Liang, L., Lu, W.: Oxygen plasma assisted enhanced photoresponse of ZnO nanowires fabricated by catalyst-free chemical vapor deposition. RSC Adv. **8**, 28928–28933 (2018). https://doi.org/10.1039/c8ra05040f
118. Korir, K.K., Catellani, A., Cicero, G.: Tailoring oxygen vacancies at ZnO(11¯00) surface: an ab initio study. J. Appl. Phys. **120**, 125301 (2016). https://doi.org/10.1063/1.4962861
119. He, S., Chong-Xin, S., Ji-Shan, L., Bing-Hui, L., Zhen-Zhong, Z., De-Zhen, S.: Stable p-type ZnO films obtained by lithium–nitrogen codoping method. Phys. Status Solidi (B) **250**, 2102–2105 (2013). https://doi.org/10.1002/pssb.201300015
120. Amrute, V., Supin, K.K., Vasundhara, M., Chanda, A.: Observation of excellent photocatalytic and antibacterial activity of Ag doped ZnO nanoparticles. RSC Adv. **14**, 32786–32801 (2024). https://doi.org/10.1039/d4ra05197a
121. Talin, A.A., Léonard, F., Katzenmeyer, A.M., Swartzentruber, B.S., Picraux, S.T.: Transport characterization in nanowires using an electrical nanoprobe. Semicond. Sci. Technol. **25**, 024015 (2010). https://doi.org/10.1088/0268-1242/25/2/024015
122. Li, Y., Li, M., Li, R., Fu, P., Wang, T., Luo, Y., Mbengue, J.M., Trevor, M.: Exact comprehensive equations for the photon management properties of silicon nanowire. Sci. Rep. **6**, 24847 (2016). https://doi.org/10.1038/srep24847

123. Zhang, X., Qiu, Y., Yang, D., Li, B., Zhang, H., Hu, L.: Enhancing performance of Ag–ZnO–Ag UV photodetector by piezo-phototronic effect. RSC Adv. **8**, 15290–15296 (2018). https://doi.org/10.1039/c8ra01189c
124. Yang, F., Zheng, M., Zhao, L., Guo, J., Zhang, B., Gu, G., Cheng, G., Du, Z.: The high-speed ultraviolet photodetector of ZnO nanowire Schottky barrier based on the triboelectric-nanogenerator-powered surface-ionic-gate. Nano Energy **60**, 680–688 (2019). https://doi.org/10.1016/j.nanoen.2019.04.015
125. Wang, Y., Chen, Y., Zhao, W., Ding, L., Wen, L., Li, H., Jiang, F., Su, J., Li, L., Liu, N., Gao, Y.: A self-powered fast-response ultraviolet detector of p–n homojunction assembled from two ZnO-based nanowires. Nano-Micro Lett. **9**, 11 (2016). https://doi.org/10.1007/s40820-016-0112-6
126. Ning, Y., Zhang, Z., Teng, F., Fang, X.: Novel transparent and self-powered UV photodetector based on crossed ZnO nanofiber array homojunction. Small **14**, 1703754 (2018). https://doi.org/10.1002/smll.201703754
127. Qiao, F., Sun, K., Chu, H., Wang, J., Xie, Y., Chen, L., Yan, T.: Design strategies of ZnO heterojunction arrays towards effective photovoltaic applications. Batter. Energy **1**, 20210008 (2022). https://doi.org/10.1002/bte2.20210008
128. Wang, L., Li, J., Wang, Y., Yu, K., Tang, X., Zhang, Y., Wang, S., Wei, C.: Construction of 1D SnO_2-coated ZnO nanowire heterojunction for their improved n-butylamine sensing performances. Sci. Rep. **6**, 35079 (2016). https://doi.org/10.1038/srep35079
129. Ganose, A.M., Scanlon, D.O., Walsh, A., Hoye, R.L.Z.: The defect challenge of wide-bandgap semiconductors for photovoltaics and beyond. Nat. Commun. **13**, 4715 (2022). https://doi.org/10.1038/s41467-022-32131-4
130. Turko, B., Nikolenko, A., Sadovyi, B., Toporovska, L., Rudko, M., Kapustianyk, V., Strelchuk, V., Panasyuk, M., Serkiz, R., Demchenko, P.: Electroluminescence from n-ZnO microdisks/p-GaN heterostructure. Opt. Quant. Electron. **51**, 135 (2019). https://doi.org/10.1007/s11082-019-1853-5
131. Tsai, S.-Y., Hon, M.-H., Lu, Y.-M.: Heterojunction photodiodes based on p-NiO/ n-ZnO for ultraviolet detection. In: 2011 6th IEEE international conference on nano/micro engineered and molecular systems 1184–1187 (2011). https://doi.org/10.1109/nems.2011.6017568
132. Li, P., Shi, H., Chen, K., Guo, D., Cui, W., Zhi, Y., Wang, S., Wu, Z., Chen, Z., Tang, W.: Construction of GaN/Ga_2O_3 p–n junction for an extremely high responsivity self-powered UV photodetector. J. Mater. Chem. C **5**, 10562–10570 (2017). https://doi.org/10.1039/c7tc03746e
133. Baviskar, P.K., Rondiya, S.R., Patil, G.P., Sankapal, B.R., Pathan, H.M., Chavan, P.G., Dzade, N.Y.: ZnO/CuSCN Nano-heterostructure as a highly efficient field emitter: a combined experimental and theoretical investigation. ACS Omega **5**, 6715–6724 (2020). https://doi.org/10.1021/acsomega.0c00006
134. Debnath, R., Xie, T., Wen, B., Li, W., Ha, J.Y., Sullivan, N.F., Nguyen, N.V., Motayed, A.: A solution-processed high-efficiency p-NiO/n-ZnO heterojunction photodetector. RSC Adv. **5**, 14646–14652 (2015). https://doi.org/10.1039/c4ra14567d
135. Liu, Z., Li, F., Li, S., Hu, C., Wang, W., Wang, F., Lin, F., Wang, H.: Fabrication of UV photodetector on TiO_2/diamond film. Sci. Rep. **5**, 14420 (2015). https://doi.org/10.1038/srep14420
136. Molina-Mendoza, A.J., Moya, A., Frisenda, R., Svatek, S.A., Gant, P., Gonzalez-Abad, S., Antolin, E., Agrait, N., Rubio-Bollinger, G., Lara, D.P., Vilatela, J.J., Castellanos-Gomez, A.: Highly responsive UV-photodetectors based on single electrospun TiO_2 nanofibres. J. Mater. Chem. C **4**, 10707–10714 (2016). https://doi.org/10.1039/c6tc02344d
137. Wang, H., Qin, P., Yi, G., Zu, X., Zhang, L., Hong, W., Chen, X.: A high-sensitive ultraviolet photodetector composed of double-layered TiO_2 nanostructure and Au nanoparticles film based on Schottky junction. Mater. Chem. Phys. **194**, 42–48 (2017). https://doi.org/10.1016/j.matchemphys.2017.03.019
138. Zheng, X., Sun, Y., Qin, H., Ji, Z., Chi, H.: Interface engineering on ZnO/Au based schottky junction for enhanced photoresponse of UV detector with TiO_2 inserting layer. J. Alloy. Compd. **816**, 152537 (2020). https://doi.org/10.1016/j.jallcom.2019.152537

139. Huynh, H.N.D., Nguyen, B.G.M., Dinh, T.M., Nguyen, N.M., Tran, C.K., Nguyen, T.T., Tran, N.H.T., La, H.P.P., Tran, T.T.V., Duy, L.T., Phan, T.B., Dang, V.Q.: Enhancing the performance of ZnO-based photodetectors by forming ZnO/(Cu:ZnO) core/shell nanorods. ACS Appl. Electron. Mater. **6**, 1894–1903 (2024). https://doi.org/10.1021/acsaelm.3c01789
140. Meng, L., Li, G., Tian, X., Bai, S., Xu, Q., Jia, X., Cui, X., Qin, Y., Wu, W.: Ultrasensitive fiber-based ZnO nanowire network ultraviolet photodetector enabled by the synergism between interface and surface gating effects. ACS Appl. Mater. Interfaces **12**, 1054–1060 (2019). https://doi.org/10.1021/acsami.9b18185
141. Yin, B., Zhang, H., Qiu, Y., Luo, Y., Zhao, Y., Hu, L.: The light-induced pyro-phototronic effect improving a ZnO/NiO/Si heterojunction photodetector for selectively detecting ultraviolet or visible illumination. Nanoscale **9**, 17199–17206 (2017). https://doi.org/10.1039/c7nr06037h
142. Yin, B., Qiu, Y., Zhang, H., Chang, Y., Yang, D., Hu, L.: Enhancing performance of ZnO/NiO UV photodetector by piezo-phototronic effect. RSC Adv. **6**, 48319–48323 (2016). https://doi.org/10.1039/c6ra07037j
143. Flemban, T.H., Haque, M.A., Ajia, I., Alwadai, N., Mitra, S., Wu, T., Roqan, I.S.: A photodetector based on p-Si/n-ZnO nanotube heterojunctions with high ultraviolet responsivity. ACS Appl. Mater. Interfaces **9**, 37120–37127 (2017). https://doi.org/10.1021/acsami.7b09645
144. Wei, C., Xu, J., Shi, S., Bu, Y., Cao, R., Chen, J., Xiang, J., Zhang, X., Li, L.: The improved photoresponse properties of self-powered NiO/ZnO heterojunction arrays UV photodetectors with designed tunable Fermi level of ZnO. J. Colloid Interface Sci. **577**, 279–289 (2020). https://doi.org/10.1016/j.jcis.2020.05.077
145. Ning, L., Jiang, T., Shao, Z., Ding, K., Zhang, X., Jie, J.: Light-trapping enhanced ZnO-MoS_2 core-shell nanopillar arrays for broadband ultraviolet-visible-near infrared photodetection. J. Mater. Chem. C **6**, 7077–7084 (2018). https://doi.org/10.1039/C8TC02139B
146. Shyam, A., Amal Kaitheri, N., Raju, R., Swaminathan, R.: Self-powered UV photodetectors based on heterojunctions composed of ZnO nanorods coated with thin films of ZnS and CuI. ACS Appl. Nano Mater. **6**, 8529–8539 (2023). https://doi.org/10.1021/acsanm.3c00894
147. Su, L., Zhu, Y., Xu, X., Chen, H., Tang, Z., Fang, X.: Back-to-back symmetric schottky type UVA photodetector based on ternary alloy BeZnO. J. Mater. Chem. C **6**, 7776–7782 (2018). https://doi.org/10.1039/C8TC02255K
148. Xie, X.H., Zhang, Z.Z., Li, B.H., Wang, S.P., Jiang, M.M., Shan, C.X., Zhao, D.X., Chen, H.Y., Shen, D.Z.: Mott-type $Mg_xZn_{1-x}O$-based visible-blind ultraviolet photodetectors with active anti-reflection layer. Appl. Phys. Lett. **102**, 231122 (2013). https://doi.org/10.1063/1.4811153
149. Zhu, Z., Wang, S., Zhu, Y., Liu, X., Zou, Y., Gu, Y., Ju, D., Zeng, H.: Fiber-shaped ZnO/graphene schottky photodetector with strain effect. Adv. Mater. Inter. **5**, 1800136 (2018). https://doi.org/10.1002/admi.201800136
150. Yi, F., Liao, Q., Huang, Y., Gu, Y., Zhang, Y.: Self-powered ultraviolet photodetector based on a single ZnO tetrapod/PEDOT: PSS heterostructure. Semicond. Sci. Technol. **28**, 105023 (2013). https://doi.org/10.1088/0268-1242/28/10/105023
151. Thyda, L., Naresh, K., Joel, K.J., Suneetha, S., Amaladass, P., Thangaraju, K.: Nitrogen-doped carbon quantum dots decorated ZnO nanorods array film for efficient UV photodetector applications. J. Mater. Sci. Mater. Electron. **64**, 1–13 (2024). https://doi.org/10.1016/j.orgel.2016.10.003
152. Nguyen, X.H., Luong, H.N., Pham, H.A., Nguyen, N.M., Dang, V.Q.: Visible photodetector based on transition metal-doped ZnO NRs/PEDOT: PSS hybrid materials. RSC Adv. **11**, 36340–36347 (2021). https://doi.org/10.1039/D1RA06315D
153. Singh, R., Choudhary, R.B., Kandulna, R.: Optical band gap tuning and thermal properties of PMMA-ZnO sensitized polymers for efficient exciton generation in solar cell application. Mater. Sci. Semicond. Process. **103**, 104623 (2019). https://doi.org/10.1016/j.mssp.2019.104623
154. Nam, M., Na, J., Shin, J., Lee, H.H., Chang, R., Ko, D.-H.: Alternative sequential deposition for optimization-free multi-component organic bulk heterojunctions. Nano Energy **74**, 104883 (2020). https://doi.org/10.1016/j.nanoen.2020.104883

155. Nozik, A.J.: Multiple exciton generation in semiconductor quantum dots. Chem. Phys. Lett. **457**, 3–11 (2008). https://doi.org/10.1016/j.cplett.2008.03.094
156. Zhao, X., Zhou, R., Hua, Q., Dong, L., Yu, R., Pan, C.: Recent progress in ohmic/schottky-contacted ZnO nanowire sensors. J. Nanomater. **2015** (2015). https://doi.org/10.1155/2015/854094
157. Ahmed, U.F., Wyatt-Moon, G.S., Flewitt, A.J.: "Nano-In-Nano" schottky diodes fabricated by combining self-aligned nanogap patterning with bottom-up ZnO nanowire growth. ACS Appl. Electron. Mater. **7**, 143–149 (2025). https://doi.org/10.1021/acsaelm.4c01609
158. Li, F., Peng, W., Pan, Z., He, Y.: Optimization of Si/ZnO/PEDOT: PSS tri-layer heterojunction photodetector by piezo-phototronic effect using both positive and negative piezoelectric charges. Nano Energy **48**, 27–34 (2018). https://doi.org/10.1016/j.nanoen.2018.03.025
159. Zhao, Y., Jiang, D., Zhao, M., Sun, Y., Li, H.: A dual-channel detection mode ZnO/organic polymer/Cu_2O heterojunction dual-band photodetector. J. Mater. Chem. A **13**, 5315–5323 (2025). https://doi.org/10.1039/d4ta07692c
160. Chen, Y., Su, L., Jiang, M., Fang, X.: Switch type PANI/ZnO core-shell microwire heterojunction for UV photodetection. J. Mater. Sci. Technol. **105**, 259–265 (2022). https://doi.org/10.1016/j.jmst.2021.07.031
161. Chandrasekhar, P.S., Dubey, A., Qiao, Q.: High efficiency perovskite solar cells using nitrogen-doped graphene/ZnO nanorod composite as an electron transport layer. Sol. Energy **197**, 78–83 (2020). https://doi.org/10.1016/j.solener.2019.12.062
162. Hoang Tran, M., Park, T., Hur, J.: Solution-processed ZnO: graphene quantum dot/Poly-TPD heterojunction for high-performance UV photodetectors. Appl. Surf. Sci. **539**, 148222 (2021). https://doi.org/10.1016/j.apsusc.2020.148222
163. Yuan, Z., Wang, W., Wu, H., Nie, F., He, J.: A solution-processed ZnO quantum dots ultraviolet photodetector with high performance driven by low operating voltage. Mater. Lett. **278**, 128413 (2020). https://doi.org/10.1016/j.matlet.2020.128413
164. Li, X., Liu, X., Li, Y., Gao, D., Cao, L.: Using novel semiconductor features to construct advanced ZnO nanowires-based ultraviolet photodetectors: a brief review. IEEE Access **9**, 11954–11973 (2021). https://doi.org/10.1109/access.2021.3051187
165. Wang, J., Chen, C., Jin, M., Zhu, Y., Zhang, N.: The carbon dots modified ZnO films photodetector with broadband and fast photoresponse. Opt. Mater. **135**, 113341 (2023). https://doi.org/10.1016/j.optmat.2022.113341
166. Hossain, F.M., Nishii, J., Takagi, S., Sugihara, T., Ohtomo, A., Fukumura, T., Koinuma, H., Ohno, H., Kawasaki, M.: Modeling of grain boundary barrier modulation in ZnO invisible thin film transistors. Phys. E Low-Dimens. Syst. Nanostructures **21**, 911–915 (2004). https://doi.org/10.1016/j.physe.2003.11.149
167. Park, W.I., Kim, J.S., Yi, G.-C., Bae, M.H., Lee, H.-J.: Fabrication and electrical characteristics of high-performance ZnO nanorod field-effect transistors. Appl. Phys. Lett. **85**, 5052–5054 (2004). https://doi.org/10.1063/1.1821648
168. Saravade, V., Feng, Z.C., Nafisa, M.T., Zhou, C., Lu, N., Klein, B., Ferguson, I.: Advances in growth, doping, and devices and applications of zinc oxide. J. Vac. Sci. Technol. A **42**, 020802 (2024). https://doi.org/10.1116/6.0003171
169. Ye, Z., He, H., Jiang, L.: Co-doping: an effective strategy for achieving stable p-type ZnO thin films. Nano Energy **52**, 527–540 (2018). https://doi.org/10.1016/j.nanoen.2018.08.001
170. Wan, Q., Xiong, Z., Dai, J., Rao, J., Jiang, F.: First-principles study of Ag-based p-type doping difficulty in ZnO. Opt. Mater. **30**, 817–821 (2008). https://doi.org/10.1016/j.optmat.2007.02.051
171. Chang, H.P., Wang, F.H., Wu, J.Y., Kung, C.Y., Liu, H.W.: Enhanced conductivity of aluminum doped ZnO films by hydrogen plasma treatment. Thin Solid Films **518**, 7445–7449 (2010). https://doi.org/10.1016/j.tsf.2010.05.020
172. Ra, H.-W., Choi, K.S., Ok, C.W., Jo, S.Y., Bai, K.H., Im, Y.H.: Ion bombardment effects on ZnO nanowires during plasma treatment. Appl. Phys. Lett. **93**, 033112 (2008). https://doi.org/10.1063/1.2965109

173. Muhammad, B.L., Cummings, F.: Nitrogen plasma treatment of ZnO and TiO_2 nanowire arrays for polymer photovoltaic applications. Surf. Interfaces **17**, 100382 (2019). https://doi.org/10.1016/j.surfin.2019.100382
174. Rakhshani, A.E.: Schottky diodes on ZnO rods grown homoepitaxially by successive chemical solution deposition. Semicond. Sci. Technol. **23**, 075037 (2008). https://doi.org/10.1088/0268-1242/23/7/075037
175. Peng, L., Zhai, J.-L., Wang, D.-J., Wang, P., Zhang, Y., Pang, S., Xie, T.-F.: Anomalous photoconductivity of cobalt-doped zinc oxide nanobelts in air. Chem. Phys. Lett. **456**, 231–235 (2008). https://doi.org/10.1016/j.cplett.2008.03.052
176. Kouklin, N.: Cu-doped ZnO nanowires for efficient and multispectral photodetection applications. Adv. Mater. **20**, 2190–2194 (2008). https://doi.org/10.1002/adma.200701071
177. Kim, W., Chu, K.S.: ZnO nanowire field-effect transistor as a UV photodetector; optimization for maximum sensitivity. Physica Status Solidi (a) **206**, 179–182 (2009). https://doi.org/10.1002/pssa.200824338
178. Lin, D.D., Wu, H., Pan, W.: Photoswitches and memories assembled by electrospinning aluminum-doped zinc oxide single nanowires. Adv. Mater. **19**, 3968–3972 (2007). https://doi.org/10.1002/adma.200602802
179. Gu, C., Shanshan, L., Huang, J., Shi, C., Liu, J.: Preferential growth of long ZnO nanowires and its application in gas sensor. Sens. Actuators B Chem. **177**, 453–459 (2013). https://doi.org/10.1016/j.snb.2012.11.044
180. Raha, S., Ahmaruzzaman, M.: ZnO nanostructured materials and their potential applications: progress, challenges and perspectives. Nanoscale Adv. **4**, 1868–1925 (2022). https://doi.org/10.1039/d1na00880c
181. Zeng, Y., Grandner, S., Oliveira, C.L.P., Thünemann, A.F., Paris, O., Pedersen, J.S., Klapp, S.H.L., von Klitzing, R.: Effect of particle size and Debye length on order parameters of colloidal silica suspensions under confinement. Soft Matter **7**, 10899 (2011). https://doi.org/10.1039/c1sm05971h
182. Shao, D., Sun, H., Gao, J., Xin, G., Anthony Aguilar, M., Yao, T., Koratkar, N., Lian, J., Sawyer, S.: Flexible, thorn-like ZnO-multiwalled carbon nanotube hybrid paper for efficient ultraviolet sensing and photocatalyst applications. Nanoscale **6**, 13630–13636 (2014). https://doi.org/10.1039/c4nr03921a
183. Hsu, Y.F., Xi, Y.Y., Yip, C.T., Djurišić, A.B., Chan, W.K.: Dye-sensitized solar cells using ZnO tetrapods. J. Appl. Phys. **103**, 083114 (2008). https://doi.org/10.1063/1.2909907
184. Abdulgafour, H.I., Abdalhadi, S.M., Mahmood, M.H.: Synthesis and characterization of ZnO tetrapod-like nanorods without catalyst by wet thermal evaporation and applied on dye-sensitized solar cells. Int. J. Mod. Phys. B **39**, 8 (2024). https://doi.org/10.1142/s0217979225500602
185. Ossai, A.N., Alabi, A.B., Ezike, S.C., Aina, A.O.: Zinc oxide-based dye-sensitized solar cells using natural and synthetic sensitizers. Curr. Res. Green Sustain. Chem. **3**, 100043 (2020). https://doi.org/10.1016/j.crgsc.2020.100043
186. Rathnasekara, R., Hari, P.: Enhancing the efficiency of dye-sensitized solar cells (DSSCs) by nanostructured Ag-doped ZnO electrodes. Chem. Sel. **7**, e202200830 (2022). https://doi.org/10.1002/slct.202200830
187. Chen, W., Qiu, Y., Yang, S.: A new ZnO nanotetrapods/SnO_2 nanoparticles composite photoanode for high efficiency flexible dye-sensitized solar cells. Phys. Chem. Chem. Phys. **12**, 9494 (2010). https://doi.org/10.1039/c000584c
188. Chen, W., Zhang, H., Hsing, I.M., Yang, S.: A new photoanode architecture of dye sensitized solar cell based on ZnO nanotetrapods with no need for calcination. Electrochem. Commun. **11**, 1057–1060 (2009). https://doi.org/10.1016/j.elecom.2009.03.013
189. Wu, C., Kim, T.W., Guo, T., Li, F.: Unique visible-light-assisted field emission of tetrapod-shaped ZnO/reduced graphene-oxide core/coating nanocomposites. Sci. Rep. **6**, 38613 (2016). https://doi.org/10.1038/srep38613
190. Gao, T., Ji, Y., Yang, Y.: Thermo-phototronic-effect-enhanced photodetectors based on porous ZnO materials. Adv. Elect. Mater. **5**, 1900776 (2019). https://doi.org/10.1002/aelm.201900776

191. Calestani, D., Zha, M., Mosca, R., Zappettini, A., Carotta, M.C., Di Natale, V., Zanotti, L.: Growth of ZnO tetrapods for nanostructure-based gas sensors. Sens. Actuators B Chem. **144**, 472–478 (2010). https://doi.org/10.1016/j.snb.2009.11.009
192. Postica, V., Paulowicz, I., Lupan, O., Schütt, F., Wolff, N., Cojocaru, A., Mishra, Y.K., Kienle, L., Adelung, R.: The effect of morphology and functionalization on UV detection properties of ZnO networked tetrapods and single nanowires. Vacuum **166**, 393–398 (2019). https://doi.org/10.1016/j.vacuum.2018.11.046
193. Alsultany, F.H., Hassan, Z., Ahmed, N.M.: Low-power UV photodetection characteristics of ZnO tetrapods grown on catalyst-free glass substrate. Sens. Actuators A **250**, 187–194 (2016). https://doi.org/10.1016/j.sna.2016.09.039
194. Gu, X., Zhang, M., Meng, F., Zhang, X., Chen, Y., Ruan, S.: Influences of different interdigital spacing on the performance of UV photodetectors based on ZnO nanofibers. Appl. Surf. Sci. **307**, 20–23 (2014). https://doi.org/10.1016/j.apsusc.2014.03.091
195. Inamdar, S.I., Ganbavle, V.V., Rajpure, K.Y.: ZnO based visible–blind UV photodetector by spray pyrolysis. Superlattices Microstruct. **76**, 253–263 (2014). https://doi.org/10.1016/j.spmi.2014.09.041
196. Park, S.J., Das, G.S., Schütt, F., Adelung, R., Mishra, Y.K., Tripathi, K.M., Kim, T.: Visible-light photocatalysis by carbon-nano-onion-functionalized ZnO tetrapods: degradation of 2,4-dinitrophenol and a plant-model-based ecological assessment. NPG Asia Mater. **11**, 8 (2019). https://doi.org/10.1038/s41427-019-0107-0
197. Mohapatra, D., Badrayyana, S., Pant, A., Parida, S.: In-situ synthesis and optical properties of CNO-ZnO nanocomposite. Mater. Lett. **183**, 56–60 (2016). https://doi.org/10.1016/j.matlet.2016.07.050
198. Newton, M.C., Shaikhaidarov, R.: ZnO tetrapod p-n junction diodes. Appl. Phys. Lett. **94**, 153112 (2009). https://doi.org/10.1063/1.3119630
199. Sun, L., Shao, Q., Zhang, Y., Jiang, H., Ge, S., Lou, S., Lin, J., Zhang, J., Wu, S., Dong, M., Guo, Z.: N self-doped ZnO derived from microwave hydrothermal synthesized zeolitic imidazolate framework-8 toward enhanced photocatalytic degradation of methylene blue. J. Colloid Interface Sci. **565**, 142–155 (2020). https://doi.org/10.1016/j.jcis.2019.12.107
200. Thongam, D.D., Gupta, J., Sahu, N.K.: Effect of induced defects on the properties of ZnO nanocrystals: surfactant role and spectroscopic analysis. SN Appl. Sci. **1**, 1030 (2019). https://doi.org/10.1007/s42452-019-1058-3
201. de Godoy, M.P.F., Gratens, X., Chitta, V.A., Mesquita, A., de Lima Jr, M.M., Cantarero, A., Rahman, G., Morbec, J.M., Carvalho, H.B.: Defect induced room temperature ferromagnetism in high quality Co-doped ZnO bulk samples. J. Alloy. Compd. **859**, 157772 (2021). https://doi.org/10.1016/j.jallcom.2020.157772
202. Shen, H., Shi, X., Wang, Z., Hou, Z., Xu, C., Duan, L., Zhao, X., Wu, H.: Defects control and origins of blue and green emissions in sol-gel ZnO thin films. Vacuum **202**, 111201 (2022). https://doi.org/10.1016/j.vacuum.2022.111201
203. Chen, Z.Q., Maekawa, M., Kawasuso, A., Sakai, S., Naramoto, H.: Electron irradiation-induced defects in ZnO studied by positron annihilation. Phys. B Condens. Matter **376–377**, 722–725 (2006). https://doi.org/10.1016/j.physb.2005.12.181
204. Karuppasamy, A., Subrahmanyam, A.: Effect of electron bombardment on the properties of ZnO thin films. Mater. Lett. **61**, 1256–1259 (2007). https://doi.org/10.1016/j.matlet.2006.07.014
205. Feng, X., Lv, B., Lu, L., Feng, X., Wang, H., Yang, Y., Zhang, F.: Role of surface oxygen vacancies in zinc oxide/graphitic carbon nitride composite for adjusting energy band structure to promote visible-light-driven photocatalytic activity. Appl. Surf. Sci. **562**, 150106 (2021). https://doi.org/10.1016/j.apsusc.2021.150106
206. Kamada, Y., Furuta, M., Hiramatsu, T., Kawaharamura, T., Wang, D., Shimakawa, S., Li, C., Fujita, S., Hirao, T.: Study on oxygen source and its effect on film properties of ZnO deposited by radio frequency magnetron sputtering. Appl. Surf. Sci. **258**, 695–699 (2011). https://doi.org/10.1016/j.apsusc.2011.07.100

207. Goswami, N., Sahai, A.: Structural transformation in nickel doped zinc oxide nanostructures. Mater. Res. Bull. **48**, 346–351 (2013). https://doi.org/10.1016/j.materresbull.2012.10.045
208. Mascarenhas, F.J., Rodney, J.D., Kim, B.C., Bhat, B.R.: Significance of transition metal (Co, Ni and Zn) doping on the nano MnSe for high-performance supercapacitor electrode. J. Alloy. Compd. **986**, 173957 (2024). https://doi.org/10.1016/j.jallcom.2024.173957
209. Asaldoust, F., Mabhouti, K., Jafari, A., Taleb-Abbasi, M.: Structural, magnetic, and optical characteristics of undoped and chromium, iron, cobalt, copper, and zinc doped nickel oxide nanopowders. Sci. Rep. **15**, 1088 (2025). https://doi.org/10.1038/s41598-025-85239-0
210. Dai, W., Pan, X., Chen, C., Chen, S., Chen, W., Zhang, H., Ye, Z.: Enhanced UV detection performance using a Cu-doped ZnO nanorod array film. RSC Adv. **4**, 31969 (2014). https://doi.org/10.1039/c4ra04249b
211. Sun, J., Liu, F.-J., Huang, H.-Q., Zhao, J.-W., Hu, Z.-F., Zhang, X.-Q., Wang, Y.-S.: Fast response ultraviolet photoconductive detectors based on Ga-doped ZnO films grown by radio-frequency magnetron sputtering. Appl. Surf. Sci. **257**, 921–924 (2010). https://doi.org/10.1016/j.apsusc.2010.07.091
212. Shinde, S.S., Rajpure, K.Y.: High-performance UV detector based on Ga-doped zinc oxide thin films. Appl. Surf. Sci. **257**, 9595–9599 (2011). https://doi.org/10.1016/j.apsusc.2011.06.073

Dr. Kuppusamy Thangaraju is currently working as Professor in Dept. of Physics, NIT-Warangal, India. He obtained his masters (2001) from Univ. of Madras, and Ph. D (2008) from Anna Univ., India. After completing his postdoctoral research (2009-2012) in South Korea (Sogang Univ., Seoul, Gyeongsang National Univ., Jinju), he worked as Assistant Professor in Dept. of Physics, Gandhigram Rural Univ., India (2012-2013). Later, he joined as Assistant Professor in Dept. of Physics, NIT Warangal (2013-2018) and Associate Professor (2018-2022). His research area includes organic semiconductors, organic optoelectronic devices such as OLEDs, OPVs, OTFTs, photodetectors, and organic/inorganic hybrid nanostructured thin films and their optoelectronic devices. He has over 50 publications in peer-reviewed international journals and 04 book chapters published. He has also received research funding worth of Rs. 85 lakhs from Govt. of India (DST-SERB/ANRF, TEQIP-III, SPARC Govt. of India) and established research facilities for the fabrication of organic optoelectronic devices.

ZnO Tetrapods for Rapid and Enhanced Photocatalytic Activity: Size, Shape and Geometrical Advantages

Yuvashree Jayavelu, Gouranga Maharana, Benjamin Hudson Baby, D. Paul Joseph, and Kovendhan Manavalan

Abstract Globally, remediation towards clean and potable water resources are a major concern due to rapid industrialization and subsequent pollution. Steady increase in the world population and enormous energy consumption has led to immense industrialization leading to persistent release of toxins and pollutants into the atmosphere and water resources that leads to various health hazards, global warming and climatic changes affecting the liveable eco-system. To address these serious issues, clean, sustainable, efficient, and cost-effective technologies are the need of the hour. This chapter discusses the application of ZnO tetrapods in photocatalytic degradation, towards environmental remediation outlining the general mechanism involved with a brief outline of various photocatalytic reactors, working principles, their design and how ZnO is integrated into it. This chapter also gives a brief overview of ZnO tetrapod's shape, arm geometry, and surface morphology aspects towards excellent catalytic performance under UV light, visible light and sunlight, highlighting the feasibilities for wastewater treatment.

1 Introduction

Globally, remediation towards clean and potable water resources are a major concern due to rapid industrialization and subsequent pollution. Steady increase in the world population and enormous energy consumption has led to immense industrialization leading to persistent release of toxins and pollutants into the atmosphere and water resources that leads to various health hazards, global warming and climatic changes affecting the liveable eco-system. To address these serious issues, clean, sustainable, efficient, and cost-effective technologies are the need of the hour. Various techniques like physical methods involving filtration, reverse osmosis, and chemical

Y. Jayavelu · G. Maharana · B. H. Baby · D. P. Joseph (✉)
Department of Physics, National Institute of Technology, Warangal, Telangana, India
e-mail: paul@nitw.ac.in

K. Manavalan
Department of Physics and Nanotechnology, SRM Institute of Science and Technology, Kattankulathur, Tamilnadu, India

K. Thangaraju et al. (eds.), *ZnO Tetrapods*, Engineering Materials,
https://doi.org/10.1007/978-981-95-1614-8_3

methods involving chlorination, disinfection, extraction, oxidation, and biological methods that involves ion-exchange, aerobic, anaerobic and anoxic degradation are predominantly utilized to tackle water pollution by these remediation methods [1]. Owing to the inherent disadvantages of the above mentioned techniques like ineffective large-scale filtration, clogging, frequent maintenance, cost-effectiveness, brittleness, membrane rupture, secondary contamination, mineral stripping etc. demands to a better and efficient system to overcome these problems worldwide. Wastewater treatment plants are unable to completely eliminate every possible toxins present in the polluted water bodies, consequently discharging them into the ecosystem further leading to biomagnification [2, 3]. Trace concentrations of these contaminants present in water bodies not only affect aquatic organisms but also every living organism which consume this water directly or indirectly.

Though wastewater treatment involves various techniques, advanced oxidation process (AOPs) that is potentially exploited for being economic, large scale and efficient water remediation. Among the various AOPs, photocatalysis is an efficient wastewater treatment and water remediation process due to its cost effective and large scale feasibility without the need for much regulation and complex equipment and its maintenance [4]. Photocatalysis is defined as a process where solar energy (or equivalent simulated low powered light) is converting into less harmful byproducts by chemical reactions in the presence of suitable catalyst. A photocatalyst accelerates the chemical reaction under the influence of incident light of appropriate energy which excites the catalyst matrix to produce radicals helping in the degradation of the pollutants to their less toxic and non-toxic forms. Exploiting natural Sunlight to degrade the pollutants in water at ambient temperature makes photocatalysis a remarkable and large-scale viable and green technique.

Predominantly photocatalysis is broadly classified into homogeneous and heterogeneous types. Homogeneous photocatalysis consists of a soluble, photosensitizer catalyst consisting of catalytically active sites for redox reactions all in a single form of state [1, 5]. Due to suitable bandgap and opto-electronic properties transition metal complex systems are mostly employed as homogeneous photocatalysts. The homogeneous photocatalytic mechanisms involve ozonisation and photo-Fenton process to produce OH radicals which helps in the generation of clean, green and accessible H_2 fuel for alternative energy related applications (Fig. 1) [6].

Heterogeneous photocatalysis remains superior to the homogeneous catalysis due to their ability to maintain greater stability, recyclability, and easy separation from waste water. Heterogenous photocatalysts are mostly inorganic semiconductors in combination with metals, other oxides, salts, and sulfides which catalytically act under the influence of light of relevant energy [5, 7]. Semiconducting materials are predominantly used as heterogeneous photocatalyst due to their optoelectronic properties including appropriate bandgap largely helping the reactions to occur on the solid surface of the semiconductor. When the photocatalyst is illuminated by light photons, electrons in the valence band (VB) of the semiconductor is photo-excited to the conduction band (CB) of the semiconductor that is separated by a bandgap, while the photoexcited electrons leave holes in the VB [8–10]. These photogenerated electrons and holes result in the oxidation and reduction process on the surface of

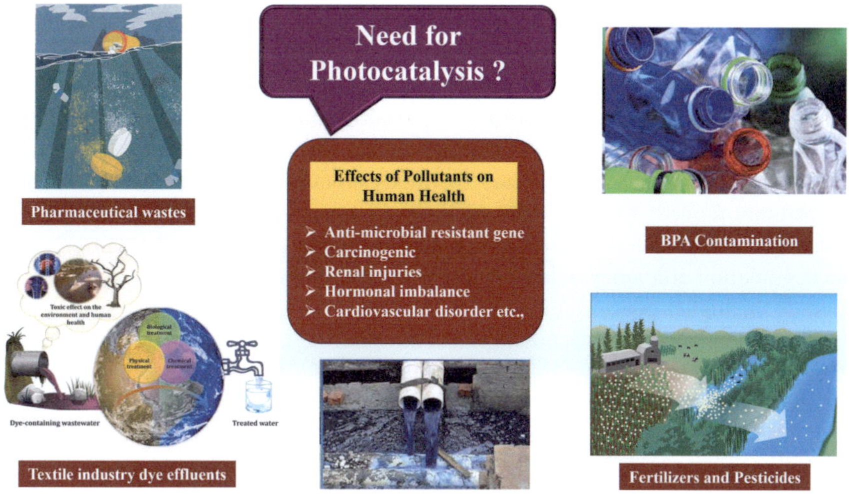

Fig. 1 Pictorial representation of the need and basic mechanism of photocatalysis using a semiconductor along with its detrimental effects on human health

the photocatalyst. The generated radicals during the oxidation and reduction process degrades the pollutants in the wastewater resulting in mineralization of the water. Among various metal oxides, TiO_2 is considered the best among the photocatalysts due to their chemical and photochemical stability, reusability, and cost effectiveness, appropriate bandgap, safety and eco-friendliness [9–11].

Photocatalytic chemical reactions depend on various parameters including the type of catalyst, size, morphology, catalyst ratio, light intensity, pH, duration of irradiation, temperature of the reaction, etc. Several types of photocatalysts like oxides of metals (TiO_2, ZnO, SnO_2, etc.) chalcogenides (CdS, ZnS, CdSe, MoS_2, WS_2, etc.) and semiconducting quantum dots have grabbed the attention for efficiently degrading various water pollutants depending on their activation under UV and/or visible light [2]. Size and surface area of the catalyst render a significant support in altering the desired reaction rate as small size and higher surface area helps in efficient pollutant-catalyst interaction due to their S/V ratio. Geometrical advantages of photocatalyst also plays a significant role as the 3D structures (flowers, sea urchins, tetrapods, snowflakes) provide increased surface area, better light adsorption and improved charge separation. Specifically, tetrapod structure has paddle-like arms providing enhanced pollutant adsorption on the surface with self-assembly and easy settling on a substrate for immobilization as well as for filtration relative to smaller spherical shaped particles. Nanomaterials show high catalytic activity compared to bulk due to their high S/V ratio and are usually engineered for their optoelectronic properties towards enhanced degradation efficiency. Spherical morphology shows high specific surface area and pore volume than other nanostructures resulting in better photocatalysis [12]. Optimum catalyst dose is a key parameter, increase in catalyst dose increases the efficiency only until the optimum value, beyond this it makes the solution turbid blocking the light catalyst interaction decreasing their efficiency. Low light intensity leads to limited production of free radicals while higher intensity provides more free radicals production in the reaction for enhancing the photocatalytic reactions for proper mineralization [13]. Change in temperature results in change in dissolved oxygen in the catalyst-pollutant solution and further it is also related to the kinetic and thermodynamic reactions indicating an ideal temperature range between 38 and 100 °C following the 1st order reaction for efficient degradation of the toxins. Higher concentration of pollutants relates inversely to the degradation efficiency while it requires other optimum parameters to efficiently degrade the higher concentration of the solution [14].

An efficient photocatalyst possess an optimum energy band gap with relevant band edge position that match with the redox potential of water. A photocatalyst should possess efficient light harvesting ability with high charge transport mobility, high excited state lifetime [15], chemical stability, anti-corrosion properties, recyclability, non-toxic and environmentally benign nature.

Fundamentally, photocatalysis are for photocatalytic water purification by oxidizing organic pollutants into less toxic or non-toxic byproducts including CO_2 and H_2O. This can also be further extended as antimicrobial photocatalyst withstanding higher chemical and mechanical stability at relatively high temperatures [16]. Self-cleaning glass with hydrophilic coating can chemically break down the adsorbed dust or pollutant in the presence of light and their antifogging property can also help in smooth driving of automobiles. Photocatalytic water splitting helps to split hydrogen and oxygen using natural sunlight or simulated light providing clean fuel and serving as renewable resources [17]. Agricultural pollutants like pesticides, herbicides, growth regulators and pharmaceutical wastes like drugs, antibiotics, are degraded under light leaving behind their un-reacted by-products in the environment. These versatility of the photocatalytic materials makes them absolutely interesting and a hot topic to further explore for its applicability in multidimensional fields including environmental, medical and energy sectors [4, 18–20]. The need for efficient photocatalysts arises from the necessity to address the critical challenges including environmental remediation, by facilitating chemical reactions using light energy making them sustainable for future requirements.

2 Photocatalytic Degradation: Strategies Using Various Materials

Predominantly for photocatalytic applications, conventional semiconductors like metal oxides, sulphides, quantum dots, 2D layered materials, and plasmonic materials having effective light absorption, charge carrier separation, and diffusion/migration, and rapid charge transfer are desired to induce the reactions. The photogenerated charge carriers are separated by vivid mechanisms corresponding to their lifetime of the photoinduced carriers varying in few orders of magnitude, from pico to microsecond in semiconducting materials in dark condition, while electrons will transit from the semiconducting matrix to the electrolyte, attaining equilibrium at the interface [21].

Eventually, the energy band is bent upwards at the solid/electrolyte interface in dark condition due to charge equilibrium. This is similar to the band structure modification when a semiconducting host comes into contact with a metal, and their Fermi levels varying like a standard Schottky junction [22]. The intrinsic in-built field in the space-charge region is by the Mott–Schottky theory for charge separation upon illumination. In this region, charge carriers are extricated either to the interface of semiconductor/electrolyte or to the bulk by the intrinsic in-built potential [23].

Unless the photocatalyst material is quantum dots or plasmonic excitation described by the Schottky theory, because of the smaller size of the semiconductor, the level of band bending is usually not very prominent and is undesired as this scenario leads to rapid recombination due to the proximity of redox reactions [22]. The concept is to enhance the reliance on the semiconductor/electrolyte interface for the Schottky junction, which eventually depend on several aspects of the interface type that are difficult or nearly not possible to tune [22, 23]. Instead, a well-formed p–n junction is created within the semiconducting host under standard conditions, the photocatalytic electrode forms ohmic contact with the liquid medium such that the band is retained flat at the semiconductor-electrolyte interface. Upon light irradiation, photogenerated carriers are isolated by the inherent in-built field due to the underlying junction. This strategy may be specifically beneficial in situations where the semiconductor/electrolyte interface is influenced by surface states such that the process of pinning is dominating.

In the case of QDs, in which a sacrificial electron donor is needed to accommodate these holes. As an alternative, photoinduced charge carriers will be gathered by the contacting semiconductor electrode, leaving the electrons for exhaustion by the sacrificial electron scavengers [25]. In every case, the intrinsic in- built energy barrier between the QD and the semiconductor electrode acts to bifurcate the charge carriers, leading to considerably enhanced lifetime. Intriguingly, the valence band of QDs remains mostly unaltered, nevertheless the CBM tends to be highly negative, leading to larger potential differences in other semiconductors [25].

The 2D layered material-based photocatalysts are quite alike to how QDs are linked to bulk semiconductors. 2D systems based photocatalysts are never subdivision of host materials with distictive aspects that are not found in other bulk semiconducting materials. The bandgap of a 2D material generally vary with layer thickness due to quantum confinement effect, and also 2D materials also show anisotropy in their characteristics primarily because of the layered structure [26, 27]. The 2D materials have rapid in-plane electron–hole bifurcation in a heterostructural nanosheet, where the carbon ring defects are identified to influence photoinduced carrier transport behaviours.

Moreover, the defects in the 2D systems may substantially influence the charge density distributions, which may result in drastic variation of charge behaviour when the layer thickness is decreased near to few atomic order [28]. The nanosheets presented significant increase in photocatalytic activity at low catalyst loading, implying increased light adsorption and effective charge carrier bifurication. Besides single-phase 2D layered systems, 2D composites are also found to possess better and facile transit of photoinduced electrons from the conduction band. Graphitic carbon nitride g-C_3N_4 is utilized not only for its semiconducting behaviour but also for its anisotropic nature in the ultrathin or single-layered configuration. Apart from this, 2D planar g-C_3N_4 can be blended with other 2D layered materials (e.g., MoS_2) to result in nano heterojunctions. MXenes as a hydrogen evolution co-catalyst was coated over CdS photocatalyst, where Schottky junction may be established between the semiconductor CdS and the conducting Ti_3C_2 MXene (Fig. 2) [29, 30].

Plasmonic material may be these most special as these are metallic like in behaviour (with the caveat of heavily doped chalcogenides). The band theory is ruled out to describe how charges are isolated within the plasmonic photocatalysts upon light irradiation. The surface plasmonic resonance (SPR) excites the electrons to the plasmonic states, leaving behind holes at lower energies on an electronic energy scale [19, 31]. As far as photocatalysis is concerned, transit of electrons may occur straight from the plasmonic states to the LUMO of the surface-adsorbed species to induce chemical interactions. As an alternative, electrons can be shifted to the LUMO of the surface species resulting in charge transfer and thermalization of hot electrons, which reverts to the Fermi level (E_F) of the plasmonic medium. As a result at E_F,

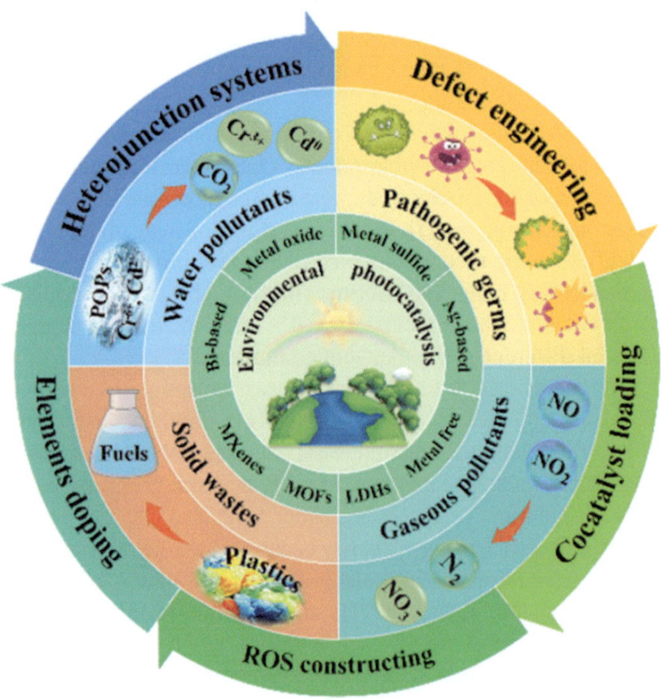

Fig. 2 General schematics of environmental photocatalysts with their modification strategies for degrading various pollutants [24]

electrons are not of enough energy to commence the targeted chemical process. In general, direct hot SPR electrons that transit tend to be of very low efficiency because of rapid cooling of hot electrons and sluggish kinetics of the chemical process [19].

Charge bifurication in plasmonic photocatalytic systems includes (a) Two-step process for indirect electron transfer. (i) Non-thermalized distribution of electron–hole pairs; (ii) Fermi–Dirac distribution of thermalized hot carriers, which are transfered to surface-adsorbed species. (b) Direct electron transfer phenomena having charge bifurication routes from metal to semiconductor. (c) Plasmon generated hot electron transfer process and (d) direct metal-to-semiconductor interfacial charge carrier transfer route. (e) Plasmon generated interfacial charge transfer route.

The charge transfer process between plasmonic photocatalyst particulates or QD are nearly similar. However, compared to other systems, plasmonic photocatalysts offer a specific mechanism by which the harvested photon energy can be transferred to the chemical species by resonant energy transfer. Electrons may be passed on from a semiconductor with larger bandgap to a metal with a low negative plasmon band energy by Föster resonance energy transfer, transferring the electronic excited state of a fluorophore process by dipole–dipole interaction. However, electrons predominantly transfer from a plasmonic metal to a semiconductor through the plasmon-induced resonance energy transfer (PIRET) process in heterojunctions [32]. In energy transfer processes, a thin insulating layer partakes to eliminate interface damping and to minimize recombination. PIRET was put forth to improve the semiconductor photocatalytic electrode to utilize excess visible and near-IR photons beyond its bandgap and increase their efficiency [32–35].

Water splitting technique is comparatively facile means to assess the efficiency of the photocatalytic medium in aspects of several performance metrics. Particularly, the absorption properties, band edge positions and flat band potentials, as well as the pros and cons of ZnO includes n-type semiconductor with 3.2 eV direct bandgap, providing high UV absorption, high and faster carrier mobility (200, 5–50 cm^2/Vs) for electron and holes, longer diffusion length (438 nm), Bohr radius of 3.2 nm, abundance, cost-effectiveness, non-toxic, appropriate CBM and VBM for water splitting, weak visible light absorption, stable in a limited pH range etc. are important for future electrochemical studies including real time solar water splitting applications [34, 35].

3 ZnO as an Effective Photocatalyst

Metal oxide semiconductors including ZrO_2, ZnO, TiO_2, CuO, Fe_2O_3 are considered best among the photocatalysts as they fulfil the criteria of being stable, reusable, and cost effective (Fig. 3). Although TiO_2 is by far the best photocatalyst considering its non-toxic nature, appropriate bandgap, stability (both chemically and photochemically), being able to maintain the water solubility and self-degradable performance.

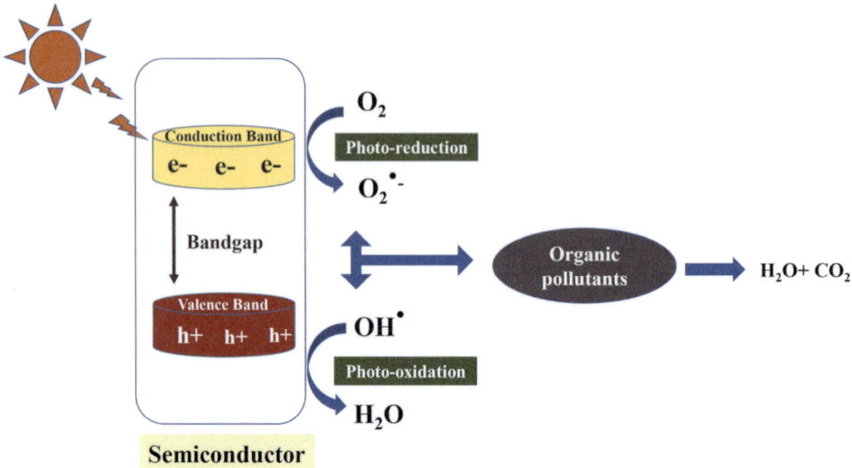

Fig. 3 General photocatalytic mechanism representing the photo-reduction and photo-oxidation reactions over a pollutant

Additionally, TiO_2 and its composites have garnered attention as environment purifying catalysts due to their high oxidative power decomposing several organic pollutants into carbon dioxide and water. TiO_2 although having a wide band gap, can only absorb a small region of the solar radiation limiting its use in UV region only. Low adsorption capacity having less affinity for organic pollutants, especially hydrophobic ones. The massive recombination of photogenerated charge carriers limits the overall photocatalytic efficacy, due to its inability to function in the visible region.

ZnO is next best photocatalyst in line after TiO_2 as it is widely explored due to its ability to absorb UV light (and in visible region due to defect levels) with its property of high photosensitivity aiding in degradation of various pollutants[10, 11, 36, 37]. TiO_2 is less effective under visible light compared to ZnO, suffers from recombination of electron–hole pairs, which can reduce its photocatalytic efficacy. Being an expensive transition metal, high-quality TiO_2 is more expensive for large scale applications. Though TiO_2 is reported to be very stable compared to ZnO, it is not as effective as ZnO in the photocatalytic disintegration of certain organic pollutants (Table 1). While TiO_2 remains one of the most widely used photocatalyst, its disadvantages in terms of visible light activity, charge recombination, and production cost and reduced efficiency in certain pollutants can make ZnO a better choice for specific applications or in combination with other materials to enhance efficiency.

In both the photocatalyst materials various strategies like doping, nanostructuring, or hybridization with other materials can enhance their photocatalytic performance and reduce some of their limitations.

Table 1 Comparison of various optoelectronic characteristics of TiO_2 and ZnO photocatalysts

Property	TiO_2	ZnO
Band gap (eV)	3.0–3.2	~ 3.3
UV absorption	Moderate (UV-A and UV-B)	Stronger in short UV-C range (200–375 nm)
Exciton binding energy (meV)	40	60
Photostability	High durability	Lower stability in acidic medium
Electron mobility (cm^2/Vs)	0.1–0.4	200–300
Electron–Hole Recombination	Lower recombination rate (200 ps)	Higher recombination rate in powdered ZnO (1.7 ps)
Cost	75% more expensive than ZnO ($4 per kg)	Inexpensive and abundant ($1–$3 per kg)
Applications	Water purification, air purification, self-cleaning, energy production etc	Organic pollutant degradation, energy generation, antifungal, antimicrobial drug delivery, etc

4 Geometrical Aspect of Nano-dimensional Photocatalyst

Though ZnO has been extensively investigated for photocatalytic activity with a variety of pollutants, the applicability of ZnO nanoparticles in water for pollutant degradation still lacks proper handling of suspended nano-dimensional photocatalysts after the process completion which is a problem of high concern [38, 39]. Lack of strategies for proper photocatalyst segregation disposal and tedious filtration process associated with the nanoparticles in water, scaling up and expanding this technique is practically limited with regard to the commercialization of this photocatalytic technique, the reuse and stability of photocatalysts with minimal effort has been given high attention. The idea of immobilizing the ZnO photocatalysts onto a relevant host surface has to be studied overcoming the shortcomings of ZnO nanoparticles.

The four-legged ZnO-T morphology has the advantage of being filtered naturally by itself when contained in a separating membrane due to its geometrical advantage (Fig. 4). Reports show that, immobilizing nanoparticles onto surfaces of glass, polymers carbon nanotubes, ceramic plates, zeolites etc., has been extensively explored for photocatalytic activity. Isolation of the nano-photocatalysts from the treatment completed water pose a severe challenge, however, the immobilization of nano-photocatalysts onto large area substrates may attribute to efficient and faster post-treatment isolation [38, 40]. Materials like glass sheets, stainless steel plates, metal wire mesh, and porous foams, etc. can be used as substrates for immobilization of nano-photocatalysts permanently in the photocatalytic reactor or materials like quartz sand, black volcanic ashes, calcium alginate hydrogen beads can be easily separated post-treatment by physical process (Fig. 5) [38].

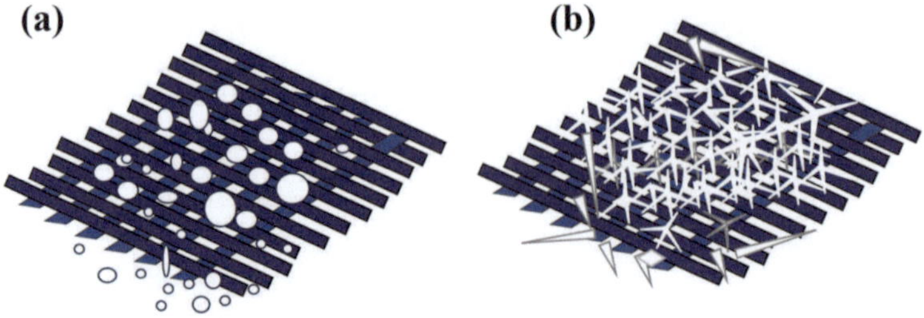

Fig. 4 Pictorial representation of envisaged morphological aspect of (**a**) spherical ZnO nanoparticles and (**b**) ZnO nanotetrapods passing through a porous membrane during the photocatalytic filtration process

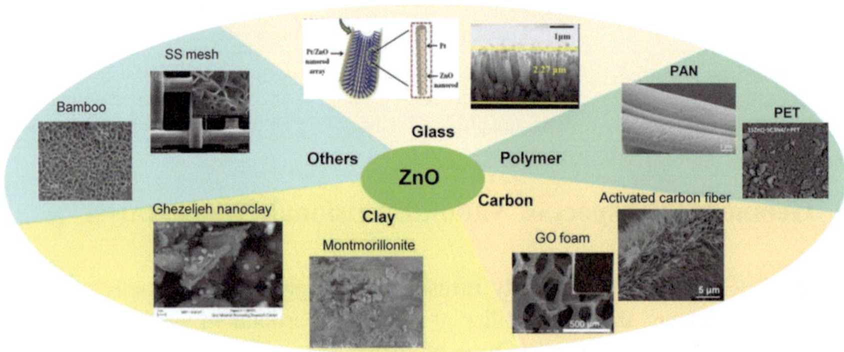

Fig. 5 Pictorial representation of ZnO photocatalyst immobilized onto various substrates [41]

Various immobilization techniques including dip-coating, electrophoretic process, sol–gel, solvothermal, spray coating, cold plasma discharge, and RF magnetron sputtering. These techniques are generally utilized for immobilization of ZnO photocatalyst onto various substrates according to the current literature listed in Table 2. The spherical shaped nano-photocatalyst percolate through the porous membrane leading to secondary contamination.

5 Immobilization Techniques of Photocatalysts

Immobilization techniques reported includes hydrothermal, electrospinning, solvent casting, thermal attachment, spin coating, dip coating and electrochemical deposition and each of which has its own characteristics specific for the substrate material [49]. Complete understanding of the immobilization process will enable to attribute desired photocatalytic efficiency for ZnO-based photocatalysts.

Table 2 Techniques for immobilization of ZnO photocatalyst for photo-degrading various pollutants from literature

Substrate	Immobilization Method/Process	Immobilization area	Pollutants	References
ITO on glass	Thermal attachment, Dip coating, Hydrothermal, Spin coating, Spray pyrolysis, Electrochemical, deposition	1–300 cm^2	p-Nitroaniline, Acid blue, Rhodamine B, Toluene, Methyl orange, Direct yellow, Diazinon, Eriochrome black, Phenol, Aniline	[41–44]
Polymers Polyurethane (PU), Polyethylene terephthalate (PET), Polydimethylsiloxane (PDMS), Poly-vinylidene fluoride (PVDF), Polyacrylonitrile (PAN), Polyethylene (PE), Polystyrene (PS), Polypropylene (PP)	Hydrothermal, Thermal attachment, Electrospinning, Solvent casting, Sonochemical, Atomic layer deposition	1–25 cm^2	Acid black, Methylene blue, Methyl orange, Rhodamine B	[41]
Carbon based materials Activated Carbon, Carbon Plate, Carbon fabric, Carbon nitride, Graphene, Carbon felt, Graphene oxide foam	Hydrothermal, Sonochemical, Precipitation, Thermal attachment, Dip-coating, Electrochemical Impregnation	0.1–20 g/L	Methylene blue, Congo red, Amoxicillin, Reactive green, Reactive orange, Reactive black, Deoxynivalenol, Ofloxacin, Black treacle, Dimethyl phenol, Brilliant green, Rhodamine B, Bisphenol	[41]

(continued)

Table 2 (continued)

Substrate	Immobilization Method/Process	Immobilization area	Pollutants	References
Clay Bentonite, Clinoptilolite, Ghezeljeh, Nanoclay, Halloysite, Tunisian clay, Montmorillonite, Zeolite, Sepiolite	Precipitation, Sono-precipitation, Precipitation, Impregnation, Sol–gel, Thermal attachment, Hydrothermal	0.1–10 g/L	Furosemide, Ibuprofen, Methylene blue, Reactive yellow, Rhodamine B, Phenol, Congo red, Metronidazole	[41]
Uncommon Substrates SS mesh, SS wire Ni foam, Ceramic plate Bamboo, Biosilica Al_2O_3	Plasma electrolytic oxidation, Hydrothermal, Dip-coating, Precipitation, Thermal attachment, CVD, Dip coating, Pyrolysis, Electrochemical	0.2–225 cm^2	Methyl orange, Rhodamine B, Ammonia, Methylene blue, Phenol, Congo red, Acid orange, Ciprofloxacin, Direct orange, 4-Nitroaniline, Methylene blue	[41]

5.1 Hydrothermal Method of Immobilization

Hydrothermal technique is a widely used immobilization method for ZnO-based photocatalysts. It enables growth of ZnO-based photocatalysts onto different types of substrates like glass, polymer, carbon, clay, and cotton fabric. Due to relatively low synthesis temperature, controlled size and geometrical aspects of several ZnO nanostructures could be achieved by hydrothermal method [41]. Predominantly a seed layer is deposited by sputtering, dip coating or spin coating before growing the ZnO nanostructures hydrothermally which will lead to a directional growth.

5.2 Electrospinning Method of Immobilization

Electro-spinning technique is predominantly adapted to immobilize ZnO photocatalysts onto a polymer substrate. It is a facile method with relatively low production cost to produce 2D nanofibers with highly effective surface area [50]. Electrospinning method results in completely embedding the ZnO-based nanoparticle photocatalysts onto a polymeric strand by using high-voltage induced electric field to generate solid fibres from a polymeric solution ejected via a sub-millimetre sized orifice (Fig. 6). The adhesion of the photocatalyst is generally not very strong.

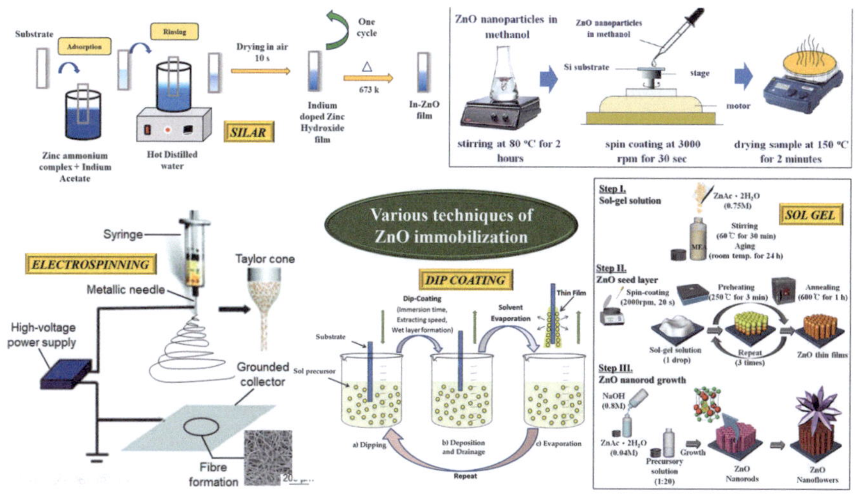

Fig. 6 Pictorial representation of various techniques utilized for immobilization of ZnO photocatalysts [45–48]

5.3 Thermal Attachment-Based Immobilization

Thermal attachment method is a facile and cost-effective method utilized for attaching ZnO photocatalysts onto various supportive hosts like glass spirals, polymeric material, carbon plate, ceramic plate, stone etc. Though ZnO photocatalysts are easily immobilized by thermal attachment, this process could minimize the surface area of photocatalyst particulates, which hamper its photocatalytic efficacy [51]. The immobilization procedure is easily performed two to three times to enhance the loading of ZnO on the surface of glass substrates. The adhesion by this method is fairly good.

5.4 Chemical Coating Methods

The chemical coating techniques like dip coating, spin coating and spray coating are suitable for immobilizing ZnO nanoparticles onto a suitable host. Utilizing these process, a thin and homogeneous ZnO film can be coated easily onto a variety of substrates [52]. Dip coating process is performed by immersing the supportive host into ZnO suspension and then the substrate is pulled out at a precise and regulated rate via a stepper motor. Dip coating is utilized to deposit a thin layer of ZnO nanoparticles over several hosts that are listed in Table 2. Spin coating method is used to coat uniform ZnO nanoparticles onto substrate via four major steps, dropping of sol, spin-up, spin-off and evaporation of the solvent medium. It is generally utilized to immobilize ZnO nanoparticles over glass and other flat host substrates (Fig. 6) [53].

In spin coating process, the sol–gel is drop-casted over a flat spinning substrate while the excess fluid is pushed out during the spinning process. A fixed speed of rotation permits homogeneous spreading over the substrate with a particular film thickness including other variables like the spinning period and spinning speed which also influence the quality of coating. Upon the thermal annealing solvent evaporates and leaves behind a dry solid nanoparticle film onto the substrate by drying and heating (Fig. 6) [54, 55].

In the dip-coating technique, a substrate is submerged into a sol–gel medium and is pulled out at a constant rate. To achieve a uniform layer, the coated sample is subjected to same kind of steps and further annealing of the spin coated sample. In general, by dip- or spin-coating films are prepared over substrate from newly prepared sols or aged sols at ambient temperature or about 60 °C. As a first step, a pre-heat step is performed for a short period for evaporation of solvent and organic species removal. In the next step, a post-heat processing is performed so as to obtain a well-crystallized films and the final conversion of organic by-products by treating in-between 250 to 900 °C [53–55].

Since the preferred growth can happen via nucleation at the film/substrate interface, it is possible to induce unidirectional formation of ZnO microcrystals can lead to a dense matrix than a porous one, and the evolved organic substances generated by pyrolysis can evaporate out of the film [56]. The number of coatings and the variation in precursor concentration led to increase/decrease in film thickness along with the speed of spin-coating and the post-heat treatment process. Film thickness also increases almost linearly with the withdrawal rate in the case of dip-coating [53]. It is also expected that the XRD peak intensity may increase proportionally with respect to film thickness inducing crystallinity upon annealing [56]. The adhesion strength by these two methods depend on the temperature and duration of annealing and the type of substrate.

6 General Photocatalytic Mechanism of ZnO

The photo degradation mechanism in ZnO semiconducting photocatalyst depends upon the irradiation of light, where the semiconducting matrix absorbs the photon to create electron–hole pairs. In retrospect to the excitation duration and the carrier recombination, photon induced charges are adhered at the pollutant sites. These electrons and holes react with the surface-adsorbed species from end to end along with the interfacial charge redistribution leading to numerous other degradation reactions [57].

Intrinsic valence and conduction band potentials determine the capability for radical generation of the photocatalyst with respect to the redox prospective of the process. It is not conceivable to catalyze a reduction process whose potential is highly negative than the intrinsic conduction band. Likewise, to catalyze an oxidative process where the reduction potential is highly positive than the intrinsic valance band is equally not feasible. Photocatalysts with higher oxidation potential, like WO_3, TiO_2,

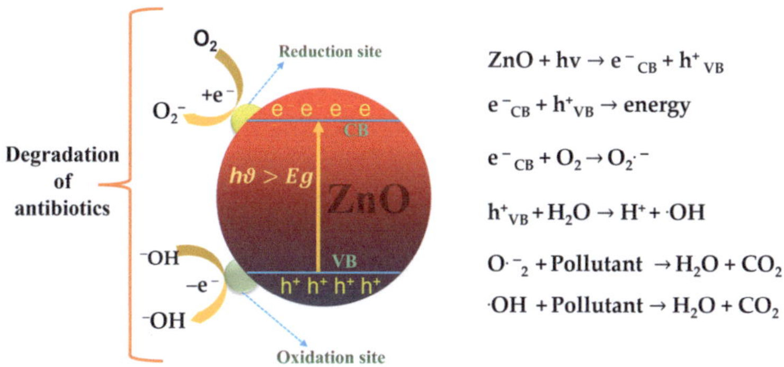

Fig. 7 General heterogenous photocatalytic degradation mechanism of ZnO for a pollutant [59]

$BiVO_4$, and ZnO are also with the limitations of including charge carriers recombination, incompatible band edge potentials, and low light absorption properties [58].

Due to efficient photodegradation of versatile pollutants ZnO has been used for heterogeneous photocatalysis (Fig. 7). The heterogeneous photocatalytic oxidation steps include:

1. Organic contaminants diffused from the liquid medium to the surface of ZnO photocatalyst.
2. Adsorption of organic contaminnats over the surface of ZnO photocatalyst.
3. Redox reactions in the adsorbed phase between the pollutant and ZnO photocatalyst.
4. Desorption of the by-products from the surface of the ZnO photocatalyst.
5. Elimination of the by-products from the interface region.

When ZnO is photo-excited by solar or visible light with energy (hv) equal to or higher than the band gap (E_g), electrons from the valence band (VB) reach the conduction band (CB) predominantly and also defect states in-between them. This process results in production of electron–hole (e^-/h^+) pairs, which can diffuse to the ZnO surface and involve in redox process (Eqs. 1–4). The H^+ interact with H_2O and OH^- ions to form hydroxyl radicals while e^- interact with O_2 to induce superoxide radical anions and then H_2O_2 (Eq. 5) [60, 61]. The produced H_2O_2 will then interact with superoxide radicals to result in hydroxyl radicals (Eqs. 7–9). Then, the formed hydroxyl radicals, which are intense oxidizing species, will degrade the adsorbed contaminants swiftly generate intermediate products which will finally be transformed into green species like CO_2, H_2O and mineral acids as indicated in (Eq. 11) [8]. The process behind the photodegradation of organic pollutants upon solar light irradiation via the redox process can be represented as given below,

$$ZnO \rightarrow ZnO\,(e) + (h) \quad (1)$$

$$ZnO(h^+(VB)) + H_2O \rightarrow ZnO + H^+ + OH^\bullet \quad (2)$$

$$ZnO(h^+(VB)) + OH^- \rightarrow ZnO + OH^\bullet \quad (3)$$

$$ZnO(e^-(CB)) + O_2 \rightarrow ZnO + O_2^{\bullet-} \quad (4)$$

$$O_2^{\bullet-} + H^+ \rightarrow HO_2^\bullet \quad (5)$$

$$HO_2^\bullet + HO_2^\bullet \rightarrow H_2O_2 + O_2 \quad (6)$$

$$ZnO(e^-(CB)) + H_2O_2 \rightarrow OH^\bullet + OH^- \quad (7)$$

$$H_2O_2 + O_2^{\bullet-} \rightarrow OH^\bullet + OH^- + O_2 \quad (8)$$

$$H_2O_2 + hv \rightarrow 2OH^\bullet \quad (9)$$

$$\text{Organic pollutant mixture} + OH^\bullet \rightarrow \text{Intermediate compounds} \quad (10)$$

$$\text{Intermediate compounds} \rightarrow CO_2 + H_2O \quad (11)$$

As discussed previously, ZnO is found to show higher absorption coefficient across a larger part of the solar spectrum than TiO_2. ZnO as a photocatalyst, the swift recombination rate of photogenerated electron–hole pairs, disturbs the photodegradation process. In addition to this, it is also observed that the solar energy conversion efficacy of ZnO is influenced by its optical absorption capability, which has been related to its wide band gap energy [2, 57]. Hence, extensive attempts have been devoted to enhance the optical characteristics of ZnO by reducing the band gap energy by predominantly doping and other techniques that leads to inhibiting the annihilation of photogenerated electron–hole pairs.

7 Photocatalytic Reactors

Photocatalytic wastewater plant comprises of an appropriate light source and a reactor setup being the major units of operation. The experimental arrangement of the reactor is the key for the efficacy of the reaction. Photocatalyst reactor involves series of steps for degradation of pollutants involving the mass transfer of the concerned pollutants to the catalyst surface including adsorption or desorption, and eventually photodecomposes the pollutants. A distinctive photocatalytic unit usually demonstrates high mass transfer and kinetic rate with high reaction surface area [62] (Table 3).

Predominantly, majority of the photocatalytic reactors explored for lab scale are successful on a limited level only. Design phase of the reactor involves significant limitations and criteria that have been come across. The main drawback is that exceedingly low catalyst concentration can lead to a low photodegradation rate as the contaminants generally are in trace amounts in water. The reaction duration may be extended to disintegrate the pollutants fully to result into less toxic by-products. The configuration of the reactor for the photodegradation process are the essential features which involve either suspended or fixed bed catalysts [71]. Nevertheless the accessible surface area per unit mass of the photocatalyst for slurry type reactors is relatively large compared to the immobilized reactors, yet the immobilized catalysts are preferred due to functional convenience including (1) additional isolation step in the downstream to recollect the photocatalyst, which can be expensive; (2) agglomeration of catalyst due to higher quantity of catalyst loading; (3) suspended catalyst particles pose other technical challenges like large scale synthesis and their process design; (4) light scattering in undesirable direction by catalyst particulates, minimizes the efficiency [62, 71]. The light flux and homogeneity and the mass transfer restrictions are crucial aspects influencing the performance and leading to significant variation in the two types of reactors. To circumvent these issues, the particle size, catalyst thickness, band gap, and the vicinity of the reactant to the photocatalyst surface are to be optimized [72].

Predominantly photocatalytic membrane reactor (PMR) is used as the system for water remediation by integrating both photocatalysis and membrane filtration techniques (Fig. 8). Photocatalytic system includes photocatalysts under light illumination to generate free radicals to degrade the pollutants. Membranes are utilized to separate catalysts and pollutants contained in the contaminated mixture after degrading by photocatalysis, and acts as a barrier, creating a chance for its reuse. Pressure-driven filtration techniques involve usage of micro, nano and ultra-filtration membranes for the normal isolation process adapted in PMR [73].

7.1 Slurry Photocatalytic Membrane Reactor

Slurry PMR is very commonly used photocatalytic unit with photocatalysts suspended freely inside of it. The catalyst dosage determines the efficiency of the photocatalytic membrane while fabricating a slurry reactor, and this has high specific surface area with catalyst suspension in the pollutant mixture. Photocatalytic efficiency depends predominantly on mass transfer rate among the catalysts and reactants. Concentration of the suspension plays the key role for efficiency as the penetration of the light photons through the reactor will be compromised as it influences the spreading of light and eventually, reducing the disintegration rate [74]. In addition to this, photocatalysts dispersed inside the reactor requires to be isolated by

Table 3 Photocatalytic reactors utilizing ZnO as the photo-catalyst to disintegrate standard methylene blue dye by pseudo 1st order reaction

Reactor type	Light source	Catalysts used	Dye Conc. (ppm)	pH range	Optimal photocatalyst dose	Degradation efficiency (%)	Reaction duration (min)	References
Flat plate continuous reactor	UV-A (4 W)	ZnO on glass	5	–	39.67 μm	~ 82	210	[63, 64]
Tubular reactor	Sunlight	ZnO	5–30	3–12	0.025 g/L	97	140	[65]
Hollow pyrex glass reactor	15 W Hg lamp	ZnO: Fe	10–50	1–10	1 g/L	88	180	[66, 67]
Aluminium foil covered wooden box	Fluorescent lamp (24 W)	ZnO/ LTO /TiO_2	5–30	2–12	0.075	95	240	[68]
Immersed lamp Jacketed photocatalytic reactor	Hg vapor lamp of (250 W) (275 nm)	ZnO	4.98–20.02	4–11	1.5 g/L	90.5 ± 2.8	180	[68, 69]
Hybrid reactor with nano filtration unit	250 W UV lamp	ZnO	250–1000	1–10	0.2 g/L	94	95	[70]

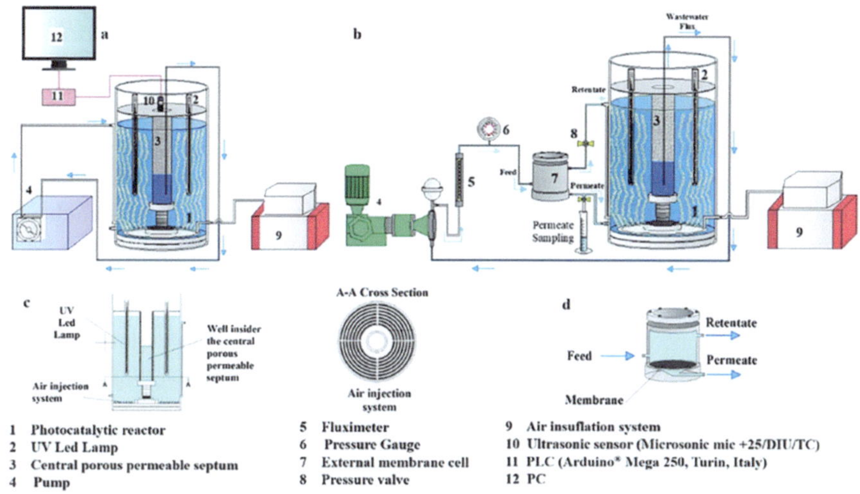

Fig. 8 Schematic representation of photocatalytic membrane reactor (PMR) and its integral working parts [73]

filtration prior to releasing the treated water. Photocatalysts made up of inorganic semiconductor may contribute to critical pollution in the environment by a significant amount. Membrane units in the slurry PMR is used to minimize the washout of the catalyst. The slurry PMR can be typically divided into two types which are separated PMR (SPMR) and incorporated PMR (IPMR) [2]. The activated photocatalysts in suspension under light illumination induce a remarkable number of oxidizing radicals that are the reason for disintegrating the contaminants. Additionally, cooling jacket is covered around the unit to retain the optimal reaction temperature and smooth working of the reactor for prolonged duration. In the SMPR system, once the organic pollutants get degraded and mineralized into less toxic by-products, and the processed water is channelled into a separate membrane unit. Then the separated photocatalysts are filtered and recycled again to the unit [75]. Nevertheless, photocatalysts gets easily segregated at the corner of the pipeline during constant flow. To 95resolve this clogging, IMPR filters are introduced to the treated mixture prior to collecting the water from the reactor. This additional filtration unit do not need any piping to forward the processed water. Ceramic membranes are preferred which have a high thermal compatibility and stability while polymeric membranes are best avoided [75].

7.2 *Parabolic Trough Reactor*

A parabolic trough reactor (PTR) consists of transparent trough collectors which has conventional photoreactor for the flow of polluted water. The collector's surface is reflective and parabolic with features enabling the sunlight to be focussed on the collectors along the parabolic focal line [68]. Several of the PTR function in sealed configuration with high flow rates to minimize the loss of the low evaporation point organic contaminants into the surrounding. High flow rate of the fluid can lead to turbulent flow that increases the mass transfer rate between the pollutants and the photocatalysts (Fig. 9). Additionally, the collectors are loaded on a movable platform with a solar tracker which rotates and lead the collectors towards the sunlight. In general, there are two variants of PTR: one-axis and two-axis parabolic troughs with the presence of one or two motors respectively where polished aluminium is usually used as the material for trough surface as it possesses high reflectivity [76]. These aspects will enhance the photodegradation efficiency and the concentration factor. The PTR degrades the pollutants by the highest flux of incident natural sunlight. Here, the amount of catalyst can be minimized against to other photoreactors because of their smaller diameter of the collector. Operating pressure and cost of construction can be reduced due to smaller size and direct usage of natural sunlight beam to degrade the pollutants [77].

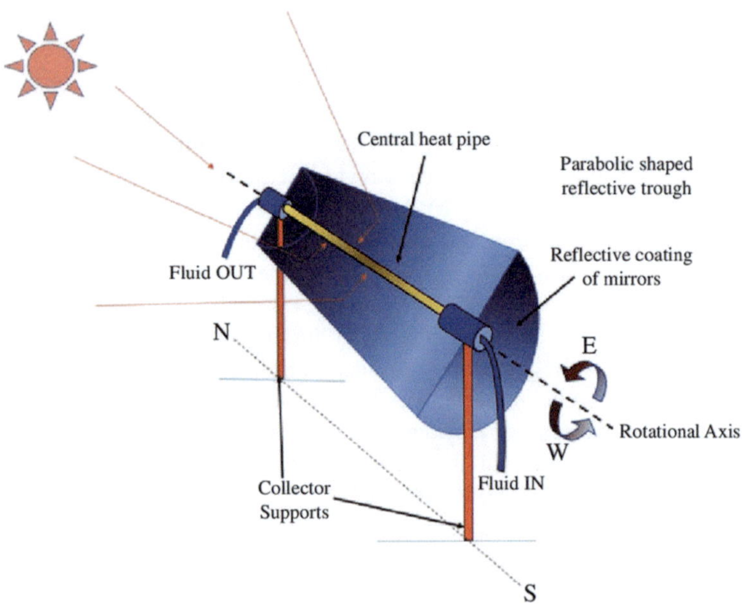

Fig. 9 Marked representation of parts of parabolic trough reactor under natural sunlight [78]

7.3 Compound Parabolic Concentrator

Compound parabolic collector (CPC) unit is designed overcoming the drawbacks of PTR and possessing both the aspects of parabolic concentrator and static flat system. CPC unit is composed of numerous cylindrical tubes and static collectors without solar tracker and rotor unit. Nevertheless, their collectors possess two halved cylinders of parabolic profiles positioned with each other that are different from the PTR [77, 79]. The parabolic design of the collectors capture both direct and diffuse radiation, reflecting and irradiating onto the receiver tube. This ensures that the complete circumference of the tube is illuminated by light, since it can function without direct sunlight, the CPC unit can be operated in a cloudy day also. Without direct heating, the quantum efficiency is enhanced because of the minimal electron–hole recombination when the photon flux is large. Although there is no other major short-coming of this design, the key interest is the availability of oxygen content in the reaction tubes for redox phenomena. Since it is a sealed unit, the oxygen content must be monitored periodically [79]. Sufficient quantity of oxygen can be rendered by infusing it directly or by mixing oxidants into the reactor tube.

7.4 Artificial Light Reactor

Other than photoreactors operating on solar energy, there are also various categories of photoreactors that use synthetic light as source of photons. The most usual artificial source of light is mercury lamp that can be sub-divided into low-pressure lamps (254 nm), fluorescent lamps (~ 365 nm), medium-pressure lamp (UV–visible range), and high-pressure lamp (UV). Among these, fluorescent lamp is cost-effective without utilizing high power and yield desired efficiency of disintegration (Fig. 10) [80].

There are two key light source configurations for the synthetic light reactors, one is positioning the source near the reactor and the emitted photons are illuminated into the unit by reflecting and focusing. Nevertheless, this geometry is not very effective as there is a loss of incident photons reflected off by the catalyst particles. The second configuration is positioning the lamp such that it surrounds the reactor which will eliminate the issues faced in the earlier configuration. In brief, the scattered photons will be reflected back and re-enter the unit with minimum loss.

The benefit of using synthetic light compared to sunlight is because of the consistency, reliability, tunability, and availability on demand. The sunlight flux is highly dependent on the climate and hard to be ascertained accurately while the flux of synthetic light is stable and can be tuned as per the need. Hence, most of the photocatalysis studies utilize artificial light of optimum flux for the photodegradation process. Nevertheless, use of artificial light source may not be economically feasible for industrial applications. This is because of the fact that artificial light system may require more power than solar light-based photosystem leading to high operational

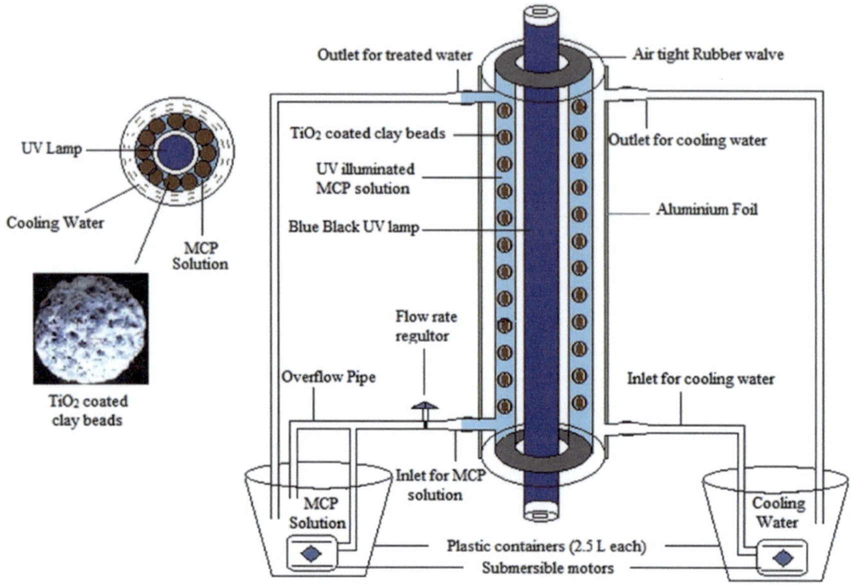

Fig. 10 Pictorial representation of artificial light reactor and its corresponding working parts [81]

cost. On the contrary, the positioning of the light source is of major concern while fabricating a synthetic light based photoreactor. This is because of the fact that the light may emit heat, varying the process temperature over prolonged use especially at higher intensity. Hence, a cooling unit is also needed which is of additional cost.

7.5 *Flat Plate Continuous Reactor*

Flat Plate Continuous Reactor is a type of photocatalytic reactor that is commonly used for processes involving continuous flow, where the reactor is designed with a flat plate or panel. This reactor is often employed with ZnO photocatalyst due to their simplicity in design, scalability, and effectiveness in promoting high mass transfer rates and efficient light utilization (Fig. 11) [82].

Design of a Flat Plate Continuous Reactor Involving Various Parts:

Flat Plate Configuration: The reactor consists of flat, parallel plates (often composed of materials such as glass, stainless steel, or other UV-transparent materials) creating a thin, uniform flow channel between them by placing them close together. These flat plates actually enhance the surface area for the photocatalytic reactions with the ZnO catalyst typically coated or immobilized on the surface of the plates. Water pollutants typically flow continuously through the reactor between the plates with either laminar or turbulent, depending on the flow rate and reactor design making the

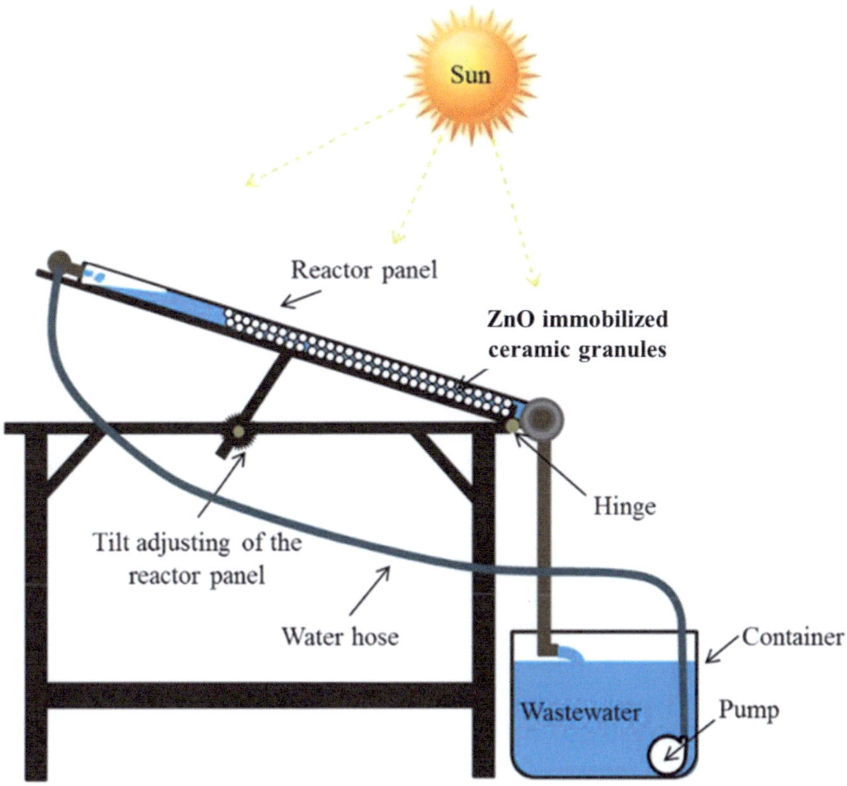

Fig. 11 Pictorial representation of flat plate continuous reactor and its working methodology [83]

reactor ideal for large-scale applications [82]. The reactor is typically illuminated by UV light (since ZnO is activated under UV) to induce photocatalytic reactions. Light is often illuminated from either of the side, top or bottom of the reactor, depending on the design. The flat plate configuration ensures that the light can penetrate the entire reactor space, maximizing the exposure of the photocatalyst to the light source. Immobilizing the photocatalyst onto a solid surface ensures it can remain adhered throughout the process, allowing for easier catalyst recovery and reuse. In a flat plate continuous reactor, the fluid flows in close proximity to the photocatalyst surface, ensuring efficient mass transfer of reactants to the catalyst and products away from the catalyst surface. The design typically focuses on laminar flow, where the fluid moves in smooth layers, allowing for better control over the reaction time and light exposure. The geometry of the flat plate reactor helps distribute light more evenly across the photocatalyst surface. The plates can be designed to maximize light penetration, promoting effective photocatalytic degradation of the pollutants or enhancement of reactions such as water splitting. Flat plate reactors are relatively easy to scale up for industrial applications. Their design allows for the addition of more plates to increase the reactor volume, making them viable for large-scale operations in wastewater treatment or hydrogen production. As a continuous flow reactor, it allows for continuous remediation of effluents, making it ideal for applications requiring steady, uninterrupted processes, such as industrial wastewater treatment [68]. Since ZnO is immobilized on the flat plates, it can be easily recovered and reused without the need for complicated filtration systems (except for marginal erosion), which is an advantage over reactors using suspended photocatalysts. Flat plate reactors are relatively simple in construction and operation. This simplicity often leads to lower costs in comparison to more complex reactor designs, making them an attractive option for large-scale or long-term applications.

One of the challenges in flat plate reactors is ensuring uniform light distribution. Since the reactor relies on UV light to activate ZnO, achieving uniform light penetration and minimizing light loss due to reflection or absorption that is crucial to maximize the degradation efficiency. ZnO can experience deactivation due to surface passivation and photo-corrosion. Regular cleaning or regeneration of the catalyst is necessary to maintain optimal performance. Achieving a uniform flow distribution across the entire surface of the flat plates can be challenging. Non-uniform flow can lead to inefficient utilization of the photocatalyst and light exposure, reducing the reactor's overall efficiency. While flat plate reactors are scalable, achieving consistent performance across large volumes of liquid or gas can be difficult [82]. Proper engineering is required to ensure light penetration, mass transfer, and catalyst recovery to be effective at large scales.

7.6 Tubular Reactors

A tubular reactor is another type of continuous flow reactor characterized by their cylindrical design, where the reactants flow through a long tube or pipe, and they are often used in processes like photodegradation. A tubular reactor typically consists of a long, cylindrical tube through which the fluid (liquid or gas) flows. The tube is often made from materials that allow light to pass through, such as quartz or transparent polymers, to allow photocatalytic reactions to take place when illuminated [68, 80] (Fig. 12). ZnO photocatalysts are immobilized either on the inner walls of the tube or in the form of suspended particles within the fluid phase. Tubular reactors are usually designed with co-current or counter-current flow arrangements. Fluid flows along the length of the tube, which ensures contact with the catalyst throughout the reactor. To activate the ZnO photocatalyst, the tubular reactor is illuminated with UV light that can be provided using UV lamps positioned along the length of the tube or external light sources that illuminate the reactor [84]. These are often designed with light guides or reflectors to maximize light utilization and ensure that the photocatalysts are sufficiently activated along the entire length of the reactor.

ZnO is usually immobilized onto the reactor surface to prevent catalyst loss during continuous flow and ensure that the catalyst is efficiently exposed to the reactants rather than suspending in the reaction fluid, but this requires additional techniques for catalyst separation, like filtration or centrifugation [41]. The continuous flow

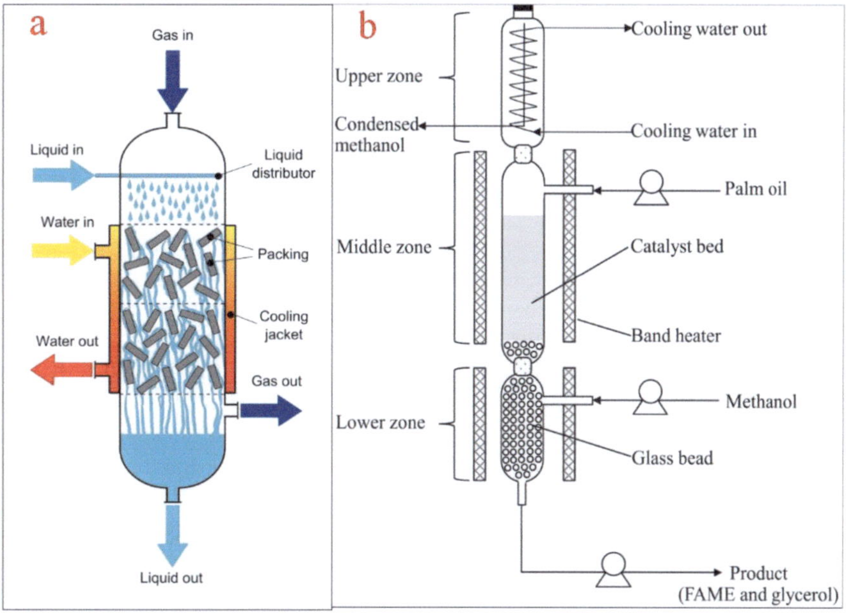

Fig. 12 Schematics of tubular reactor and their corresponding working parts [84]

and tubular geometry ensure high mass transfer rates between the reactants and the ZnO photocatalyst. Tubular reactors can be designed to optimize light penetration along the length of the reactor from multiple angles or through the tube walls, ensuring effective activation of the ZnO photocatalyst. Light guides or reflectors can be accommodated into the reactor design to further boost light irradiation on the catalyst, ensuring better photocatalytic performance. Tubular reactors are highly scalable and can be used for both small laboratory-scale experiments and large industrial applications [84]. For large-scale processes tubular reactors can be easily scaled by increasing the length or diameter of the reactor, or by connecting multiple reactors in parallel. Continuous flow nature of tubular reactors makes them ideal for long-term, uninterrupted processes. This is specifically beneficial for industrial applications where large volumes of effluent need to be treated continuously [82]. ZnO photocatalyst is immobilized onto the inner walls of the tube, recovery is simple and efficient. The catalyst remains fixed in place, reducing the need for complicated and time-consuming separation methods and minimizing catalyst loss during the process.

A key challenge in tubular reactors is ensuring uniform light distribution along the entire length of the reactor. Inadequate light exposure can lead to inefficient photocatalytic reactions. Regular maintenance, catalyst regeneration, or the use of more stable catalysts may be needed to ensure long-term performance. The rate of flow of the fluid through the reactor must be optimized to ensure sufficient residence time for the reactants to interact with the photocatalyst. While tubular reactors can be scaled by adjusting their length or diameter, very long reactors may suffer from pressure drops and uneven light distribution [84]. To address this, multiple shorter reactors can be connected in parallel to achieve higher throughput.

7.7 Hollow Pyrex Reactors

A hollow pyrex glass reactor is a type of photocatalytic reactor commonly used for reactions involving light-sensitive processes, such as those that utilize ZnO photocatalysts. The reactor is typically designed with a hollow cylindrical or spherical shape made from Pyrex glass, a type of borosilicate glass known for its high resistance to heat, chemicals, and thermal shock [85] (Fig. 13). Hollow structure refers to the design where the reactor has an empty internal volume, which is used to house the fluid or gas reactants. The hollow structure can be cylindrical or spherical, providing space for the continuous or batch processing of fluids. It is also crucial for creating a higher internal surface area for interaction with the catalyst and the light source. Pyrex glass is often chosen for its thermal stability, chemical resistance, and optical transparency in the UV range, where UV light needs to pass through the reactor walls to activate the photocatalyst, like nanostructured ZnO [85]. The high melting point and resistance to thermal shock make pyrex an ideal material for reactors that may experience varying temperatures during the process. The prime property of hollow pyrex glass reactors is their ability to transmit UV light through the glass walls. Since ZnO is excited by UV light, the reactor is often illuminated from the outside with UV lamps, or the reactor can be placed in an environment where sunlight can pass through the glass (Fig. 13). Some hollow reactors are designed to maximize light penetration by adding reflectors or light guides to ensure that the photocatalyst is sufficiently activated along the entire volume of the reactor [85]. These reactors can be used for loading a fixed quantity of reactants into the hollow reactor, letting the reaction proceed, and then removing the products or continuous flow operations, depending on the design. Pyrex glass allows for excellent light transmittance, especially in the UV range along with their high resistance to chemicals and thermal fluctuations, which makes it durable in harsh environments. There is a need for working with aggressive chemicals or when the reactor undergoes temperature changes during operation, and pyrex is chemically inert and does not leach substances and ensures that the reactor's material doesn't interfere with the photocatalytic process [86]. Pyrex glass also provides uniform heat distribution when external heating is applied. This is important when specific temperatures are needed to maintain optimal reaction conditions. They can also be easily modified in terms of size, light intensity, and flow configuration, permitting flexibility in the experimental design. Scaling up of hollow pyrex reactors can be more challenging than scaling batch processes, as maintaining uniform light exposure and mass transfer in larger systems can be difficult. Design modifications may be required to optimize performance on a larger scale.

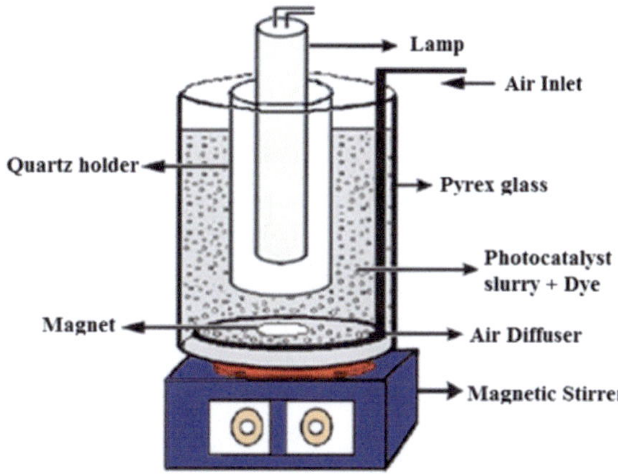

Fig. 13 Schematic representation of the pyrex glass photocatalytic reactor [87]

7.8 Jacketed Photocatalytic Reactor

A hybrid immersed lamp jacketed photocatalytic reactor is a specialized design that combines the principles of immersion lamps and jacketed reactors to optimize the efficiency of photocatalytic reactions, highly effective for processes that require controlled light exposure, temperature management, and high mass transfer (Fig. 14). Immersed lamps are light sources placed directly into the reactor, submerged in the water to be remediated [80]. The immersed lamp design ensures that the light intensity is distributed more evenly throughout the reactor, directly activating the photocatalyst in the liquid medium. By immersing the lamp in the reactor, the system minimizes light loss due to reflection or absorption by the reactor walls. A jacketed reactor refers to a reactor that has an outer jacket surrounding the main chamber with temperature control which maintain a specific thermal environment inside the reactor. The jacket can be filled with a coolant that circulates around the reactor, helping to remove excess heat generated by the photocatalytic reaction, which can be important for maintaining the optimal conditions for the catalyst and reactants [80, 88]. The integration of internal UV light sources and external temperature control via the jacket creates an efficient system that maximizes both light activation and thermal management. The light intensity is needed to activate the catalyst, but also when thermal management is crucial to prevent catalyst degradation or overheating, which could negatively impact the reaction or catalyst performance. This reactor design can help enhance mass transfer within the system along with the immersed lamp with uniform light distribution, while the jacket ensures uniform thermal management [88]. The combination of these factors leads to a more homogeneous reaction environment, promoting efficient interaction between the photocatalyst and the pollutants. The cost and lifetime of UV lamps must be considered, as they are expensive and degrade over time and may need to be replaced periodically. The combination of immersed UV lamps and jacketed systems may be more expensive compared to simpler reactor designs. Designing and maintaining a hybrid reactor system with

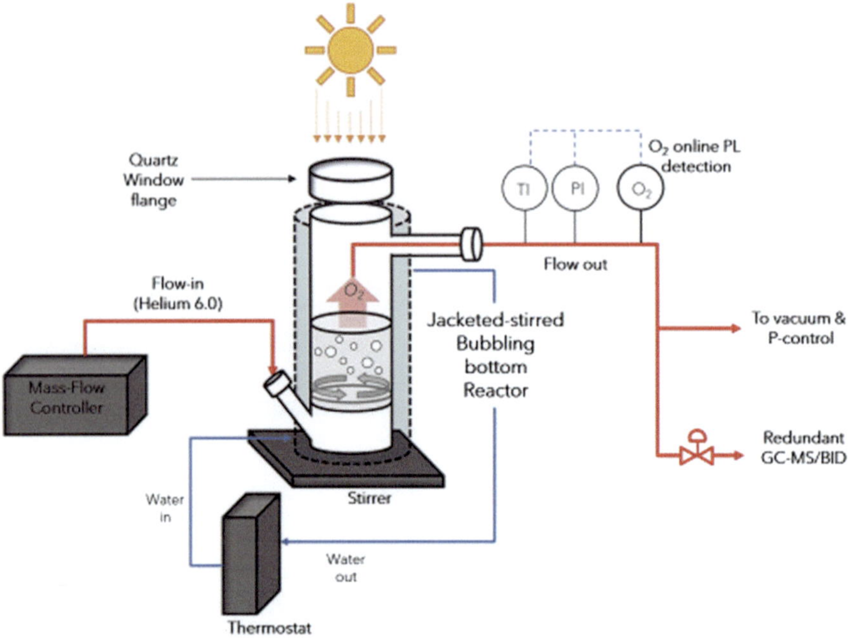

Fig. 14 Schematics of jacketed photocatalytic reactor with its respective working parts [81]

immersed lamps and jacketed thermal control may require more technical expertise compared to simpler reactors. Proper optimization of flow, temperature, and light intensity is crucial for ensuring efficient operation.

7.9 Hybrid Photocatalytic Reactors

A hybrid photocatalytic reactor with a nanofiltration unit combines the aspects of photocatalysis with the advanced filtration capabilities of nanofiltration (NF). This system is highly effective for applications such as wastewater treatment, pollutant degradation, and water purification, as it involves both the strengths of photocatalytic degradation and selective filtration [89, 90]. The photocatalytic reactor can be designed in various forms, such as batch reactors, continuous flow reactors, or jacketed reactors with temperature control for liquid or gas-phase reactions. Nanofiltration is a type of membrane filtration that operates between reverse osmosis (RO) and ultrafiltration (UF) which typically allow the passage of divalent ions and larger organic molecules but reject monovalent ions and small molecules. The NF unit in the hybrid reactor is used to selectively filter out fine particles, organic contaminants, and soluble ions that may remain in the water after the photocatalytic degradation [89, 90]. The treated fluid enters the nanofiltration unit, where residual pollutants are filtered out, leaving clean, purified water or air with residual contaminants through selective size exclusion and electrostatic repulsion. The hybrid reactor system ensures that the photocatalysis step significantly reduces the volume of contaminants or breaks them into less harmful compounds with subsequent nanofiltration step ensuring remaining harmful components, including by-products, are removed, leaving the output cleaner and safer for discharge or reuse [70]. Although photocatalysis reduces organic fouling, nanofiltration membranes can still suffer from inorganic fouling or scale formation, especially when dealing with high concentrations of salts or metal ions. Periodic cleaning and maintenance of the membrane are necessary. The initial setup cost for a hybrid photocatalytic and nanofiltration system can be relatively high due to the need for photocatalysts, UV lamps, and membranes. However, the system's long-term operational efficiency may justify the investment, particularly in high-demand industrial applications.

(i) ***Batch Reactor*** have photocatalyst (ZnO) is suspended in a liquid phase along with the pollutants or reactants. The reactor is illuminated with a light source (typically UV) during the reaction. These ZnO catalyst is mixed with the reaction mixture, and the system is exposed to light in a controlled environment [95] (Fig. 15). After the reaction time, the liquid is removed, and the products are analyzed. They are simple and easy to design, inexpensive, and useful for small-scale lab studies. Although they have their own advantages, limited scalability, as it is not efficient for large volumes or continuous reactions and the photocatalyst may not be easily recovered after the reaction.

(ii) ***Continuous Flow Reactors*** allow for a continuous flow of reactants through a reactor chamber containing ZnO photocatalyst. These reactors are often equipped with light sources (UV lamps or LED arrays) that irradiate the reactor chamber. The solution containing ZnO is continuously circulated through the reactor under UV light exposure, and reactants are continuously supplied. The products are continuously removed, maintaining a steady-state reaction. High throughput and better scalability for large-scale applications (e.g., industrial

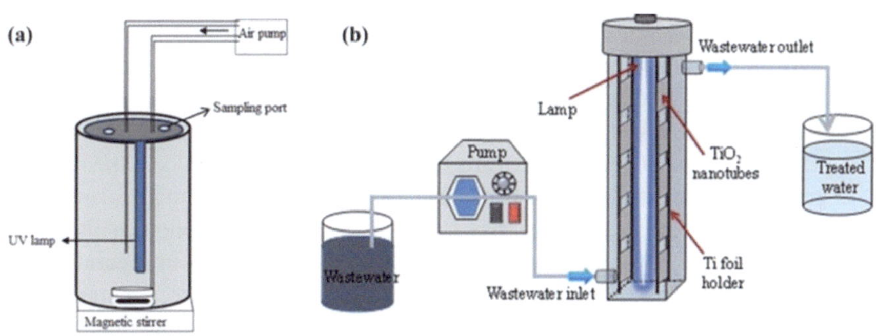

Fig. 15 Schematics of batch reactors used for photocatalytic degradation **(a)** representing the sample on magnetic stirrer, **(b)** photocatalytic filtration process with TiO_2 nanotubes [91]

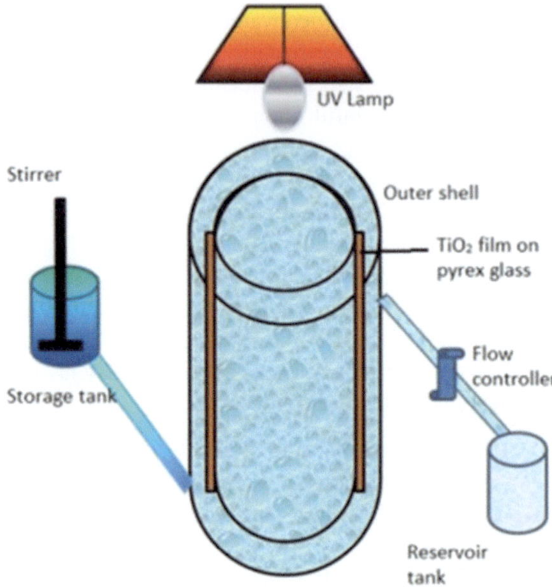

Fig. 16 Schematics of continuous flow reactor for photodegradation of pollutants [68]

water treatment) (Fig. 16). They provide more efficient use of the photocatalyst; however, it has a complex design and higher operational costs compared to batch reactors [95]. The flow rate and light intensity must be optimized to ensure maximum efficiency.

(iii) ***Fixed-Bed Photocatalytic Reactors*** have ZnO particles (often in the form of powders, granules, or coatings) are packed into a reactor vessel. The liquid (or gas) phase containing pollutants or reactants flows through the bed, where it interacts with the ZnO catalyst. UV light is usually directed onto the reactor, and the reactants flow through the bed where they are treated by the ZnO photocatalyst (Fig. 17). Simple and effective for treating large volumes of wastewater. It provides easy isolation of the catalyst and degraded products. The flow of reactants may be limited by the porosity of the bed, and the catalyst may undergo deactivation over prolonged usage [68, 95, 96]. In these reactors, ZnO is suspended in the liquid medium, often as fine nanoparticles. The light source (usually UV lamp) is used to irradiate the reactor and catalyze the reaction. The ZnO particulates are suspended in the solution, and the reaction proceeds while the mixture is continuously agitated to maintain suspension. High surface area exposure, easy mixing of reactants and photocatalyst, and high efficiency for smaller-scale operations [97]. Difficulty in separating the ZnO catalyst from the treated solution, requires additional porous membrane like filtration (Fig. 4) or centrifugation steps with considerable amount of time.

(iv) ***Annular Photocatalytic Reactors*** consist of a cylindrical design where the catalyst (ZnO) is typically placed on the inner surface of the cylinder, and the reactants flow through the annular space between the two concentric cylinders.

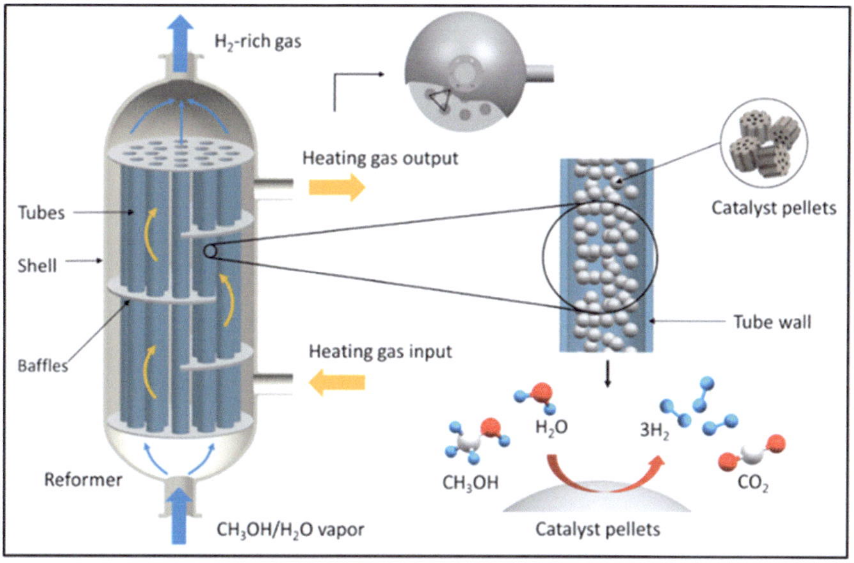

Fig. 17 Schematics of fixed bed reactor and associated process [92]

UV light is generally used to irradiate ZnO residing on the inner surface, where the photocatalytic reaction occurs as the reactants pass through it [98, 99] (Fig. 18). This reactor design can ensure good light penetration and contact between reactants and the catalyst. It also provides high surface area for the photocatalytic reaction. These reactors can be complex to design and scale up for large sized applications.

(v) ***Photocatalytic Fluidized-Bed Reactors*** have ZnO particles that are suspended in a fluidized state (usually by air or liquid flow). The photocatalyst particles are fluidized, which enhances the contact between reactants and catalyst. The

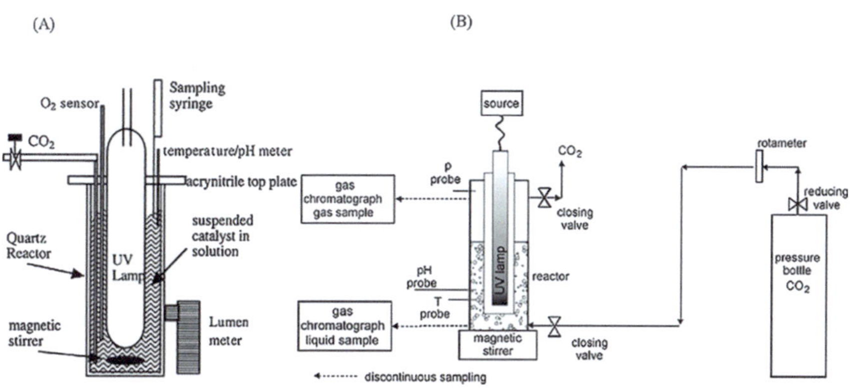

Fig. 18 Schematics of working mechanism of annular photocatalytic reactor [93]

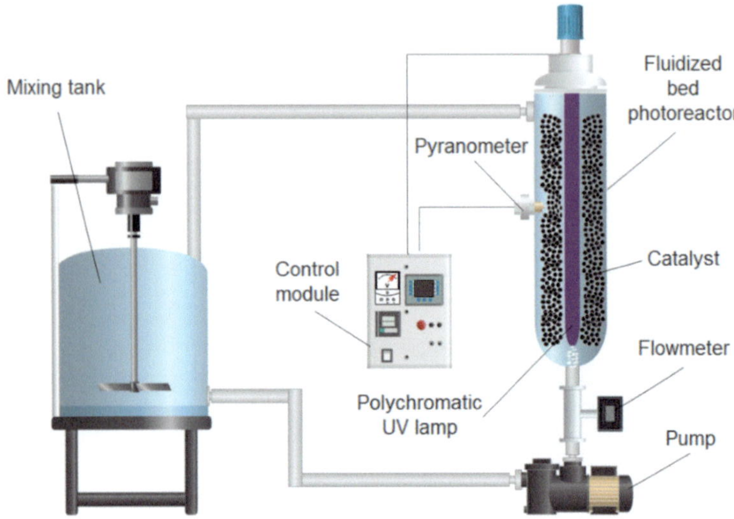

Fig. 19 Pictorial representation of fluidized bed photocatalytic reactor unit [94]

reaction mixture flows upward through the bed of ZnO particles while being irradiated by UV light (Fig. 19). The fluidization ensures constant movement of the particles, maintaining high contact efficiency [100]. Improved heat and mass transfer, and high contact between the reactants and the photocatalyst. It is ideal for continuous processing. Complexity in controlling fluidization conditions and catalyst separation after the reaction [101].

(vi) ***Photocatalytic Solar Reactors*** use solar energy to activate the photocatalytic process. ZnO can be used as the catalyst to degrade pollutants or generate hydrogen (upon proper heterostructure interface formation) in the presence of sunlight. These reactors are designed to maximize sunlight exposure, with reflectors or lenses utilized to focus light on the catalyst. It may operate similar to batch or continuous reactors. Sustainable and energy-efficient, as they use solar energy instead of artificial UV light sources [102]. Dependent on the availability of sunlight and intensity make them less efficient in regions with limited sunlight and during night time. Efficient photocatalytic reactors must ensure that light is uniformly distributed and that the ZnO photocatalyst is exposed to adequate light intensity. This can involve optimizing reactor design or using reflectors or light guides to enhance light penetration. The contact period between the reactants (e.g., pollutants, water) and the ZnO catalyst is crucial for efficient photocatalysis. Reactors should be designed to maximize the surface area exposure of the catalyst while ensuring effective mixing or flow of reactants. The ability to recover and reuse ZnO catalyst is important, especially for reactors like batch or suspended-bed systems where the catalyst is dispersed in the solution. Techniques like centrifugation, porous membrane filtration (Fig. 4), or using a fixed-bed design can help separate and reuse the catalyst. For industrial applications, reactor design must be scalable to handle large volumes of water or gas and maintain high photocatalytic efficiency. Some photocatalytic processes using ZnO may require specific temperature or pH conditions for optimal activity.

Photocatalytic reactors of optimal design using ZnO are integral in harnessing the photocatalytic properties of ZnO for applications like environmental cleanup, renewable energy generation, and sustainable chemical production. The choice of reactor type (batch, continuous flow, fixed-bed, etc.) depend on the choice of application, scale, budget and operational efficiency that is desired.

8 Zinc Oxide Tetrapods for Photocatalytic Degradation

8.1 Physico-chemical Advantages of ZnO Tetrapods

8.1.1 Optical Characteristics of ZnO Tetrapods

Photoluminescence (PL) experiments confirm the presence of two emission maxima related to ultraviolet (UV) emission at 380 nm found around 3.2 eV and green emission at 495 nm related to the near band-edge emission due to inter-band transitions and the green emission has been found to be a resultant of singly ionized oxygen vacancies [58, 103]. Photoemission is a result of annihilation of photo-induced holes with the electrons residing at the oxygen vacancy sites. Predominantly the UV emission for the bulk ZnO is observed only at very low temperature whereas for the ZnO-T it was evident even at ambient temperature. The major reasons could be being of the high quality or single crystallinity and low impurity and structural defects in ZnO-T relative to the bulk counterparts and it can also be limited to the quantum confinement effects in nanostructured form. The room temperature photoluminescence spectra indicates a significant variation in response from a collection of arbitrarily positioned tetrapods and an individual tetrapod aligned in a fashion such that one of its arm is parallel to the incident laser beam and behaved like a Fabry–Perot resonant cavity.

8.1.2 Electronic Properties of ZnO Tetrapods

The limitations of establishing proper electrical contacts to the nanocrystals makes it difficult to measure their electronic properties. Reports show metallo-organic coating supported process by focused ion beam (FIB) technique is made for the single tetrapod diode structure to test the electrical contacts. Among the three electrical contacts of the tetrapod morphology, two were W contacts that are ohmic in nature and one is a Pt contact which behaved as a Schottky contact [104, 105]. The rectification aspect and electrical output of the ZnO-T Schottky diode was tested using a continuous wave laser irradiation along with a boost in laser power density, where the device drifts from a rectifying to more ohmic nature. Similar results are obtained for measurements with a single nanorod Schottky diode. The ZnO-T by direct thermal evaporation was explored for field emission and reported to have an extremely low 'turn on' field of 1.6 V μm^{-1} at a current density of 1 $\mu A/cm^2$ because of the enhanced aspect ratio of the ZnO-T [106, 107].

8.1.3 Multifunctional Characteristics of ZnO Tetrapods

The utility of ZnO-T are multifunctional due to its distinctive combination of physico-chemical aspects along with a specific morphology allowing it to be employed for a various requirement. The optical response of tetrapods is used for the optimization of self-reporting materials dependent on their photoluminescence emission of ZnO-T which varies with stress. ZnO-T/polydimethylsiloxane (PDMS) elastomeric composite was fabricated unlike the arbitrarily formed short fillers that orient along the stress direction and have no impeding effect. The tetrapods have arms that are non-parallel to the applied stress direction and therefore have an impeding effect by an interlocking process [52]. When stress increases, the elastic energy is initially stored in the ZnO-T network thus stiffening the composite and when the critical stress allows this framework to break and the stored energy is liberated. The witnessed optical output is because of the defects induced by stress during the rupturing of the tetrapods. This behaviour permits for the non-destructive assessment of the induced internal damage in the lattice and therefore unlocks new potential for its application in several technological requirements.

ZnO-T was recently explored for low surface energy polymers by utilizing it as the linker of polytetrafluoroethylene (PTFE) and polydimethylsiloxane (PDMS) that were combined with enhanced peeling resistant strengths [105, 108]. The concave part of the tetrapods acts as micro/nano velcro filler at the interface of the polymers unlike the convex shorter fibres. This results in adhesion among the polymer layers merely on the basis of mechanical effects with chemical reaction. The geometry of the tetrapods is like if it is attempted to pull out, it leads to a considerable bending of its arms or deformation of the polymer in several embedded circumstances. This process of combining of materials can be supplemented with other linking procedure like plasma process to enhance the adhesive force among the materials and are specifically essential in situation where the materials cannot be linked by the usual process.

8.1.4 Biological Advantages of ZnO Tetrapods

The antiviral response of the ZnO-T against the HSV-1 (Herpes simplex virus-1) and HSV-2 (Herpes simplex virus-2) viruses show destruction of the cell by the process of their linking to the surface of ZnO-T. The innate oxygen vacancy defects hold the virus to bind over the surface of the ZnO-T arms as these vacancies are the reason for the anchoring of the viruses [108]. The efficiency can be boosted by irradiation with UV light which further enhances the O_2 vacancy density in the ZnO-T. The cytotoxicity of the ZnO-T is little when compared to other nanostructures that makes it a promising futuristic material for biomedical applications.

ZnO-T templates to design newer light in weight material called aero graphite making it a key low-density matrix in the materials world. The ZnO structures utilized by sacrificial templates by flame transport synthesis process in continuation by compaction of the tetrapod powder and ensuing thermal treatment to result in inter-linking [108]. Chemical vapor deposition (CVD) is executed on this template using toluene precursor with relevant gas purging. This template is eliminated by converting ZnO to metal Zn by H_2 which is then isolated out by the exhaust unit. This material has extended utility in sectors where electrically conductive, porous and physically stable networks are a necessity for applications such as Li-ion batteries and supercapacitors [109].

The ZnO-T has also been effectively utilized for sensors as an in-situ lift out process. ZnO-T produced in an aqueous medium to fabricate a single ZnO-T UV sensor presented lowering of resistance when it was irradiated with UV light. The reduced resistance is ascribed to the desorption of the chemisorbed O_2 from the surface and increase in charge carrier density due to electron–hole pair production. Interpenetrated framework of ZnO-T in a gap induced by rupture of thin film of Au with fabrication approaches allows a direct integration of the electrical contacts on chips [110]. The UV sensor exhibited rapid recovery and response period associated with a large signal ratio for a dried drop of ZnO-T solution between the two single walled carbon nanotubes (CNT) serving as electrical contacts on a polyethylene terephthalate (PET) substrate.

ZnO-T gas sensor showed an improved performance for ethanol in comparison to other ZnO sensors with a value as high as 20 at 100 ppm of ethanol. ZnO-T has a extremely porous framework allowing a high S/V ratio for gas sensing [111]. Silica-coated amino-modified ZnO-T were employed as novel carriers for plasmid DNA delivery with their distinctive morphology bound to the cell membranes with their tetrapodal arms linked to the cell for DNA delivery but their entangled network do not allow to get across the cell membrane [112]. ZnO-T also carries plasmid DNA into the cells without entering them into the cells and which in turn minimizes cytotoxic effects where phages carry genes to the cells without invading into them. The photocatalytic performance of ZnO-T is also being explored in specific as it has scope for future requirements such as waste water treatment and cleansing. ZnO-T has unique 3D morphology preventing its aggregation in fluid medium, thus offering a higher effective surface area compared to other nanostructures for photocatalytic degradation to occur.

The field emission aspects of the ZnO-T by depositing it on a nickel and doped-silicon substrates with a low 'turn on' field [108]. Low disintegration rate of the cathode has been published which make ZnO-T a perfect host for cold cathodes and vacuum electron appliances in field emission displays. ZnO-lead–zinc-borosilicate glass varistors using ZnO-T which needed low annealing temperatures for making varistors due to enhanced performance of the ZnO-T powders and showed a large non-linear coefficient α of 35 and associated with leakage current lower than 5 μA.

8.2 Special Aspects of ZnO Tetrapods for Photocatalytic Degradation

ZnO-T with their unique nanostructure exhibit distinct physico-chemical, opto-electronic properties, making them highly promising for versatile applications, including photocatalysis. The key parameters distinguishing ZnO-T as photocatalyst in comparison to other morphologies are discussed below.

ZnO-T with four-armed structure significantly increases its overall surface area compared to regular near spherical shaped ZnO particles [113]. This high surface area intentionally yields more active sites for photocatalytic process to commence, enhancing the overall efficiency. The tetrapod structure also facilitates better light absorption because of its 3D geometry and multiple reflection paths with angles and electron–hole pair separation, improving the photocatalytic activity, particularly for applications like degradation of organic pollutants or hydrogen production [109].

2D materials such as graphene, MXenes, graphene oxide (GO), and transition metal dichalcogenides (TMDs) have garnered attention because of their remarkable characteristics including high surface area, improved charge carrier mobility, tuneable electronic properties and ability to facilitate electron transfer and inhibit charge recombination. When ZnO is combined with 2D materials, several synergistic effects improve photocatalytic performance as they act as electron acceptors, improving the separation of excitons in ZnO. This minimizes recombination rates and enhances photodegradation performance. Some 2D materials, such as graphene, can absorb visible light and extend the light absorption range of ZnO, improving its efficiency in utilizing the solar spectrum. The large surface area of 2D materials yieldds more active spots for photocatalytic reactions, increasing the overall catalytic activity. The incorporation of 2D materials may also improve the stability of ZnO under reaction conditions, preventing photo-corrosion and maintaining long-term efficiency. The ZnO-T combined with graphitic carbon nitride gC_3N_4 shows improved photocatalytic activity due to the properties mentioned above. The degradation efficiency increased by 3% for ZnO/gC_3N_4 photocatalyst under sunlight within 30 min relative to the pure ZnO-T due their increase in more active sites available for photodegradation. The ZnO/gC_3N_4 sample has highest rate of photodegradation with least ratio of change in concentration indicating improvement in its efficiency (Fig. 20).

ZnO has an optimum bandgap of around 3.3 eV, allowing the absorption of ultraviolet (UV-C) light. Effective photocatalytic reactions under UV light irradiation can generate electron–hole pairs that drive the photocatalytic processes. The tetrapodal morphology may enhance the light harvesting capability by scattering and trapping light, even extending the absorption into the visible range to some extent when modified with other dopants or incorporates. Their unique morphology

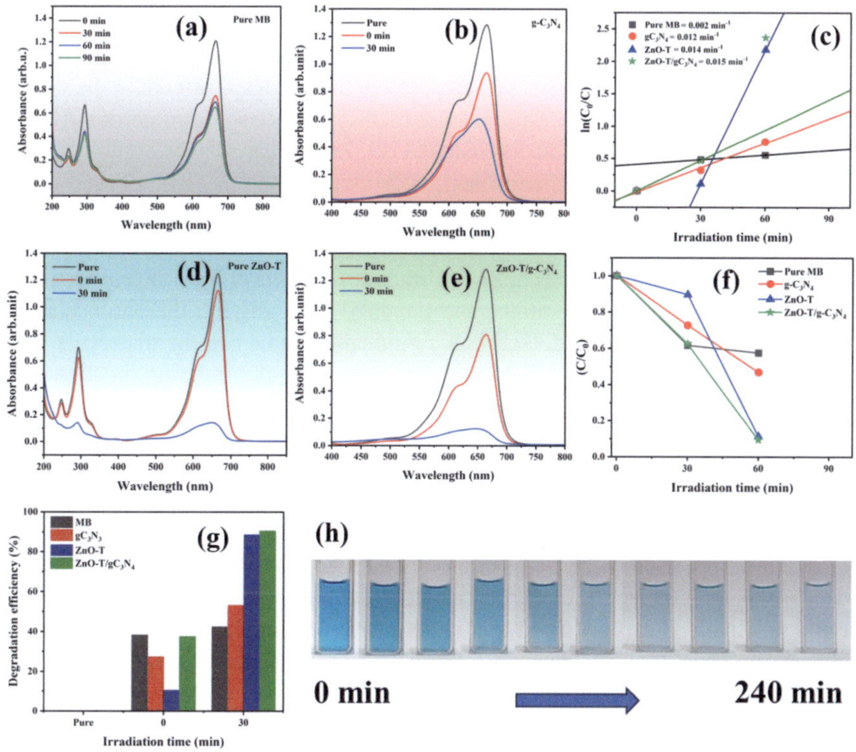

Fig. 20 Absorption spectra of photodegraded pure **(a)** methylene blue dye with photocatalysts **(b)** gC_3N_4, **(c)** change in concentration plot of methylene blue dye **(d)** ZnO tetrapods, **(e)** ZnO/ gC_3N_4, **(f)** rate constant plot, **(g)** bar graph representing photodegradation efficiency within 30 min, and **(h)** photograph representing methylene blue photodegradation over an irradiation time of 240 min under white LED light of 200 W

helps in minimizing the ratio of recombination of photo-induced electron–hole pairs. The tetrapodal structure provides multiple directions for electron movement, potentially improving charge carrier separation and prolonging their lifetime for effective photocatalysis [113]. ZnO-T often have surface defects, oxygen vacancies, or Zn interstitials, which act as trap states for charge carriers, minimizing recombination and boosting photocatalytic efficiency [112, 113]. ZnO is predominantly chemically stable in a wide range of conditions, making ZnO-T mostly durable during the photocatalytic processes. They are physically stable and can withstand mechanical pressure, while maintaining high catalytic performance [106]. The tetrapodal shape enhances electron transport by offering continuous path for charge carriers making the electron transport system efficient. This increases the photocatalytic performance by boosting the mobility of electrons and holes to the surface where the photocatalytic reactions occur predominantly [106].

ZnO-T can be doped with various metals including some transition metals or non-metals which can further enhance their photocatalytic activity. Doping or incorporating other metals and non-metals can narrow the bandgap, extending the visible region photon absorption also. This will also modify the electronic properties of ZnO-T to improve the photocatalytic efficiency [112]. ZnO-T can also be combined with other materials (of varying dimensions), such as graphene or carbon-based materials, to form composites. These composites can further boost the photocatalytic efficacy by increasing charge transfer and expanding the light absorption range. Defect-mediated photocatalysis in ZnO-T includes the role of intrinsic and extrinsic defects (like vacancies, interstitials, and dopants) in modifying the photocatalytic properties of ZnO-T. These defects can significantly influence the material's light absorption, charge separation, and reactivity, enhancing or hindering the photocatalytic process. ZnO, naturally possesses various types of defects that can be engineered during synthesis to improve its photocatalytic activity. These defects include oxygen vacancies (V_O) as one of the most common intrinsic defects in ZnO are typically formed under reduced oxygen conditions during synthesis or by relevant thermal treatments. Oxygen vacancies act as electron traps, capturing the electrons generated during photocatalytic reactions. These trapped electrons can interact with adsorbed oxygen molecules to form reactive superoxide radicals (O_2•−), which are highly reactive and can degrade organic pollutants during photocatalytic reactions. This charge separation mechanism improves the overall photocatalytic efficiency since it prevents the recombination of electrons and holes, which would otherwise decrease the efficiency of the photocatalyst. Zinc vacancies (V_{Zn}) also play a key role in the photocatalytic process, though they are less common than oxygen vacancies. Zinc vacancies can also act as electron traps, leading to the formation of photo-induced holes. These holes can interact with water molecules or organic substrates, generating hydroxyl radicals (•OH), which are also highly reactive and can contribute to the disintegration of pollutants in the photocatalytic process. The balance between oxygen and zinc vacancies can help optimize the material's photocatalytic efficiency by ensuring the right amount of charge carrier separation and radical generation. The surface of ZnO-T is rich in defects, especially at the tips and edges of the arms. These surface defects increase the density of active sites for reactions, enhancing the photocatalytic process. Surface defects can create localized electronic states in the band gap of ZnO, which can capture photo-generated charge carriers and prevent recombination. These defect states can also serve as active sites for the adsorption of reactants (such as water, oxygen, or organic pollutants), further enhancing the photocatalytic activity.

Doping ZnO-T with certain dopants (such as transition metals like Fe, Cu, Ag, or non-metals like N, S, or C) can create controlled point and surface defects in the material. These dopants can introduce mid-gap states or band bending, leading to a modification of the electronic structure and photocatalytic behaviour. Doped defects can improve light absorption by narrowing the band gap, extending absorption into the visible light region, and increasing the number of active sites for photocatalysis. For example, nitrogen doping can decrease the ZnO band gap and enhance photocatalytic activity under visible light. Electron trapping at oxygen vacancies helps in the formation of superoxide radicals, while holes can be captured at zinc vacancies or surface defects, promoting hydroxyl radical formation. The efficient isolation of charge carriers in ZnO-T due to their unique 3D morphology combined with defect states significantly improves photocatalytic performance by enhancing charge transfer and minimizing recombination. Defects on ZnO-T can lead to the generation of hydroxyl radicals (•OH) and superoxide anion radicals (O_2•−), which are key species responsible for breaking down of the organic compounds.

During water splitting, oxygen vacancies can assist in the generation of reactive oxygen species (ROS), which aid in the oxygen evolution reaction (OER), and zinc vacancies may help generate hydrogen by facilitating the hydrogen evolution reaction (HER). CO_2 reduction is another area where defect-mediated photocatalysis in ZnO-T can be advantageous. Oxygen vacancies may help in CO_2 adsorption and activation, facilitating its reduction to valuable products such as methane or methanol. Defects can also improve the photostability of ZnO-T, however, excessive defects, especially at the surface, can lead to photocorrosion or the disintegration of the photocatalyst over time. Therefore, controlling the density and distribution of defects is essential for maintaining long-term photocatalytic activity. Additionally, defect engineering can be used to enhance the structural stability of ZnO-T by reducing the agglomeration of nanoparticles during repeated photocatalytic cycles (Fig. 21). Defect-mediated photocatalysis in ZnO-T plays a vital role in enhancing their photocatalytic performance by carefully tuning the type, concentration, and distribution of defects, the performance of ZnO-T can be optimized for photocatalytic degradation of a variety of dyes.

ZnO-T are used to degrade organic pollutants in wastewater by generating hydroxyl radicals (•OH) and superoxide anions (•O_{2-}) upon UV irradiation. The

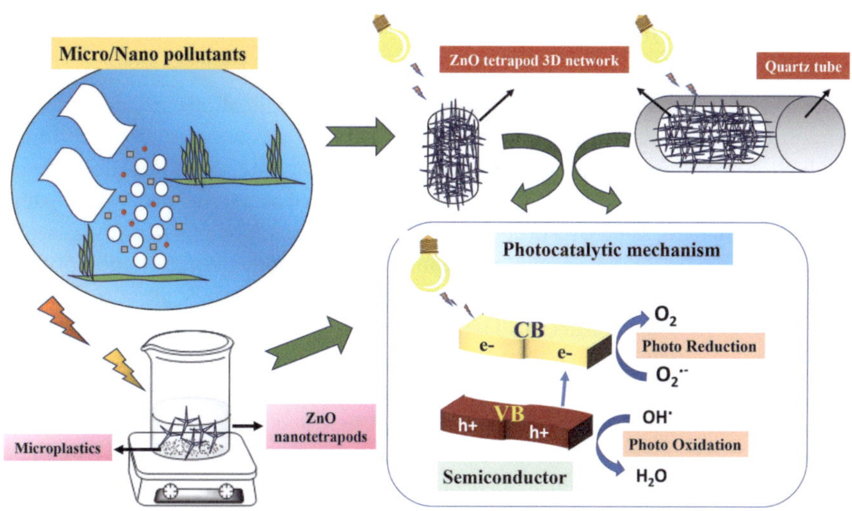

Fig. 21 ZnO tetrapod-based prototype photocatalytic filtration device for pollutant filtration and degradation

reactive species attack and break down the organic molecules, making them less toxic [114, 115]. Their cyclic study without altering their physico-chemical stability makes them industrially viable. ZnO-T can generate electron–hole pairs that drive the water-splitting reaction, converting water into beneficial hydrogen and oxygen. ZnO-T generally exhibit good photostability, which renders them a viable means for long-term photocatalytic utility when compared to spherical shaped particles and other morphologies. They do not degrade quickly (only gradually) under prolonged UV exposure, as they are a combination of both roughly spherical centroid and nanorods ensuring a significant sustainable photocatalytic performance [14, 115]. Preliminary characterization of single crystalline nanostructured tetrapods by Brunauer–Emmett–Teller (BET) study revealed unanticipated results with ~ 0.077 m^2/g surface area as higher surface area provides more active sites and thus increases the adsorption capacity and photocatalytic activity. Photodegradation results of synthesized ZnO-T using modified catalyst free rapid thermal sublimation technique showed rapid degradation of various dyes, antibiotics and effluents with negligible surface area, which raises curiosity about their surface properties and series of key surface chemistry questions leading to specific aims and experiments that can lead to the fundamental understanding of the physico-chemical aspects of ZnO-T. The role played by oxygen species and surface oxygen vacancies in the catalytic oxidation, may be linked to the improvement in the mobility of O_2 and the easy formation of ROS (reactive oxygen species). There are four major types of oxygen species physically adsorbed oxygen (O_2(ad)), superoxide species (O_2–), monoatomic oxygen (O–), and lattice oxygen (O_{latt2}–). As a consequence, the catalytic activity is expected to show a direct relation with the surface O_2 vacancy concentration and its types.

9 Future Perspectives

ZnO nano tetrapods are significant towards various applications because of their extraordinary features constituting abundant Lewis's acidic/basic sites, optimum bandgap, defect sites, charge separation and electron trapping, and compatibility with various metals. Photocatalytic, thermocatalytic, electrocatalytic processes extensively exploit their intensive properties for applications like photocatalytic water purification by removing various pollutants. Though other morphologies of ZnO also exhibit significant potentials, the challenge lies toward large-scale cost-effective production of ZnO nanostructures [108]. High surface accessibility and re-usability of ZnO nanostructures play significant role in photocatalytic and other electrocatalytic applications. Though the conventional spherical ZnO has fully exposed surface area, and ZnO nanorods have high S/V ratio but they lack the additional capabilities of forming intertwined network of highly porous 3D structures (Fig. 22). Combination of both the near spherical shaped ZnO as the centroid and each of their arms as separate nanorods, ZnO-T exhibit both their exceptional properties together. Thus, due to its exceptional surface sensitivity, the ZnO-T are highly porous and their 3D

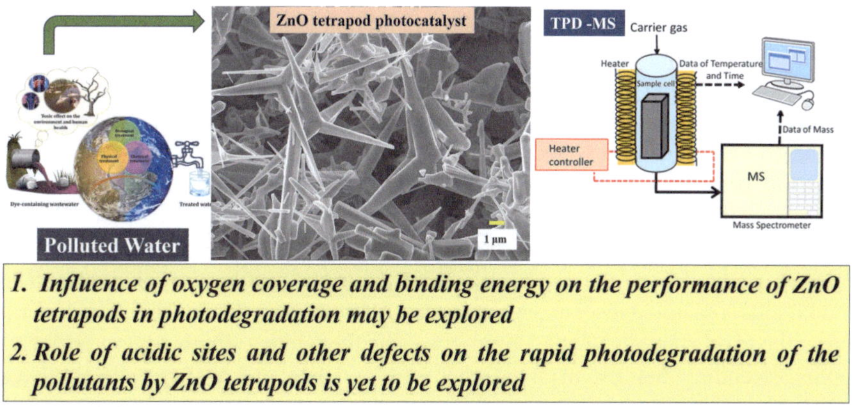

Fig. 22 Graphical representation of ZnO-T for futuristic exploration on the photocatalytic activity with TPD-MS technique

interconnected frameworks extend their capability for photocatalytic requirements [112, 113, 116].

The ZnO-T with gradually varying arm dimensions and morphology have manifested their photocatalytic efficacy against many dyes including methylene blue, rhodamine-B, methyl orange and some antibiotics including ciprofloxacin, microplastics etc. Complex shaped ZnO structures with high S/V ratio and further enhanced surface area has led to enhanced photocatalytic disintegration. The performance of ZnO-T can be further tuned by surface sensitization of various other metals, nanoparticles, quantum dots and 2D materials, incorporating a variety of defects. Functionalizing the slender arms of ZnO-T can make them more robust for photocatalytic process by selectively engineering with materials for their utilization in versatile applications. Additionally, as an intrinsic functionality, the wettability of ZnO-T can be modified by irradiating it with UV photons. Combination of ZnO-T's UV-induced wettability, easily engineerable surface and their networked porous structure may be efficiently exploited for several bio-chemistry based applications. Due to their morphological advantage, these ZnO-T can be used as an ideal model for filtering device requirements and can be quite easily upscaled to an industrial level as they are efficient, cost effective, non-toxic, and can be produced in mass scale [109].

Typically, filtering and reusing spherical shaped ZnO particles for photocatalytic or filtration techniques is a huge task. Overcoming this, some bigger sized species out of liquids is a comparatively easier task using sieving techniques, but ZnO-T can be used in the place of spherical particles to achieve this straight forward in a porous membrane. ZnO-T can be a game changer in most filtration processes including photocatalytic filtration techniques, biological filtration techniques involving bacteria and viruses in blood is a critical job due to bigger size variations in the bio-entities. The diameter of viruses are generally a few hundred nanometres while blood cells are a few micrometres, removing viruses from blood is not a facile technique as blood cells can also be removed during the blood dialysis. Viruses can be easily trapped with UV-illuminated ZnO-T as their surface are highly sensitive with anti-viral and anti-bacterial properties [108, 112]. Reports suggest that the 3D linked porous frameworks can be efficiently made use for separating viruses from the blood having pore size much bigger than the blood cells. The micron sized arm dimensions exhibit important role in identifying wettability of different liquids as extremely porous ZnO-T framework may be used effectively as filters for separating out the required fluids. Polluted water, drugs, chemicals, oils, defected blood, antibiotics, pharmaceutical compounds, organic dyes, etc., can be easily filtered using the multiple functionalities of the nanostructured ZnO-T by engineering them into a simple prototype device model, extending their use in environmental, biomedical and pharmaceutical related applications.

The free radical study on photocatalytic activity of the textile effluents in the presence of ZnO-T revealed the significant task of hydroxyl radicals. To further investigate the influence of change in concentration of surface hydroxyl species on the performance of the pristine and dye degraded catalysts. The surface acidity and basicity can be investigated by performing NH_3 and CO_2-based Temperature programmed desorption (TPD) measurements. Brønsted (BA) and Lewis's acid (LA) sites are essential for photocatalytic activity indicating the presence of hydroxyl groups which can trap holes leading to formation of hydroxyl radicals, minimizing recombination of electron and hole, and thereby enhancing photocatalysis (Fig. 22).

The TPD apparatus, may be used to examine the role of different surface adsorbed oxygen species and along with determining surface acidity and basicity on the ZnO-T surface. Oxygen isotope exchange at high temperature with O_2 TPD analysis can be performed for O_2 adsorption and to identify various different types of surface oxygen species and vacancies may be detected by Mass spectrometer (MS). The surface acidity and basicity can be investigated by Ammonia Temperature-programmed Desorption (NH_3-TPD) technique using He as a carrier gas. Prior to TPD experiments, the catalysts shall be reduced at atmospheric pressure and purged with He gas until achieving a constant baseline, and then the evolved NH_3 or CO_2 may be identified by the MS. The proposed technique may enable to understand the unexplored fundamental behaviour of the synthesised single crystalline ZnO nano tetrapods and its efficient photodegradation aspects by exploring the surface properties using TPD–MS and oxygen isotope exchange measurements. ZnO-T represent a promising host material with advantageous geometry for photocatalytic applications because of their unique structural characteristics and photocatalytic efficacy. Their morphology, stability, surface binding aspects and charge transport properties enable them to efficiently degrade pollutants or produce clean energy like hydrogen under suitable light irradiation.

10 Conclusion

ZnO-T are a unique and promising form of nanostructures that have demonstrated significant potential in photocatalysis due to their distinctive shape, which enhances the material's photocatalytic activity. These tetrapod shape inherently provides a high surface-to-volume ratio, increasing surface area which permits for highly active sites for photocatalytic reactions, leading to enhanced light absorption and reaction efficiency. The 3D structure of ZnO-T increase their light-harvesting ability since it creates multiple light paths and reflections within the material, allowing for better utilization of light energy. Their branching structure of the tetrapods can facilitate the isolation of the photo-generated electron–hole pairs helping in charge transport. This geometry also helps to control recombination, enhancing the collective performance of the photocatalytic process. ZnO-T, like other ZnO-based photocatalysts, can be easily recovered after use in liquid-phase reactions, allowing them to be reused in subsequent cycles without significant loss of performance. ZnO is an environmentally friendly and cost-effective photocatalyst which can contribute to sustainable processes and green chemistry. Although, ZnO is a proven efficient photocatalyst and is most effective under UV light since its bandgap being large enough to absorb UV light but not significantly in visible region. ZnO nanoparticles synthesized by various traditional techniques are relatively expensive, time consuming (in terms of annealing and processing) and produces larger sized particles and size distribution. ZnO synthesized by modifieds catalyst free thermal sublimation method resulted in highly single crystalline nano structured ZnO-T. Green emission of the synthesised ZnO-T confirmed the presence of defects leading to the prominent electron–hole radiative recombination. The ZnO-T manifest these as an efficient photocatalyst in disintegrating both the cationic and anionic dyes individually or even when mixed together under white LED light irradiation by defect mediated mechanism.

To tackle the limitation in visible light absorption, dye-sensitized ZnO-T or incorporate/dope them with other materials or surface sensitize to extend their activity to visible light. ZnO-T may tend to agglomerate or aggregate over time, reducing the available surface area and photocatalytic efficiency. This issue can be minimized by using stabilizing agents to minimize aggregation. While ZnO-T are relatively durable than other photocatalyst, they may still experience catalyst deactivation over time due to the accumulation of degraded by-products or surface contamination. Periodic regeneration or cleaning may be necessary to maintain high photocatalytic performance. Immobilization of ZnO-T offer several benefits in the photocatalytic processes including enhanced stability without aggregation, leaching, photocorrosion or sedimentation. It is also relatively easy to retrieve and reuse the photocatalyst after every cycle leading to saving the catalyst cost. These ZnO-T not only represent a highly effective photocatalyst with a unique 3D structure that enhances light absorption, charge separation, and surface area. These features make them highly efficient for various other applications, including pollutant degradation, hydrogen production, air purification, and CO_2 reduction. Their high stability, environmental friendliness, and reusability make them a promising candidate for sustainable technologies aimed at addressing environmental challenges. However, to fully capitalize on their potential, ongoing research is focused on expanding their application in relevant sectors and industrializing it for the larger benefit of the global society and the environment.

References

1. Jayavelu, Y., Maharana, G., Rajender, G., Muniramaiah, R., Divyadharshini, S., et al.: Defect-mediated time-efficient photocatalytic degradation of methylene blue and ciprofloxacin using tungsten-incorporated ternary perovskite $BaSnO_3$ nanoparticles. Chemosphere **351**, 141128 (2024). https://doi.org/10.1016/j.chemosphere.2024.141128
2. Koe, W.S., Lee, J.W., Chong, W.C., Pang, Y.L., Sim, L.C.: An overview of photocatalytic degradation: photocatalysts, mechanisms, and development of photocatalytic membrane. Environ. Sci. Pollut. Res. **27**, 2522–2565 (2019). https://doi.org/10.1007/s11356-019-07193-5
3. Szynkowska, M.I., Pawlaczyk, A., Maćkiewicz, E.: Bioaccumulation and biomagnification of trace elements in the environment. In: Recent Advances in Trace Elements, pp. 251–276. (2018). https://doi.org/10.1002/9781119133780.ch13
4. Oturan, M.A., Aaron, J.-J.: Advanced oxidation processes in water/wastewater treatment: principles and applications. A Rev., Critical Rev. Environ. Sci. Technol. **44**, 2577–2641 (2014). https://doi.org/10.1080/10643389.2013.829765
5. Ibhadon, A., Fitzpatrick, P.: Heterogeneous photocatalysis: recent advances and applications. Catalysts **3**, 189–218 (2013). https://doi.org/10.3390/catal3010189
6. Jayavelu, Y., Maharana, G., Joseph, D.P., Divyadharshini, S., Sakthinathan, V., et al.: Stable and efficient hydrogen evolution activity of tungsten-incorporated nanocrystalline perovskite $BaSnO_3$ electrocatalyst. Electrochim. Acta **502**, 144777 (2024). https://doi.org/10.1016/j.electacta.2024.144777
7. Zhuang, H., Xu, W., Lin, L., Huang, M., Xu, M., et al.: Construction of one dimensional $ZnWO_4$@$SnWO_4$ core-shell heterostructure for boosted photocatalytic performance. J. Mater. Sci. Technol. **35**, 2312–2318 (2019). https://doi.org/10.1016/j.jmst.2019.05.036
8. Maharana, G., Jayavelu, Y., Muniramaiah, R., Paul Joseph, D., Baskar. L., Manavalan, K.: Visible light driven defect mediated photocatalytic activity of spray pyrolyzed Ga and Al doped ZnO thin film electrodes. ChemistrySelect **9** (2024). https://doi.org/10.1002/slct.202402590
9. Rajender, G., Kumar, J., Giri, P.K.: Interfacial charge transfer in oxygen deficient TiO_2-graphene quantum dot hybrid and its influence on the enhanced visible light photocatalysis. Appl. Catal. B **224**, 960–972 (2018). https://doi.org/10.1016/j.apcatb.2017.11.042
10. Wang, T.M., Zheng, S.K., Hao, W.C., Wang, C.: Studies on photocatalytic activity and transmittance spectra of TiO_2 thin films prepared by R. F. magnetron sputtering method. Surf. Coat. Technol. **155** 141–145 (2002). https://doi.org/10.1016/s0257-8972(02)00004-x
11. Kim, M.G., Lee, J.E., Kim, K.S., Kang, J.M., Lee, J.H., Kim, K.H., Cho, M., Lee, S.G.: Photocatalytic degradation of methylene blue under UV and visible light by brookite–rutile bi-crystalline phase of TiO_2. New J. Chem. **45**, 3485–3497 (2021). https://doi.org/10.1039/d0nj05162d
12. Guan, S., Hao, L., Murayama, M., Xie, X., Komuro, S., Zhao, X.: Influence of anneal temperature in air on surface morphology and photoluminescence of ZnO thin films. In: IOP Conference Series: Materials Science and Engineering vol. 522, p. 012004. (2019). https://doi.org/10.1088/1757-899x/522/1/012004
13. Maharana, G., Muniramaiah, R., Yuvashree, J., Mandal, D., Mondal, S., et al.: Tungsten and fluorine co-doping induced morphology change and textured growth of spray-pyrolyzed SnO_2 thin films viable for photocatalytic application. Surfaces and Interfaces **42**, 103413 (2023). https://doi.org/10.1016/j.surfin.2023.103413

14. He, Y., Rehman, A.U., Xu, M., Not, C.A., Ng, A.M.C., Djurišić, A.B.: Photocatalytic degradation of different types of microplastics by TiOx/ZnO tetrapod photocatalysts. Heliyon **9**, e22562 (2023). https://doi.org/10.1016/j.heliyon.2023.e22562
15. Mukherjee, I., Cilamkoti, V., Dutta, R.K.: Sunlight-driven photocatalytic degradation of ciprofloxacin by carbon dots embedded in ZnO nanostructures. ACS Appl. Nano Mater. **4**, 7686–7697 (2021). https://doi.org/10.1021/acsanm.1c00883
16. Yang, X., Wang, D.: Photocatalysis: from fundamental principles to materials and applications. ACS Appl. Energy Mater. **1**, 6657–6693 (2018). https://doi.org/10.1021/acsaem.8b01345
17. Zhu, S., Wang, D.: Photocatalysis: basic principles, diverse forms of implementations and emerging scientific opportunities. Adv. Energy Mater. **7** (2017). https://doi.org/10.1002/aenm.201700841
18. Ahmed, S.N., Haider, W.: Heterogeneous photocatalysis and its potential applications in water and wastewater treatment: a review. Nanotechnology **29**, 342001 (2018). https://doi.org/10.1088/1361-6528/aac6ea
19. Hassaan, M.A., El-Nemr, M.A., Elkatory, M.R., Ragab, S., Niculescu, V.-C., El Nemr, A.: Principles of photocatalysts and their different applications: a review. Top. Curr. Chem. (Z) **381** (2023). https://doi.org/10.1007/s41061-023-00444-7
20. Raha, S., Ahmaruzzaman, Md.: ZnO nanostructured materials and their potential applications: progress, challenges and perspectives, Nanoscale Adv. **4**, 1868–1925 (2022). https://doi.org/10.1039/d1na00880c
21. Jain, S., Shah, J., Negi, N.S., Sharma, C., Kotnala, R.K.: Significance of interface barrier at electrode of hematite hydroelectric cell for generating ecopower by water splitting. Int. J. Energy Res. **43**, 4743–4755 (2019). https://doi.org/10.1002/er.4613
22. Zhang, Z., Yates, J.T., Jr.: Band bending in semiconductors: chemical and physical consequences at surfaces and interfaces. Chem. Rev. **112**, 5520–5551 (2012). https://doi.org/10.1021/cr3000626
23. Schleuning, M., Ahmet, I.Y., van de Krol, R., May, M.M.: The role of selective contacts and built-in field for charge separation and transport in photoelectrochemical devices. Sustain. Energy Fuels **6** 3701–3716 (2022). https://doi.org/10.1039/d2se00562j
24. Wang, H., Li, X., Zhao, X., Li, C., Song, X., et al.: A review on heterogeneous photocatalysis for environmental remediation: from semiconductors to modification strategies. Chin. J. Catal. **43**, 178–214 (2022). https://doi.org/10.1016/s1872-2067(21)63910-4
25. Janbandhu, S.Y., Munishwar, S.R., Gedam, R.S.: Synthesis, characterization and photocatalytic degradation efficiency of CdS quantum dots embedded in sodium borosilicate glasses. Appl. Surf. Sci. **449**, 221–227 (2018). https://doi.org/10.1016/j.apsusc.2018.02.065
26. Kumbhakar, P., Jana, I.D., Basu, S., Mandal, S., Banerjee, S., et al.: Utilization of DNA and 2D metal oxide interaction for an optical biosensor. Phys. Chem. Chem. Phys. **25**, 17143–17153 (2023). https://doi.org/10.1039/d3cp01402a
27. Shahabuddin, S., Mehmood, S., Ahmad, I., Sridewi, N.: Synthesis and characterization of 2D-WS2 incorporated polyaniline nanocomposites as photo catalyst for methylene blue degradation. Nanomaterials **12**, 2090 (2022). https://doi.org/10.3390/nano12122090
28. Wang, Y., Chhowalla, M.: Making clean electrical contacts on 2D transition metal dichalcogenides. Nat Rev Phys **4**, 101–112 (2021). https://doi.org/10.1038/s42254-021-00389-0
29. Zeng, W., Zhang, H., Yuan, X., Chen, T., Pei, Z., Ji, X.: Two-Dimensional nanomaterial-based catalytic medicine: theories, advanced catalyst and system design. Adv. Drug Deliv. Rev. **184**, 114241 (2022). https://doi.org/10.1016/j.addr.2022.114241
30. Chen, Y., Fan, Z., Zhang, Z., Niu, W., Li, C., Yang, N., Chen, B., Zhang, H.: Two-Dimensional metal nanomaterials: synthesis, properties, and applications. Chem. Rev. **118**, 6409–6455 (2018). https://doi.org/10.1021/acs.chemrev.7b00727
31. Nagella, S.R., Vijitha, R., Ramesh Naidu, B., Krishna Rao, K.S.V., Ha, C.-S., Venkateswarlu, K.: Benchmarking recent advances in hydrogen production using g-C_3N_4-based photocatalysts, Nano Energy **111** 108402 (2023). https://doi.org/10.1016/j.nanoen.2023.108402
32. Gieseking, R.L., Ratner, M.A., Schatz, G.C.: Review of plasmon-induced hot-electron dynamics and related SERS chemical effects. ACS Symp. Ser. 1–22 (2016). https://doi.org/10.1021/bk-2016-1245.ch001

33. Ma, X.-C., Dai, Y., Yu, L., Huang, B.-B.: Energy transfer in plasmonic photocatalytic composites. Light Sci Appl **5**, e16017–e16017 (2016). https://doi.org/10.1038/lsa.2016.17
34. Villa, K., Galán-Mascarós, J.R., López, N., Palomares, E.: Photocatalytic water splitting: advantages and challenges, Sustainable. Energy Fuels **5**, 4560–4569 (2021). https://doi.org/10.1039/d1se00808k
35. Kamaruzaman, N.A., Wan Mohamed Zin, W.M.K., Kamarudin, K.H., Saleh, N.M., Yusoff, F.: Recent advances in transition metals- based materials as electrocatalysts for water splitting. Int. J. Electrochem. Sci. **18** 100187 (2023). https://doi.org/10.1016/j.ijoes.2023.100187
36. Li, D., Song, H., Meng, X., Shen, T., Sun, J., Han, W., Wang, X.: Effects of particle size on the structure and photocatalytic performance by alkali-treated TiO_2. Nanomaterials **10**, 546 (2020). https://doi.org/10.3390/nano10030546
37. Zheng, X., Xu, S., Wang, Y., Sun, X., Gao, Y., Gao, B.: Enhanced degradation of ciprofloxacin by graphitized mesoporous carbon (GMC)-TiO_2 nanocomposite: strong synergy of adsorption-photocatalysis and antibiotics degradation mechanism. J. Colloid Interface Sci. **527**, 202–213 (2018). https://doi.org/10.1016/j.jcis.2018.05.054
38. Kulis-Kapuscinska, A., Kwoka, M., Borysiewicz, M.A., Wojciechowski, T., Licciardello, N., Sgarzi, M., Cuniberti, G.: Photocatalytic degradation of methylene blue at nanostructured ZnO thin films. Nanotechnology **34**, 155702 (2023). https://doi.org/10.1088/1361-6528/aca910
39. Sathya, M., Selvan, G., Karunakaran, M., Kasirajan, K., Usha, S., et al.: Synthesis and characterization of cadmium doped on ZnO thin films prepared by SILAR method for photocatalytic degradation properties of MB under UV irradiation. Eur. Phys. J. Plus **138** (2023). https://doi.org/10.1140/epjp/s13360-023-03667-1
40. Gopinath, C.S., Nalajala, N.: A scalable and thin film approach for solar hydrogen generation: a review on enhanced photocatalytic water splitting. J. Mater. Chem. A **9**, 1353–1371 (2021). https://doi.org/10.1039/d0ta09619a
41. Le, A.T., Le, T.D.H., Cheong, K.-Y., Pung, S.-Y.: Immobilization of zinc oxide-based photocatalysts for organic pollutant degradation: a review. J. Environ. Chem. Eng. **10**, 108505 (2022). https://doi.org/10.1016/j.jece.2022.108505
42. Liu, Y., Zhao, N., Gao, W.: Microstructure, growth process and enhanced photocatalytic activity of immobilized hierarchical ZnO nanostructures. RSC Adv. **3**, 21666 (2013). https://doi.org/10.1039/c3ra42902d
43. Vaiano, V., Iervolino, G.: Facile method to immobilize ZnO particles on glass spheres for the photocatalytic treatment of tannery wastewater. J. Colloid Interface Sci. **518**, 192–199 (2018). https://doi.org/10.1016/j.jcis.2018.02.033
44. Malakootian, M., Khatami, M., Mahdizadeh, H., Nasiri, A., Amiri Gharaghani, M.: A study on the photocatalytic degradation of p-Nitroaniline on glass plates by Thermo-Immobilized ZnO nanoparticle, Inorg. Nano–Met. Chem. **50** 124–135 (2019). https://doi.org/10.1080/24701556.2019.1662807
45. Preda, M.D., Popa, M.L., Neacşu, I.A., Grumezescu, A.M., Ginghină, O.: Antimicrobial clothing based on electrospun fibers with ZnO nanoparticles. IJMS **24**, 1629 (2023). https://doi.org/10.3390/ijms24021629
46. Jung, H.J., Lee, S., Yu, Y., Hong, S.M., Choi, H.C., Choi, M.Y.: Low-temperature hydrothermal growth of ZnO nanorods on sol–gel prepared ZnO seed layers: optimal growth conditions. Thin Solid Films **524**, 144–150 (2012). https://doi.org/10.1016/j.tsf.2012.10.007
47. Kathwate, L.H.: In-doped ZnO films deposited by modified SILAR method for enhanced ethanol gas sensor application. Ceram. Int. **50**, 48462–48473 (2024). https://doi.org/10.1016/j.ceramint.2024.09.195
48. Jin, X., Götz, M., Wille, S., Mishra, Y.K., Adelung, R., Zollfrank, C.: A Novel concept for self-reporting materials: stress sensitive photoluminescence in ZnO tetrapod filled elastomers. Adv. Mater. **25**, 1342–1347 (2012). https://doi.org/10.1002/adma.201203849
49. Zakria, H.S., Othman, M.H.D., Kamaludin, R., Sheikh Abdul Kadir, S.H., Kurniawan, T.A., Jilani, A.: Immobilization techniques of a photocatalyst into and onto a polymer membrane for photocatalytic activity, RSC Adv. **11** 6985–7014 (2021). https://doi.org/10.1039/d0ra10964a

50. Rather, A.H., Khan, R.S., Wani, T.U., Beigh, M.A., Sheikh, F.A.: Overview on immobilization of enzymes on synthetic polymeric nanofibers fabricated by electrospinning. Biotech Bioeng. **119**, 9–33 (2021). https://doi.org/10.1002/bit.27963
51. Ashkan, Z., Hemmati, R., Homaei, A., Dinari, A., Jamlidoost, M., Tashakor, A.: Immobilization of enzymes on nanoinorganic support materials: an update. Int. J. Biol. Macromol. **168**, 708–721 (2021). https://doi.org/10.1016/j.ijbiomac.2020.11.127
52. Sha, R., Basak, A., Maity, P.C., Badhulika, S.: ZnO nano-structured based devices for chemical and optical sensing applications. Sens. Actuators Rep. **4**, 100098 (2022). https://doi.org/10.1016/j.snr.2022.100098
53. Marouf, S., Beniaiche, A., Guessas, H., Azizi, A.: Morphological, structural and optical properties of ZnO thin films deposited by dip coating method. Mat. Res. **20**, 88–95 (2016). https://doi.org/10.1590/1980-5373-mr-2015-0751
54. Benrezgua, E., Deghfel, B., Zoukel, A., Basirun, W.J., Amari, R., Boukhari, A., Yaakob, M.K., Kheawhom, S., Mohamad, A.A.: Synthesis and properties of copper doped zinc oxide thin films by sol-gel, spin coating and dipping: a characterization review. J. Mol. Struct. **1267**, 133639 (2022). https://doi.org/10.1016/j.molstruc.2022.133639
55. Shafi, M.A., Bouich, A., Fradi, K., Guaita, J.M., Khan, L., Mari, B.: Effect of deposition cycles on the properties of ZnO thin films deposited by spin coating method for CZTS-based solar cells. Optik **258**, 168854 (2022). https://doi.org/10.1016/j.ijleo.2022.168854
56. Athare, A.E.: Synthesis and characterisation of ZnO nanoparticals by XRD, EDX, SEM, FTIR and UV-DRS. Int. J. Sci. Res. Sci. Eng. Technol **4** 627–631 (2018)
57. Zhou, H., Wang, H., Yue, C., He, L., Li, H., Zhang, H., Yang, S., Ma, T.: Photocatalytic degradation by TiO_2-conjugated/coordination polymer heterojunction: preparation, mechanisms, and prospects. Appl. Catal. B **344**, 123605 (2024). https://doi.org/10.1016/j.apcatb.2023.123605
58. Das, A., Liu, D., Wu, Y., Abzakh, B.A., Madhumitha, R., et al.: Origin of the improved photoelectrochemical and photocatalytic activity in a ZnO-TiO_2 nanohybrid revealed by experimental and density functional theory studies. J. Phys. Chem. Lett. **15**, 7524–7532 (2024). https://doi.org/10.1021/acs.jpclett.4c01641
59. Mohamed, K.M., Benitto, J.J., Vijaya, J.J., Bououdina, M.: Recent Advances in ZnO-based nanostructures for the photocatalytic degradation of hazardous, non-biodegradable medicines. Crystals **13** 329 (2023). https://doi.org/10.3390/cryst13020329
60. Ong, C.B., Ng, L.Y., Mohammad, A.W.: A review of ZnO nanoparticles as solar photocatalysts: synthesis, mechanisms and applications. Renew. Sustain. Energy Rev. **81**, 536–551 (2018). https://doi.org/10.1016/j.rser.2017.08.020
61. Ghamarpoor, R., Fallah, A., Jamshidi, M.: A review of synthesis methods, modifications, and mechanisms of ZnO/TiO_2-based photocatalysts for photodegradation of contaminants. ACS Omega **9**, 25457–25492 (2024). https://doi.org/10.1021/acsomega.3c08717
62. Sabouni, R., Gomaa, H.: Photocatalytic degradation of pharmaceutical micro-pollutants using ZnO. Environ. Sci. Pollut. Res. **26**, 5372–5380 (2019). https://doi.org/10.1007/s11356-018-4051-2
63. Ramos, B., Carneiro, J.G.M., Nagamati, L.I., Teixeira, A.C.S.C.: Development of intensified flat-plate packed-bed solar reactors for heterogeneous photocatalysis. Environ. Sci. Pollut. Res. **28**, 24023–24033 (2021). https://doi.org/10.1007/s11356-020-11806-9
64. Baghbani Ghatar, S., Allahyari, S., Rahemi, N., Tasbihi, M.: Response surface methodology optimization for photodegradation of methylene blue in a ZnO coated flat plate continuous photoreactor. Int. J. Chem. React. Eng. **16** (2018). https://doi.org/10.1515/ijcre-2017-0221
65. Olivo-Alanís, D., García-González, A., Mueses, M.A., García-Reyes, R.B.: Generalized kinetic model for the photocatalytic degradation processes: validation for dye wastewater treatment in a visible-LED tubular reactor. Appl. Catal. B **317**, 121804 (2022). https://doi.org/10.1016/j.apcatb.2022.121804
66. Parvaz, S., Rabbani, M., Rahimi, R.: Fabrication of novel magnetic ZnO hollow spheres/pumice nanocomposites for photodegradation of Rhodamine B under visible light irradiation. Mater. Sci. Eng. B **263**, 114863 (2021). https://doi.org/10.1016/j.mseb.2020.114863

67. Sayed, M., Xu, F., Kuang, P., Low, J., Wang, S., Zhang, L., Yu, J.: Sustained CO_2-photoreduction activity and high selectivity over Mn, C-codoped ZnO core-triple shell hollow spheres. Nat Commun **12** (2021). https://doi.org/10.1038/s41467-021-25007-6
68. Binjhade, R., Mondal, R., Mondal, S.: Continuous photocatalytic reactor: critical review on the design and performance. J. Environ. Chem. Eng. **10**, 107746 (2022). https://doi.org/10.1016/j.jece.2022.107746
69. Tadkar, P.S., Borkar, P.K., Gholap, S.N., Gole, V.L.: Treatment of methylene blue dye using immersed lamp photocatalytic reactor: 5 L scale study. J. Inst. Eng. India Ser. E **100**, 199–204 (2019). https://doi.org/10.1007/s40034-019-00150-4
70. Theodorakopoulos, G.V., Arfanis, M.K., Sánchez Pérez, J.A., Agüera, A., Cadena Aponte, F.X., Markellou, E., Romanos G. E., Falaras, P.: Novel pilot-scale photocatalytic nanofiltration reactor for agricultural wastewater treatment. Membranes **13**, 202 (2023). https://doi.org/10.3390/membranes13020202
71. Visan, A., van Ommen, J.R., Kreutzer, M.T., Lammertink, R.G.H.: Photocatalytic reactor design: guidelines for kinetic investigation. Ind. Eng. Chem. Res. **58**, 5349–5357 (2019). https://doi.org/10.1021/acs.iecr.9b00381
72. Akyol, A., Bayramoglu, M.: Performance comparison of ZnO photocatalyst in various reactor systems. J of Chemical Tech Biotech 85 1455–1462 (2010). https://doi.org/10.1002/jctb.2449
73. Molinari, R., Argurio, P., Szymański, K., Darowna, D., Mozia, S.: Photocatalytic membrane reactors for wastewater treatment. Curr. Trends Futur. Dev. (Bio-) Membr. 83–116 (2020). https://doi.org/10.1016/b978-0-12-816823-3.00004-6
74. Rani, C.N., Karthikeyan, S.: Performance of an indigenous integrated slurry photocatalytic membrane reactor (PMR) on the removal of aqueous phenanthrene (PHE). Water Sci. Technol. **77**, 2642–2656 (2018). https://doi.org/10.2166/wst.2018.220
75. Ariyanti, D., Wicaksana, F., Gao, W.: Submerged membrane photo reactor (SMPR) with simultaneous photo degradation and TiO_2 catalyst recovery for efficient dyes removal. ASEAN J. Chem. Eng., **21** 225–240 (2021). https://doi.org/10.22146/ajche.65952
76. Shi, Y., Toghraie, D., Nadi, F., Ahmadi, G., Alizadeh, A., Zhang, L.: The effect of the pitch angle, two-axis tracking system, and wind velocity on the parabolic trough solar collector thermal performance. Environ. Dev. Sustain. **23**, 17329–17348 (2021). https://doi.org/10.1007/s10668-021-01368-2
77. Malan, A., Ravi Kumar, K.: A comprehensive review on optical analysis of parabolic trough solar collector. Sustain. Energy Technol. Assess. **46** 101305 (2021). https://doi.org/10.1016/j.seta.2021.101305
78. Kumar, L., Ahmed, J., El Haj Assad, M., Hasanuzzaman, M.: Prospects and challenges of solar thermal for process heating: a comprehensive review. Energies **15** 8501 (2022). https://doi.org/10.3390/en15228501
79. Masood, F., Nor, N.B.M., Elamvazuthi, I., Saidur, R., Alam, M.A., et al.: The compound parabolic concentrators for solar photovoltaic applications: opportunities and challenges. Energy Rep. **8**, 13558–13584 (2022). https://doi.org/10.1016/j.egyr.2022.10.018
80. Wang, D., Mueses, M.A., Márquez, J.A.C., Machuca-Martínez, F., Grčić, I., et al.: Engineering and modeling perspectives on photocatalytic reactors for water treatment. Water Res. **202**, 117421 (2021). https://doi.org/10.1016/j.watres.2021.117421
81. Sraw, A., Kaur, T., Pandey, Y., Sobti, A., Wanchoo, R.K., Toor, A.P.: Fixed bed recirculation type photocatalytic reactor with TiO_2 immobilized clay beads for the degradation of pesticide polluted water. J. Environ. Chem. Eng. **6**, 7035–7043 (2018). https://doi.org/10.1016/j.jece.2018.10.062
82. Khositanon, C., Saothayanun, T., Sirimongkhol, W., Sukpancharoen, S., Ogawa, M., et al.: Continuous flow hydroxylation of benzene to phenol in a photocatalytic millistructured flat-plate reactor. ACS Sustainable Chem. Eng. **12**, 13998–14008 (2024). https://doi.org/10.1021/acssuschemeng.4c04834
83. Sutisna, S., Rokhmat, M., Wibowo, E., Khairurrijal, K., Abdullah, M.: Prototype of a flat-panel photoreactor using TiO_2 nanoparticles coated on transparent granules for the degradation of Methylene Blue under solar illumination. Sustain. Environ. Res. **27** 172–180 (2017). https://doi.org/10.1016/j.serj.2017.04.002

84. Awogbemi, O., Kallon, D.V.V.: Application of tubular reactor technologies for the acceleration of biodiesel production. Bioengineering **9**, 347 (2022). https://doi.org/10.3390/bioengineering9080347
85. Wang, G., Zhang, J., Zhong, L., Tang, Z.: Extending the light transmission in the continuous flow photocatalytic process by 3D printed silica packing. Ceram. Int. **50**, 38532–38541 (2024). https://doi.org/10.1016/j.ceramint.2024.07.221
86. Molinari, R., Limonti, C., Lavorato, C., Siciliano, A., Argurio, P.: Upgrade of a slurry photocatalytic membrane reactor based on a vertical filter and an external membrane and testing in the photodegradation of a model pollutant in water. Chem. Eng. J. **451**, 138577 (2023). https://doi.org/10.1016/j.cej.2022.138577
87. Soltani, N., Saion, E., Hussein, M.Z., Erfani, M., Abedini, A., Bahmanrokh, G., Navasery, M., Vaziri, P.: Visible light-induced degradation of methylene blue in the presence of photocatalytic ZnS and CdS nanoparticles. IJMS **13**, 12242–12258 (2012). https://doi.org/10.3390/ijms131012242
88. Iyyappan, J., Gaddala, B., Gnanasekaran, R., Gopinath, M., Yuvaraj, D., Kumar, V.: Critical review on wastewater treatment using photo catalytic advanced oxidation process: role of photocatalytic materials, reactor design and kinetics. Case Stud. Chem. Environ. Eng. **9**, 100599 (2024). https://doi.org/10.1016/j.cscee.2023.100599
89. Molinari, R., Argurio, P., Giorno, L., Palmisano, L., Drioli, E.: Photocatalytic nanofiltration reactors. Nanofiltration 707–757 (2021). https://doi.org/10.1002/9783527824984.ch17
90. Molinari, R., Pirillo, F., Loddo, V., Palmisano, L.: Heterogeneous photocatalytic degradation of pharmaceuticals in water by using polycrystalline TiO_2 and a nanofiltration membrane reactor. Catal. Today **118**, 205–213 (2006). https://doi.org/10.1016/j.cattod.2005.11.091
91. Saygı, G., Kap, Ö., Özkan, F.Ç., Varlikli, C.: Photocatalytic reactors design and operating parameters on the wastewater organic pollutants removal. In: Green chemistry and sustainable technology, pp. 103–151. (2023). https://doi.org/10.1007/978-3-031-27707-8_5
92. Zhu, J., Araya, S.S., Cui, X., Sahlin, S.L., Kær, S.K.: Modeling and design of a multi-tubular packed-bed reactor for methanol steam reforming over a $Cu/ZnO/Al_2O_3$ catalyst. Energies **13**, 610 (2020). https://doi.org/10.3390/en13030610
93. Khan, A.A., Tahir, M.: Recent advancements in engineering approach towards design of photo-reactors for selective photocatalytic CO_2 reduction to renewable fuels. J. CO_2 Util. **29** 205–239 (2019). https://doi.org/10.1016/j.jcou.2018.12.008
94. Ferreiro, C., Villota, N., Lombraña, J.I., Rivero, M.J., Zúñiga, V., Rituerto, J.M.: Removal of aniline and benzothiazole wastewaters using an efficient MnO_2/GAC catalyst in a photocatalytic fluidised bed reactor. Materials **14**, 5207 (2021). https://doi.org/10.3390/ma14185207
95. Suhadolnik, L., Pohar, A., Novak, U., Likozar, B., Mihelič, A., Čeh, M.: Continuous photocatalytic, electrocatalytic and photo-electrocatalytic degradation of a reactive textile dye for wastewater-treatment processes: batch, microreactor and scaled-up operation. J. Ind. Eng. Chem. **72**, 178–188 (2019). https://doi.org/10.1016/j.jiec.2018.12.017
96. Alexiadis, A., Mazzarino, I.: Design guidelines for fixed-bed photocatalytic reactors. Chem. Eng. Process. **44**, 453–459 (2005). https://doi.org/10.1016/j.cep.2004.06.009
97. Vaiano, V., Sacco, O., Pisano, D., Sannino, D., Ciambelli, P.: From the design to the development of a continuous fixed bed photoreactor for photocatalytic degradation of organic pollutants in wastewater. Chem. Eng. Sci. **137**, 152–160 (2015). https://doi.org/10.1016/j.ces.2015.06.023
98. Mondal, R.: Micro-segregated two-dimensional fluid and mass transport modelling in unsteady rotating annular photocatalytic reactor. Results Eng. **16**, 100752 (2022). https://doi.org/10.1016/j.rineng.2022.100752
99. Imoberdorf, G.E., Cassano, A.E., Irazoqui, H.A., Alfano, O.M.: Optimal design and modeling of annular photocatalytic wall reactors. Catal. Today **129**, 118–126 (2007). https://doi.org/10.1016/j.cattod.2007.06.057
100. Yang, Z., Liu, M., Lin, C.: Photocatalytic activity and scale-up effect in liquid–solid mini-fluidized bed reactor. Chem. Eng. J. **291**, 254–268 (2016). https://doi.org/10.1016/j.cej.2016.01.050

101. Gusmão, C., Palharim, P.H., Diniz, L.A., de Assis, G.C., de Carvalho e Souza, T., et al.: Advances in fluidized bed photocatalysis: bridging gaps, standardizing metrics, and shaping sustainable solutions for environmental challenges. Ind. Eng. Chem. Res. **63**, 14967–14982 (2024). https://doi.org/10.1021/acs.iecr.4c00821
102. Abdel-Maksoud, Y., Imam, E., Ramadan, A.: TiO_2 solar photocatalytic reactor systems: selection of reactor design for scale-up and commercialization—analytical review. Catalysts **6**, 138 (2016). https://doi.org/10.3390/catal6090138
103. Schmidt-Mende, L., MacManus-Driscoll, J.L.: ZnO – nanostructures, defects, and devices. Mater. Today **10**, 40–48 (2007). https://doi.org/10.1016/s1369-7021(07)70078-0
104. Sharma, D.K., Shukla, S., Sharma, K.K., Kumar, V.: A review on ZnO: fundamental properties and applications. Mater. Today: Proc. **49**, 3028–3035 (2022). https://doi.org/10.1016/j.matpr.2020.10.238
105. Yan, L., Uddin, A., Wang, H.: ZnO tetrapods: synthesis and applications in solar cells. Nanomater. Nanotechnol. **5**, 19 (2015). https://doi.org/10.5772/60939
106. Maiti, S., Maiti, U.N., Behera, B.C., Pal, S., Chattopadhyay, K.K.: Controlling the sharpness of ZnO tetrapods by restricted zinc oxidation in the open air: a low turn-on field emitter stabilized by graphene. J. Mater. Chem. C **1**, 4940 (2013). https://doi.org/10.1039/c3tc30775a
107. Orudzhev, F., Muslimov, A., Selimov, D., Gulakhmedov, R.R., Lavrikov, A., et al.: Oxygen vacancies and surface wettability: key factors in activating and enhancing the solar photocatalytic activity of ZnO tetrapods. IJMS **24**, 16338 (2023). https://doi.org/10.3390/ijms242216338
108. Mishra, Y.K., Adelung, R.: ZnO tetrapod materials for functional applications. Mater. Today **21**, 631–651 (2018). https://doi.org/10.1016/j.mattod.2017.11.003
109. Mishra, N., Wu, W.-Y., Srinivasan, B.M., Hariharaputran, R., Zhang, Y.-W., Chan, Y.: Continuous shape tuning of nanotetrapods: toward shape-mediated self-assembly. Chem. Mater. **28**, 1187–1195 (2016). https://doi.org/10.1021/acs.chemmater.5b04803
110. Zhang, X., Han, Y., Liu, W., Pan, N., Li, D., Chai, J.: A novel synthesis of hexagonal cylinder-like ZnO with an excellent performance by a surfactant-free microemulsion-hydrothermal method. J. Ind. Eng. Chem. **97**, 326–336 (2021). https://doi.org/10.1016/j.jiec.2021.02.019
111. Hu, J., Ma, H., Zhou, Y., Ma, L., Zhao, S., Shi, S., Li, J., Chang, Y.: Gas-Sensing properties and mechanisms of 3d networks composed of ZnO tetrapod micro-nano structures at room temperature. Materials **17**, 203 (2023). https://doi.org/10.3390/ma17010203
112. Terracciano, M., Račkauskas, S., Falanga, A.P., Martino, S., Chianese, G., et al.: ZnO tetrapods for label-free optical biosensing: physicochemical characterization and functionalization strategies. IJMS **24**, 4449 (2023). https://doi.org/10.3390/ijms24054449
113. Modi, G.: Zinc oxide tetrapod: a morphology with multifunctional applications. Adv. Nat. Sci: Nanosci. Nanotechnol. **6**, 033002 (2015). https://doi.org/10.1088/2043-6262/6/3/033002
114. Kumar, S., Kaushik, R.D., Purohit, L.P.: Novel ZnO tetrapod-reduced graphene oxide nanocomposites for enhanced photocatalytic degradation of phenolic compounds and MB dye. J. Mol. Liq. **327**, 114814 (2021). https://doi.org/10.1016/j.molliq.2020.114814
115. Chamoli, P., Shukla, R.K., Bezbaruah, A.N., Kar, K.K., Raina, K.K.: Microwave-assisted rapid synthesis of honeycomb core-ZnO tetrapods nanocomposites for excellent photocatalytic activity against different organic dyes. Appl. Surf. Sci. **555**, 149663 (2021). https://doi.org/10.1016/j.apsusc.2021.149663
116. Baruah, S., Afre, R.A., Pugliese, D.: Effect of size and morphology of different ZnO nanostructures on the performance of dye-sensitized solar cells. Energies **17**, 2076 (2024). https://doi.org/10.3390/en17092076

D. Paul Joseph, presently serving as an Associate Professor, Dept. of Physics, NIT-Warangal completed M.Sc (Loyola College, Chennai), B.Ed, M.Phil, and Ph.D. from the Univ. of Madras, India. He was a post-doctoral fellow at National Taiwan Univ., Taipei (2010–2013). His research areas are alternative transparent conducting thin film electrodes, Proximity effect between Superconductor/Half-metallic Ferromagnet, alternative components for dye-sensitized solar cells, nanomaterials for batteries, swift heavy-ion-beam irradiation effects and photocatalytic activity for pollutant degradation. He has 07 contributed book chapters, 80 articles in journals of repute. Five of his X-Ray diffraction data has been included in the ICDD database by invitation. As a PI/Co-PI he has completed several funded research projects (SERB/ANRF, TEQIP-III, NPIU, etc.) and a SPARC project is ongoing. He has several academic laurels to his credit; a notable one is 'Young Scientist Award' (2017) by 'Academy of Sciences-Chennai'. He has co-organized 01 International conference, 09 workshops/FDPs, out of which 02 are National level programs each of one-month duration. He has guided 03 Ph.D. scholars and 28 students for their M.Sc (Tech) and M.Tech projects. He has designed several indigenous equipments (automated spray pyrolysis unit, standalone dip-coating unit, high temperature VSM rod adaptor import substitute, Bending-Twisting unit to study flexible films) and real time YBCO and bismuth levitation kits as a teaching aid for students.

ZnO Tetrapods in Biosensing, Drug Delivery, and Antimicrobial Strategies

Saumya Ranjan Pradhan, Krishnakanth Chithari, Ramadevi Suguru Pathinti, Vattikondala Ganesh, and Jayalakshmi Vallamkondu

1 Introduction

Nanostructured Zinc Oxide particles (ZnO-NPs) have gained significant attention in research owing to their excellent biocompatibility, minimal toxicity, environmental friendliness, and affordability. As a widely utilized metal oxide nanoparticles, ZnO-NPs possess unique and tunable optical, electrical, and chemical properties, making them promising for various applications. Their exceptional characteristics have enabled their use in optical and biomedical fields, including drug delivery, antimicrobial coatings, biosensors, and antibacterial applications. Ongoing research continues to investigate its potential in biosensing and antibacterial activity, with a focus on optimizing synthesis methods, enhancing the functionality, and addressing the associated health concerns [1–9]. It has a variety of morphologies for enhancing surface to volume (S/V) ratio, isoelectric point, charge carrier transport, biocompatibility etc. [10, 11].

When it comes to the biological uses of nanostructured ZnO based materials, zinc is one of the most essential elements for the human body, since it regulates over 300 metabolic activities [12] for which the body acquires zinc through various sources, including meals containing plant-based foods, fruits, and meats, as well as through dietary zinc supplements. In fact, the zinc-based nanomaterials are being used since ancient times, even though the term "nanotechnology" was not known by that time. Zinc based materials were widely used in medicine in ancient times to treat a variety of deadly diseases, and it is considered to be quite effective. In terms of

S. R. Pradhan · K. Chithari · R. S. Pathinti · J. Vallamkondu (✉)
Department of Physics, National Institute of Technology, Warangal, Telangana, India
e-mail: jayalakshmiv@nitw.ac.in

V. Ganesh (✉)
Department of Physics and Nanotechnology, College of Engineering and Technology, SRM Institute of Science & Technology, Kattankulathur, Tamilnadu, India
e-mail: ganeshv3@srmist.edu.in

K. Thangaraju et al. (eds.), *ZnO Tetrapods*, Engineering Materials,
https://doi.org/10.1007/978-981-95-1614-8_4

scientific view point, ancient zinc-based medications were not well understood. Zinc oxide-based materials are being explored in terms of their potential applications in biomedical engineering, for example in the biomedical field for treating nano range cytotoxicities on cancer cells [13]. There is a vast biological significance in featuring the novel functionalities of ZnO nanostructures, in this research field [13].

Unlike conventional 1D and 2D nanostructures, 3D engineered materials are designed using micro- to nanoscale components such as ZnO tetrapods (ZnO-T), offering enhanced structural complexity and functionality. They display excellent stability and large free volume to satisfy many of the technological criteria. ZnO-T feature four distinct arms radiating from a central core, each extending outward at angles approximately between 105° and 110°, creating a unique and symmetrical three-dimensional architecture. This unique 3D positioning of arms allows them to access a wide range of characteristics like the surface area of individual arms. The high stability of these materials results from the transmission of forces down one arm to the others [14, 15]. The geometry has applications, including microscopic concrete tetrapod structures that are widely used in coastal engineering for excellent stability. The nano to microscale ceramic tetrapods rather exhibits unique potential towards advanced nanotechnology. Zinc-based materials equipped with unique abilities for making nanostructures, the aspect of this structural engineering of tetrapods can be easily grown by following well known synthesis strategies, including the hydrothermal method, Vapour-Liquid–Solid (VLS) mechanism, Chemical Vapour Deposition (CVD), Flame Transport Synthesis (FTS) approach etc. [14, 15]. The distinctive 3D geometry of ZnO tetrapods, formed by the integration of four one-dimensional nanorods connected at a central junction, enables the exploration of both physical and chemical characteristics typically associated with micro-nanorod structures. This spatial configuration offers exceptional properties and enhanced surface accessibility, setting it apart from conventional nanostructures.

ZnO-T are micro-nano scale structures composed of four monocrystalline ZnO arms, each exhibiting a microstructure surface morphology. They are extensively studied in scientific domains like electronics, sensing devices, composites or surface coatings, energy harvesting, and biomedical applications [16–21]. ZnO-T are significant factor in the development of biosensing among ZnO nanostructures due to their larger active surface area, which is supplied by the huge pods, and their extremely additional sites for analyte adsorption and, thus, increased sensitivity to the detection of tiny molecular weight substances, thanks to the porous 3D linked networks [22, 23]. All these features of tetrapods make them the most important structure to be applicable in a wide range of applications. Furthermore, recent studies highlight the superior catalytic efficiency of ZnO-T, showing two to three times higher antibacterial activity in sulphonamide breakdown compared to ZnO nanoparticles or nanoflowers [24]. This exceptional performance is further reinforced by the unique semiconductor and photoactive properties of ZnO-T, stemming from their monocrystalline arms, which are absent in polycrystalline and amorphous ZnO nanoparticles [25]. These distinctive features can be engaged for bio-sensing, drug delivery and anti-microbial applications.

This chapter briefly outlines the on key synthesis techniques for ZnO-T, which include the vapor–liquid-solid (VLS) method, chemical vapor deposition (CVD), and the hydrothermal method. Additionally, the growth mechanisms of ZnO-T and their specific biomedical applications and their role in biosensing and antibacterial activity, highlighting their potential to address critical challenges in these fields are discussed. By understanding the synthesis methods and functional properties of ZnO-T, this chapter aims to give an insight into their optimization for advanced biomedical and technological applications.

2 Synthesis Methods

Several synthesis techniques (Fig. 1) have been developed for the preparation of ZnO-T, which are broadly classified into two categories: the vapour transport method and the wet chemical method. This section focuses on the VLS technique and CVD, both of which fall under the vapour transport category, as well as the hydrothermal method, which belongs to the wet chemical category. These methods are particularly important as they provide effective routes to address the challenges associated with achieving precise control over the size, shape, and yield of ZnO-T. Selecting the appropriate synthesis method is essential for optimizing the physicochemical characteristics of ZnO-T to meet the specific requirements of diverse applications.

2.1 *Vapor–Liquid-Solid Technique*

In Vapor–Liquid-Solid VLS Technique method, ZnO powder was mixed with deionized water (DI) to form a suspension, which was then grinded for 20 min using a mortar grinder to ensure homogeneity. The resulting suspension was poured into a

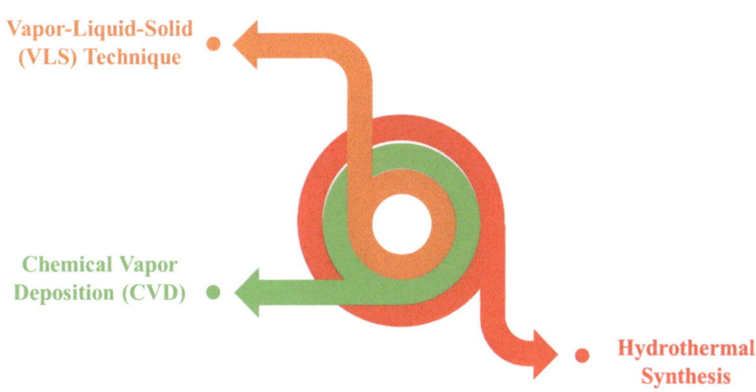

Fig. 1 Few methods for synthesis of ZnO-tetrapods

flat panel, left uncovered, and allowed to dry for 72 h at room temperature exceeding 20 °C. Following this, the mixture was dried at 100 °C for another six hours. The dried powder was subsequently placed in a muffle furnace and heated to 920 °C under an oxygen atmosphere, with an airflow rate ranging from 3.5 to 7.5 L/min. Upon completion of the reaction, the crucible contained the synthesized ZnO-T (Fig. 2a and b) [14, 26, 27].

The vapor transport method is a preferred approach for synthesizing ZnO-T due to its simplicity, cost-effectiveness, high yield, and minimal particle agglomeration. In this high- temperature process (> 600 °C), zinc metal undergoes evaporation, oxidation, nucleation, and subsequent growth. Under these conditions, tetrapods nucleate rapidly and can grow into significantly large structures. ZnO-T produced by this technique typically feature arm diameters exceeding 100 nm and lengths over 2 μm.

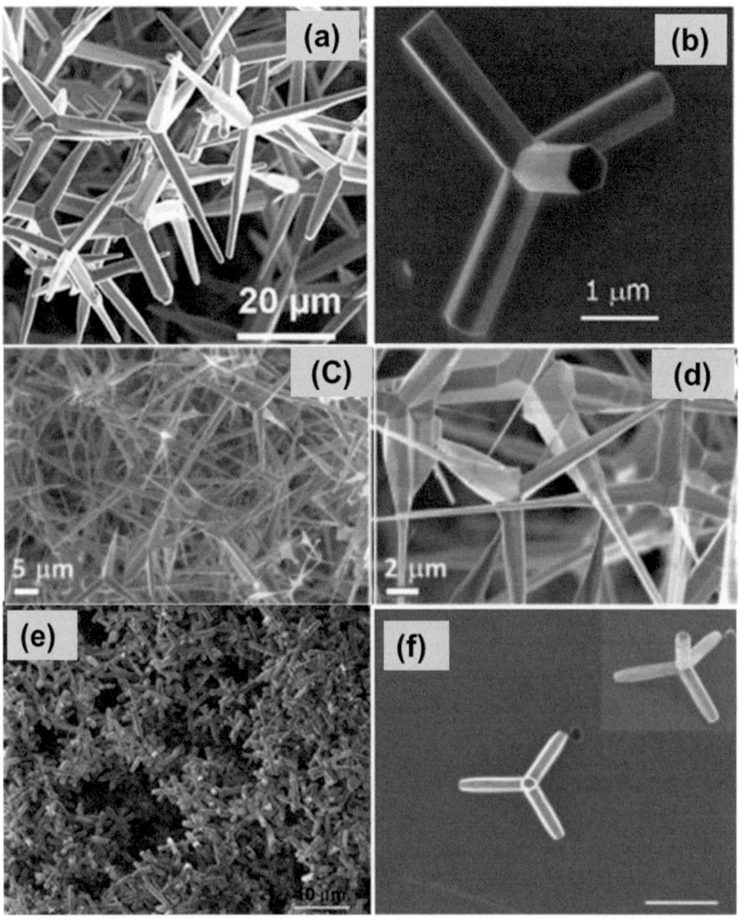

Fig. 2 FESEM images of ZnO tetrapods prepared by (**a, b**) VLS technique [33], (**c, d**) CVD method [34] and (**e, f**) hydrothermal method [32]

Key factors influencing the vapor transport synthesis include reagent vapor pressure, reaction temperature, and the resulting particle size [28, 29]. It has advantages like facileness, high yield and low cost but this process needs high temperature furnace to produce ZnO-T.

2.2 Chemical Vapor Deposition Process

In CVD method, ZnO powder was loaded into the base of an alumina boat, which was subsequently placed at the center of a horizontally aligned quartz tube inside a tubular furnace. A silicon substrate was then positioned face-down directly above the alumina boat to facilitate material deposition. The furnace was heated to 630 °C, with an airflow rate of 300 SCCM and an O_2 flow rate of 25 SCCM, maintained for 30 min. After this duration, the flow rates were shut down, and the alumina boat was allowed to cool under ambient air protection. The resulting ZnO-T (Fig. 2c and d) was collected from the substrate [30, 31]. This process requires high-temperature furnace operating above 630 °C, along with the presence of oxygen and air. The resulting tetrapods exhibit asymmetric nature, and its formation occurs at a very low rate.

2.3 Hydrothermal Method

In general, in hydrothermal process, to prepare ZnO nanomaterials, zinc sulphide and ammonia were used as precursor in hydrothermal reactor, with a glass substrate mounted inside. The system was heated to a temperature between 90 and 98 °C, then allowed to cool naturally without stirring. Once the reaction is over, the substrate was washed with DI and dried using hot air at 150 °C for 5 min. This process resulted in the formation of ZnO with various architectures (Fig. 2e and f) on the substrate [32].

The synthesis of ZnO-T involves significant challenges associated with wet chemistry and physical methods. The VLS, CVD, and hydrothermal synthesis methods presents unique advantages and limitations of their own. The continuous advancements in these methodologies have improved the ability to tailor the physicochemical characteristics of ZnO-T to meet specific requirements. These innovations are critical for realizing the full potential of ZnO-T in biomedical and technological applications, paving the way for future breakthroughs in materials science. Hydrothermal synthesis is a multi-step process that results in the formation of uniform tetrapods, though with a poor yield.

2.4 *Mechanism of Growing ZnO Tetrapods*

The morphology and growth of ZnO-T are significantly influenced by various factors, which are determined by the type of synthesis method employed and the corresponding synthesis conditions. Three main mechanisms have been proposed for the growth of ZnO-T through catalytic-free vapor deposition. Shiojiri et al. prepared ZnO-T by mixing zinc metal in air (80%) with oxygen (20%) through the burning of zinc metal [35]. Initially, ZnO nanoparticles with diameter less than 20 nm were formed, exhibiting a zinc blende (ZB) structure. The ZnO-T structure developed as four arms grew along the < 111 > direction and perpendicular to the < 111 > direction or zinc faces. Subsequently, a wurtzite (WZ) crystal formed on the {111} faces due to the introduction of stacking faults, leading to growth along the [0001] direction and resulting in the ZnO-T structure. Nishio et al. putforth a growth model suggesting that ZnO-T originate from multiple wurtzite (WZ) twin structures emerging within a zinc blende (ZB) nucleus, which is stable only at elevated temperatures [36]. As the temperature decreases to room temperature, the ZnO-T degenerates into multiple twins, leading to the formation of the final structure. Iwanaga et al. reported that ZnO-T is composed of a core-leg structure with the same wurtzite structure [37, 38]. An octa-twin growth model was suggested, in which the formation begins with octahedral inversion twins composed of eight interconnected tetragonal pyramidal units. The eight surfaces of the octahedron alternated between [0001] plus (+c) and $[000\bar{1}]$ minus (− c) orientations. The + c surface, which terminates with Zn, was a favorable site for the attachment of Zn vapor species. During growth, some twin sides cracked, allowing four legs to extend along the + c direction, ultimately resulting in the formation of ZnO-T. Throughout the experiment, Dai et al. [39] provided direct evidence for the existence of the octa-twin nucleus, strongly supporting the octa-twin model. Other researchers believe that the growth mechanism of ZnO-T and multi-mode nanostructures can follow a self-catalyzed vapor–solid process (VSP), which generally involves two key steps: initial nucleation followed by epitaxial growth. In this process, zinc or ZnO suboxide vapor is generated at high temperatures and transported to a lower-temperature region, where it condenses into liquid droplets on the substrate. Consequently, incoming Zn and O atoms are absorbed, facilitating nucleation and the subsequent growth of ZnO nanostructures. Comprehensive morphology analysis, spectral measurements, and theoretical calculations are till essential to better understand and elucidate the growth mechanisms of ZnO-T. The role of synthesis parameters on the growth mechanisms of ZnO-T along with size, quality and yield is given in Table 1 in a detailed manner.

Table 1 Application of ZnO-T in biosensing and antibacterial activity

S. No.	Application	Materials	Mechanism	References
1	Glucose sensor	GOx, ZnO-T, Glucose	Oxidation of H_2O_2 generated from reactions catalyzed by GOx change the CV curve	[61]
2	Glucose sensor	Tetrapods, Chitosan, Glutaraldehyde (GA), glucose, and Glucose oxidase	Due to H_2O_2 oxidation and oxygen reduction potentials, CV curve changes	[62]
3	Glucose sensor	3-aminopropyl triethoxysilane (aptes), Mxene, tetrapods, Glucose	Due to catalytic oxidation of glucose on GOx produced electrode, charges are observed in CV	[55]
4	Pathogen sensor (streptavidin)	Streptavidin, sulfo-n-hydroxysuccinimide biotin, tetrapod	Interaction between biotinylated ZnO-T and streptavid change the PL intensities	[57]
5	Glucose sensor	GOx, ZnO-T	Collisional quenching by H_2O_2 produced from the oxidation of glucose change PL intensities	[63]
6	Sars-cov-2 virus proteins	3-aminopropyltriethoxsilan (aptes), sars-cov-rs–spike protein, sars-cov-2-rn–nucleocapsid protein	Interaction of anti-sars-cov-antibodies with the biosensor surface resulted in increase of the PL intensity	[71]
7	Helicobacter Pylori sensing (Pathogen sensor)	CagA antibody, antigen and ZnO-T	Due to immunocomplex forming an antibody and CagA antigen over Zn, it decreases the peak current	[72]
8	Tetracycline sensing	TC Polyclonal antibody, antigen, tetrapod, (3-aminopropyl) triethoxysilane (APTES)	PL intensities change due to interaction between the bio-receptor (anti-TC antibodies) and the TC molecules	[73]
9	Human chorionic gonadotropin hromes	ZnO-T, antibody, anti-hcg, hcg tetrapod	Variations in PL emission intensity following the interaction between immobilized antibodies and hCG	[74]

(continued)

Table 1 (continued)

S. No.	Application	Materials	Mechanism	References
10	Antibacterial activity	ZnO-T	Addition of ZnO-T leads to the formation of nanostructured fibrous membranes with a unique rose spike-like structure	[125]
11	Antiviral activity	ZnO-T	ZnO-T suppress the infection by binding and trapping the virus	[102]
12	Cytotoxicity (antimicrobial agents)	ZnO-T	Due to larger in size compared to ZnO nanoparticles	[78]
13	Antiviral activity	ZOTeN (ZnO-T) with oxygen vacancy)	Due to ZOTeN traps HSV-2 virion it inhabits virus growth	[114]
14	Antiviral activity	ZnO-T	Due to negatively charged ZnO MNSS trap the HSV-1to prevent virus-cell interaction	[107]
15	Antiviral activity	ZnO-T with oxygen vacancy	ZnO-T trapping HSV-2 virions neutralize it	[108]
16	Antiviral activity	ZnO-T	Due to its ability of trap the virus and neutralize it	[134]
17	Antiviral activity	ZOTeN (ZnO-T with oxygen vacancy)	ZOTeN trap the virus and inhibit the virus growth	[114]
18	Wound healing (genital infection)	ZnO-T	ZOTeN trap the virus and inhibit the virus growth	[110]
19	Wound healing	ZnO-T	Due to its ability antiproliferative and anti-inflammatory potentials of ZnO-T, can be used for wound healing in ocular surgery	[122]
20	Wound healing	ZnO-Ts nanoparticles (ZOTeN) engineered with controlled O_2 vacancies	Strong HSV-2 trapping capability of ZOTeN notably diminished the clinical symptoms of vaginal infection	[114]

(continued)

Table 1 (continued)

S. No.	Application	Materials	Mechanism	References
21	Wound healing bandage	ZnO-T incorporated with hydrogel	Low cytotoxicity and improved angiogenic properties while maintaining antibacterial functions in vitro, lead to use as wound healing bandage	[21]
22	Wound healing treatment	Polycaprolactone (PCL) polymer with a dispersion of anisotropic ZnO-T	Due to its antiviral properties against both HSV-1 and HSV-2, it is used as bandage	[125]
23	Wound healing	ZnO-Tand sodium alginate	Antibacterial activity and biocompatibility of sodium alginate can be used as bandage	[126]

3 Properties of ZnO Tetrapods

3.1 *Optical and Electronic Properties*

Sirbuly et al. [40] extensively studied the optical characteristics of ZnO nanostructures, revealing that several ZnO nanostructures exhibit optically pumped lasing properties. Huang et al. were the first to demonstrate this phenomenon in ZnO nanorods [41]. Numerous research teams have investigated the photoluminescence (PL) of ZnO nanostructures, identifying two distinct emission peaks [42–44]. The PL spectra (Fig. 3a) reveal ultraviolet (UV) emission at 380 nm, corresponding to an energy of around 3.2 eV, and green emission at 495 nm, corresponding to 2.3 eV. Vanheusden et al. and Wang et al. [45, 46] demonstrated that green emission in ZnO is associated with near-band-edge emission caused by inter-band transitions, resulting from singly ionized oxygen vacancies.

Photoemission occurs when a photogenerated hole recombines with an electron occupying an oxygen vacancy site. In tetrapod-shaped ZnO, UV emission is detectable even at ambient temperature due to two primary factors: the high quality of ZnO-T with fewer impurities and structural defects, and the quantum confinement effect in the nanostructures. In contrast, UV emission in bulk ZnO is only detectable at very low temperatures. Newton et al. observed significant variations in the optical response of ZnO-T, particularly when a single tetrapod was arranged with one of its arms parallel to the incident laser beam compared to a collection of randomly oriented ZnO-T [49–51].

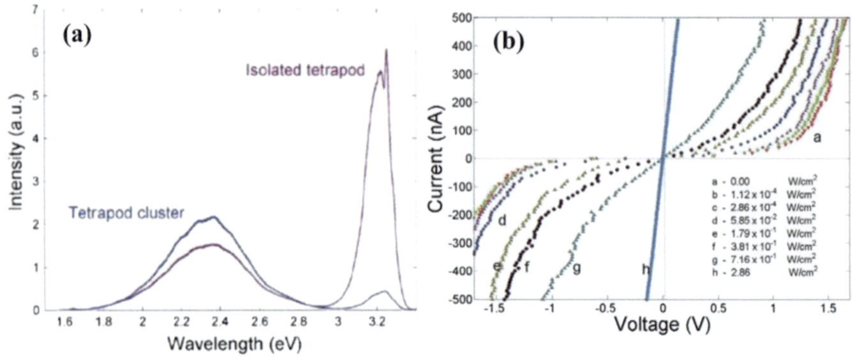

Fig. 3 (**a**) Variation in the PL intensity of isolated tetrapod and tetrapod clusters (**b**) Current -Voltage of ZnO-Tetrapods [47, 48]

Newton et al. investigated the electronic characteristics of ZnO nanocrystals and addressed the challenge in establishing electrical connections using focused ion beam (FIB) technology. They created a diode structure with a single ZnO-T and formed electrical connections through metallo-organic deposition [48, 52]. Three electrical connections were established: a Pt contact functioning as a Schottky contact and two W- contacts exhibiting Ohmic behavior. The rectification characteristics of the ZnO-T Schottky diode were studied, and its electrical response was further investigated under continuous wave laser illumination. It was found that as the laser power density increased, the device's behavior shifted from rectifying to a more Ohmic in nature. Similar results were reported by Heo et al. [53] in their experiments, which included scanning electron microscope (SEM) images of trumpet-shaped ZnO structures captured at both low and high magnification in Fig. 2.

The photoluminescence of ZnO-T clusters and isolated tetrapods at room temperature has been examined [33] as shown in Fig. 3a. Researchers investigated a single Schottky diode nanorod. Wan et al. [54] synthesized tetrapod-shaped ZnO through direct thermal evaporation and analyzed its field emission potential. The high aspect ratio of ZnO-T, an exceptionally low turn-on field of 1.6 V $\mu m^{-}1$ was observed at a current density of 1 $\mu A\ cm^{-}2$ in Fig. 3b.

3.2 *Biocompatibility and Functionalization*

The capacity of a substance to interact with live tissue without producing negative side effects like rejection or inflammation is known as biocompatibility. It plays a significant role in medical device development and approval. ZnO interacts nicely with biological surroundings without having any negative consequences since it is biocompatible [55, 56]. For use as a biosensor, ZnO-T bio-conjugates with biotin to generate a biological probe [57]. Functionalized ZnO- nanostructures can be utilized as gene delivery carriers, PCR delivery agents, and solid phase absorbents [56].

4 Biomedical Applications of Zinc Oxide Tetrapods

Zinc oxide nanostructures have emerged as a highly versatile material in biomedical science, owing to their distinctive physicochemical properties. This chapter explores the multifaceted roles of ZnO-T, focusing on its applications in biosensing, photoluminescence, pathogen detection, antimicrobial activity, wound healing, and the mitigation of cytotoxic effects (Fig. 4).

4.1 Electrochemical Biosensors

ZnO is a highly favored material in sensor and biosensor applications due to its excellent stability, biocompatibility, and increased surface-to-volume ratio, particularly when incorporated into hierarchical structures. The stability of the transducer and its biocompatibility are crucial factors in determining electrochemical biosensor performance. Among various ZnO morphologies, tetrapod-shaped ZnO offers the largest active surface area, making it an ideal choice for enhancing detection efficiency [23, 58]. Glucose is a fundamental molecule in human physiology, serving as a primary source of energy for the body's cells. The primary energy source for

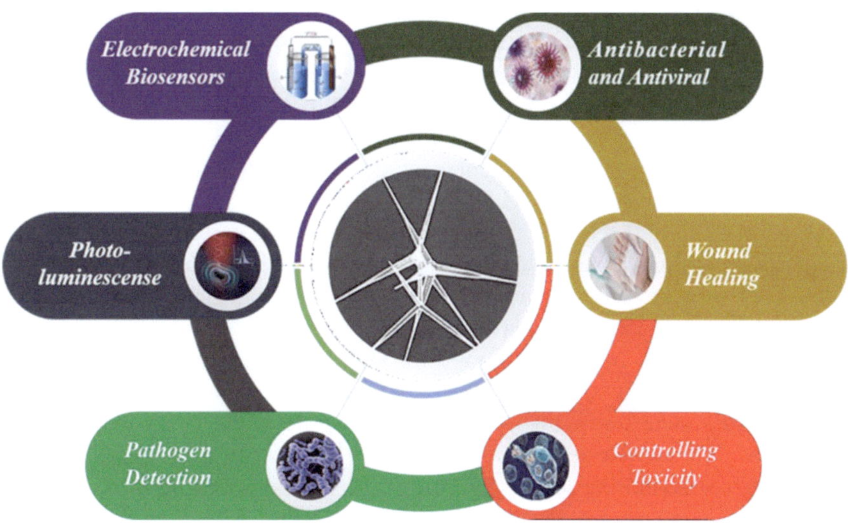

Fig. 4 Overview of ZnO tetrapods in biomedical applications

cellular respiration is glucose, which is transformed by cells into adenosine triphosphate (ATP), the chemical that drives the majority of cellular functions. Deficiency and excess amount of glucose in the body show several symptoms. High glucose (blood sugar) is frequently linked to diabetes mellitus. Low blood sugar can result in symptoms including weakness, disorientation, and lightheadedness. Consequently, the accurate and sequence- specific detection of glucose is very critical humans. In this regard, various studies have explored the application of ZnO-T sensors to detect glucose.

Yeong kim et al. [59] demonstrated the mechanism of amperometric glucose sensors that relies on the variation in free oxygen concentration around the working electrode. The reduction current is influenced by the electrode's catalytic activity in the oxygen reduction in reaction, with higher oxygen levels leading to increased reduction currents and vice versa. In this context, the oxygen concentration is regulated by the glucose oxidation process at the flavin adenine dinucleotide (FAD) site of glucose oxidase (GOx), which consumes free oxygen. This interaction underscores the significance of optimizing the surface area of ZnO nanorods to improve the efficiency and sensitivity of glucose sensors as per the following equation,

$$Glucose + O_2 \rightarrow^{GO_x} glucono - \delta - lactone + H_2O_2 \quad (1)$$

Yang Lei et al. expanded on this by demonstrating that a ZnO nano tetrapod electrochemical biosensor can quantitatively detect a very little quantity of glucose [60]. Because of their positively charged surface and excellent electron communication, the tetrapod-shaped ZnO nanostructures were able to immobilize negatively charged glucose oxidase by the interplay of electrostatics. The as-fabricated biosensor demonstrated a low detection limit of 4 M for glucose and a repeatable sensitivity of 25.3 μA/mM·cm2. The constructed Nafion/GOx/ZnO/Au electrode was tested in PBS and 1 mM glucose, respectively, to demonstrate the adsorption of GOx. The CV curve with 1 mM glucose shows a modest shoulder peak at about 0.6 V and a clear increase in oxidation current above the potential of 0.2 V as compared to the PBS without glucose. This effect results from the oxidation of the H_2O_2, and it is comparable to enzyme electrodes in hydrogen peroxide (H_2O_2) solutions resulting from the processes that GOx catalyzes (Fig. 5).

Myndrul et al. [62] reported that the glucose detection performance of ZnO-T-based electrodes functionalized with chitosan and Nafion polymers. Cyclic Voltammetry (CV) studies reveal that chitosan-modified ZnO-T facilitates glucose oxidation at a reduced applied potential. This might be because the chitosan micelles are bigger than nafion micelles, making it simpler for glucose to pass through the chitosan membrane. ZnO-T/Nafion and ZnO-T/chitosan electrodes show potential for non-invasive glucose detection in bodily fluids for glucose detection. In this related study, Myndrul et al.[55] developed and utilized ZnO-T/MXene based electrodes for qualitative glucose monitoring in sweat on the body. The fabricated sensors exhibited sensitivity of 27.87 $\mu A\ mM^{-1}\ cm^{-2}$ in PBS and 29.88 $\mu A\ mM^{-1}\ cm^{-2}$ in artificial sweat, respectively (Fig. 6).

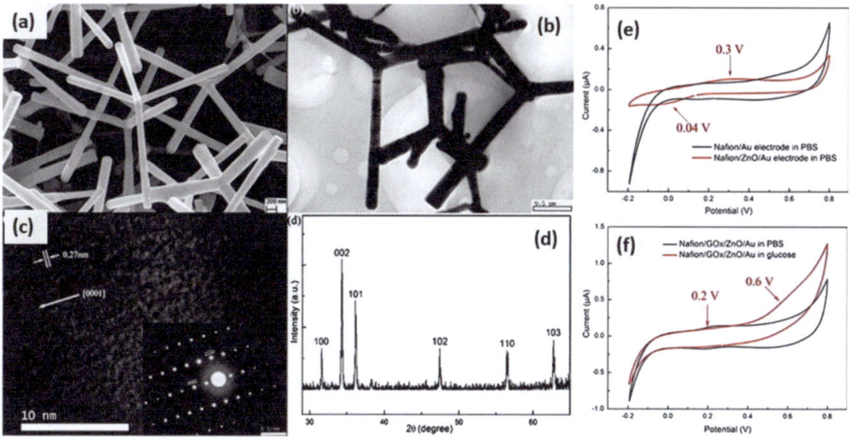

Fig. 5 (**a**) FESEM image showcasing the morphology of the synthesized ZnO-T, (**b**) TEM image illustrating ZnO-T particles, (**c**) HRTEM image of a single ZnO leg, (**d**) typical XRD pattern of ZnO-T. (**e**) Cyclic voltammograms of Nafion/Au (black line) and gold electrodes coated with ZnO (red line), (**f**) CV curves of the GOx/ZnO biosensor in PBS (black line) and 1 mM glucose solution (red line) with a scan rate of 50 mV/s [61]

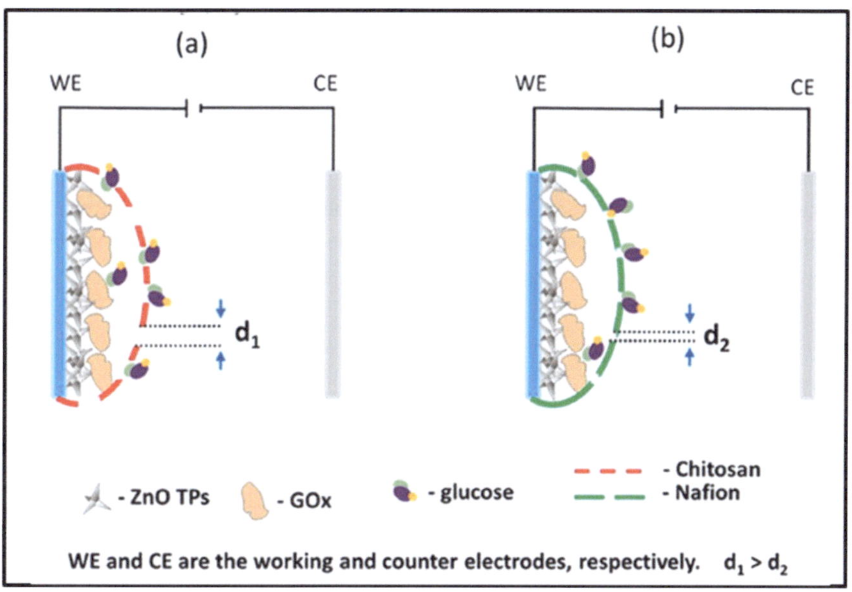

Fig. 6 Schematics of **a** ZnO-T/chitosan and **b** ZnO-T/Nafion electrodes, illustrating the effect of micelle size on glucose diffusion through the semi-permeable polymer layer [62]

4.2 Photoluminescence

4.2.1 Glucose Detection

Zhao et al. explore the photocatalytic activity and luminescence of ZnO-T in a glucose solution to simulate a biologically relevant environment [63]. The photoluminescence properties of ZnO-T were analyzed at varying glucose concentrations using a fluorescence spectrometer. It was observed that as the glucose concentration increased up to 4 mM, the PL intensity in the UV region decreased linearly. ZnO exhibits PL emission in two distinct regions, the UV region and the visible region, as shown in Fig. 7a. Only the peak intensity differs in the UV area, but the peak location and intensity are similar in the visible region for all samples. Consequently, the UV area was the primary focus of this work's study of PL intensity quenching. As a reference, the ZnO-T PL spectra in DI water was acquired, it is found that the NBE peak intensity decreases when glucose is present and further decreases as glucose is added one after the other. The reference peak intensity of the ZnO-T without glucose is used to standardize the change in the NBE peak intensity following the addition of glucose (Fig. 7b).

4.3 Pathogen Detection

Pathogen is a microbe or agent that induces illness in its host. Pathogens encompass a diverse array of species, including bacteria, viruses, fungi, and parasites, in addition to non-living infectious agents such as prions. They are accountable for a variety of illnesses in humans, animals, and plants, and are spread through many channels, including air, water, food, and direct contact.

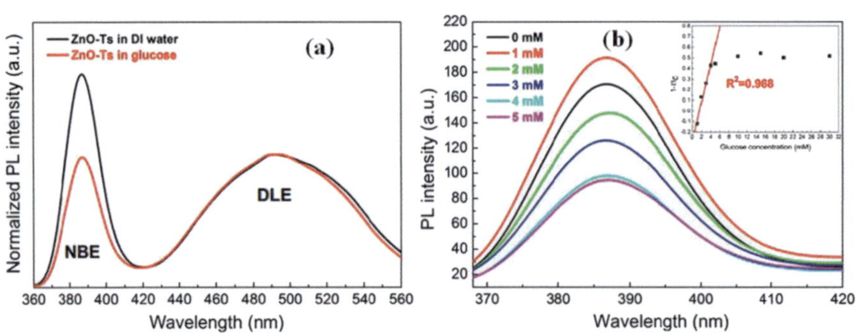

Fig. 7 PL emission spectra of **a** ZnO-T in DI water, both in the absence and presence of 4 mM glucose, and **b** ZnO-T in DI water with varying glucose concentrations. (Inset: Variation in NBE PL intensity with glucose concentration) [63]

Detection of the pathogen help accurate treatment of humans or animals to properly cure from the infection. There are several methods to detect host of the pathogen. In the creation of biosensors, the device's selectivity is still another crucial concern. Selectivity can be attained by using specific bioprobes, which may be synthesized externally and then attached to the surface through electrostatic or covalent bonding, or alternatively, incorporated directly into the PSIs matrix through in situ synthesis [64–66]. Several chemical techniques were established to retain the bioprobe's functioning for biosensing applications because this sensitive process might alter the mobility, conformation, and functions of the chosen bioprobes [67]. In this regard, ZnO-T are a novel method for pathogen detection that exploits the special qualities of ZnO nanostructures to detect microorganisms with high sensitivity and specificity, including bacteria, viruses, and fungi [68]. The three-dimensional architecture of ZnO-T offers advantages over one-dimensional nanostructures in biosensing, due to its enhanced electrical conductivity, minimized particle agglomeration, and ease of electrode and sensor device fabrication, enabling the development of a label-free, multi-parameter detection platform [69, 70]. Terracciano et al. [57] advanced the utility of ZnO nanostructures by employing a multi-step chemical modification process. The process involved a silanization step with 3-(aminopropyl) triethoxysilane (APT), followed by the attachment of biotin to enable real-time detection of streptavidin. This was achieved through optical techniques such as steady-state photoluminescence and fluorescence microscopy, as illustrated in Fig. 8a, b, c and d [57].

Building on the versatility of ZnO, Viter et al. reported a novel photoluminescence-based platform integrated with a microfluidic system was developed for detecting anti-SARS-CoV-2 antibodies in blood samples. Through anti-SARS-CoV-2 antibodies interacted with its surface which was observed through rise of PL intensity is shown in Fig. 9 [71]. Chauhan et al. developed an electrochemical immunosensor utilizing ion-irradiated ZnO-T as a biohybrid interface for the detection of helicobacter pylori shown in Fig. 10 [72]. The CagA antigen is immobilized on ZnO-T electrodeposited on the surface of SP-AuE, which is then utilized to create an immunosensor interface using the redesigned SP-AuE. As the CagA antibody concentration increased from 0.2 to 50 ng/mL, the peak current progressively dropped. CagA antibody and CagA antigen form an immunocomplex over ZnO-T/SP-AuE, which is responsible for the subsequent drop in peak current ($\phi = 10^{15}$). This was noted because the immunoelectrode's surface is covered in protein, which prevents electron flow or redox reactions as depicted in Fig. 11.

Bars et al. [73] developed a luminescent-based immunosensor to detect tetracycline by using ZnO-T as transducers. Since ZnO-T has a large surface area for analyte reaction and bioreceptor immobilization, as shown in Fig. 12. Additionally, it exhibits strong PL emission at room temperature. The PL quenching occurs as the analyte concentration increases due to interaction between the Tetracycline (TC) molecules and the bioreceptor (anti-TC antibodies) gradual reduction in NBE PL intensity with elevated concentrations of TC incubated on the transducer surface gradual reduction in NBE PL intensity with elevated concentrations of TC incubated on the transducer surface. Hence, the minimal changes observed in the PL spectra—both in

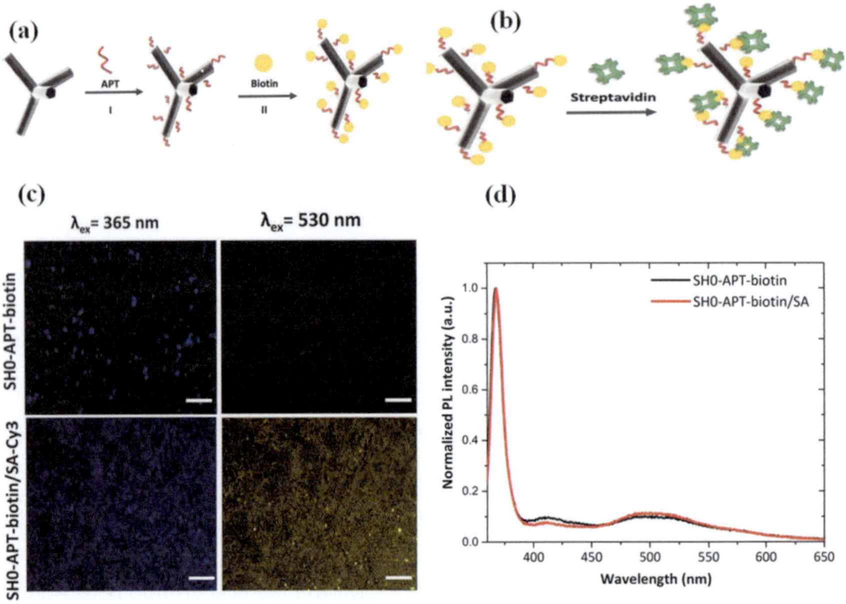

Fig. 8 **a** Schematics of the surface functionalization of ZnO-T with APT (I) and biotin (II), followed by streptavidin detection using biotinylated-ZnO-T. **b** Schematics of streptavidin detection via biotinylated ZnO-T. **c** Fluorescence microscope images of functionalized ZnO-T before and after interaction with Cy3-labeled streptavidin. Scale bar: 75 μm. **d** Normalized photoluminescence (PL) intensity spectra of functionalized ZnO-T before and after exposure to label-free streptavidin at an excitation wavelength of 325 nm [57]

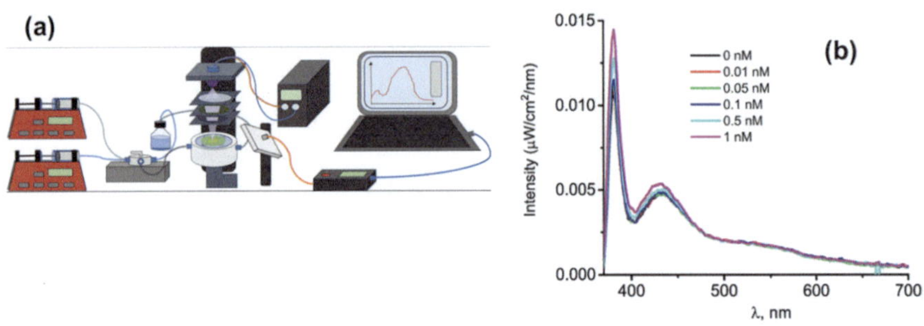

Fig. 9 **(a)** Schematic representation of a photoluminescence-based biosensor for detecting antibodies against SARS-CoV-2 virus proteins, utilizing a ZnO-T structure integrated within a microfluidic system **(b)** steady state PL spectrum [71]

terms of intensity and profile—for the control samples strongly suggest that nonspecific binding between the transducer surface and the target analyte are negligible. Rodrigues et al. [74] developed photoluminescence-based immunosensor to detect the human chorionic gonadotropin (hCG) hormone by using ZnO nanotetrapods. The

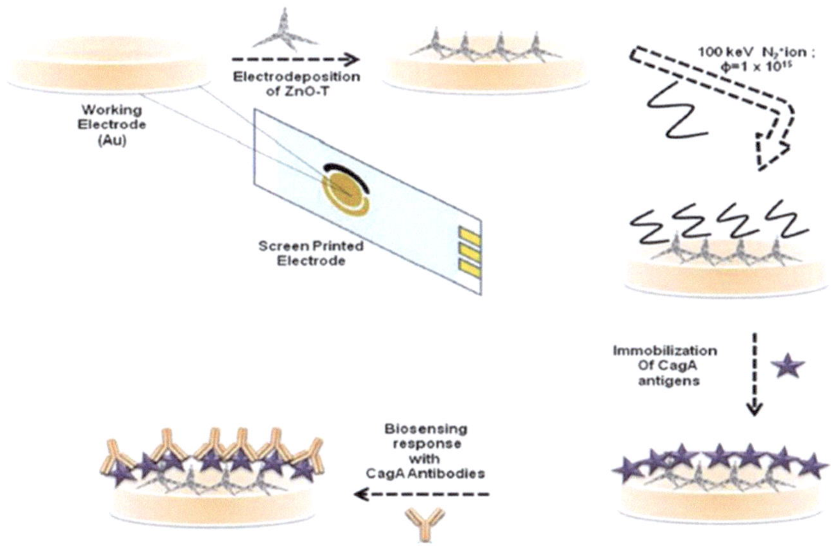

Fig. 10 Illustration of the steps involved in the formation of the CagA antigen @ZnO-T/SP- AuE immunosensor [72]

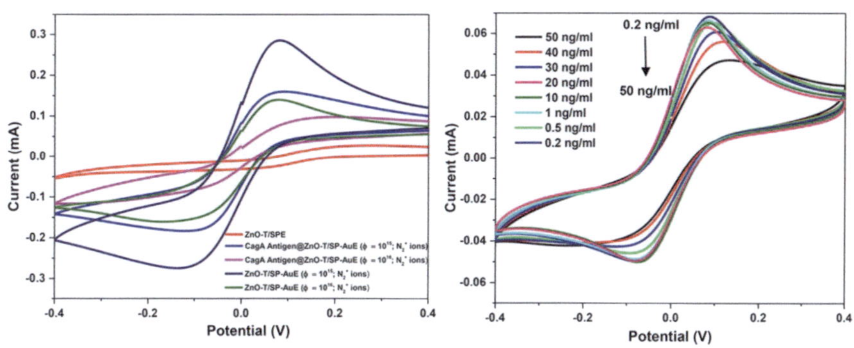

Fig. 11 **(a)** Cyclic voltammetry (CV) curves of ZnO-T/SP-AuE (unirradiated), ZnO-T/SP-AuE ($\phi = 1 \times 10^{15}$), ZnO-T/SP-AuE ($\phi = 1 \times 10^{16}$), CagA antigen@ZnO-T/SP-AuE ($\phi = 10^{15}$), and CagA antigen@ZnO-T/SP-AuE ($\phi = 10^{16}$), recorded using a 5 mM $[Fe(CN)_6]^{3-/4-}$ electrolyte in 0.5 mM sodium phosphate buffer (pH 7.5) at a scan rate of 50 mV/s. **(b)** CV curves showing the effect of different antibody concentrations on CagA antigen@ZnO-T/SP-AuE [72]

proposed sensors relied on variations in PL emission intensity as a means of signal transduction, triggered by the interaction between immobilized antibodies and hCG (Fig. 13). This antigen–antibody binding caused charge redistribution at the semiconductor interface, leading to a reduction in PL intensity. As the hCG concentration increased, the PL signal progressively decreased, demonstrating a direct relationship between the optical response and analyte concentration.

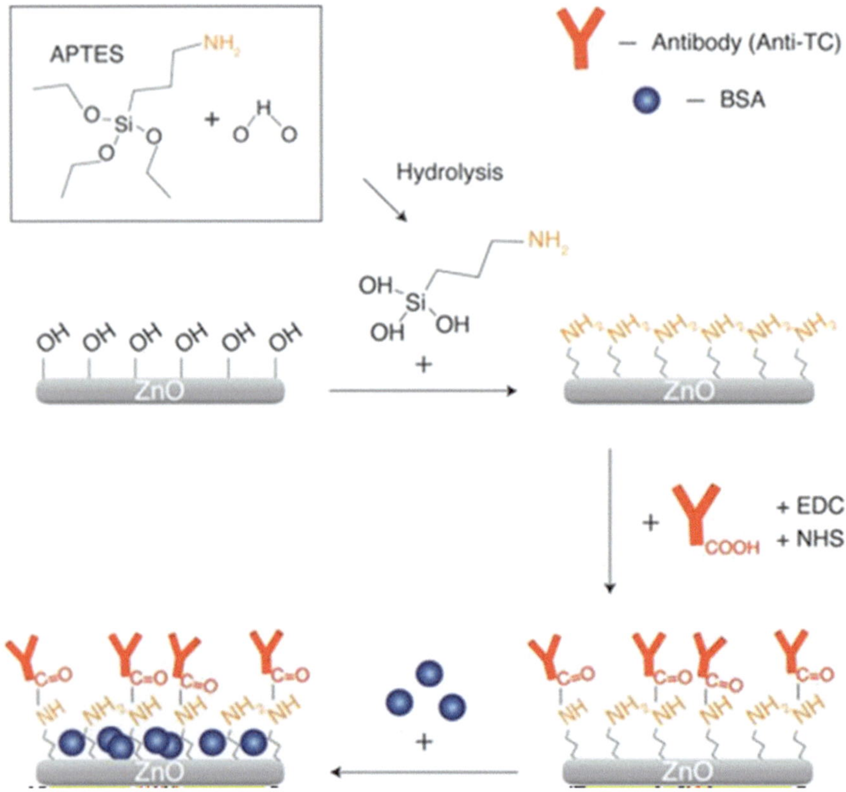

Fig. 12 Illustration of the step-by-step process used for the covalent attachment of bioreceptors onto the surface of transducer platforms [73]

4.4 *Antibacterial and Antiviral Applications*

The antibacterial properties of zinc have been mentioned for several medicinal applications [23, 75, 76]. Tetrapods of ZnO are a special type of nanostructure that have a characteristic three-dimensional tetrapodal shape [77]. This structure makes to highly adaptable in biological applications, including antibacterial and antiviral capabilities. By employing their vast surface area, sharp tips, and controllable growth mechanism [78]. It has some special properties toward antibacterial activity that it has delayed the antibacterial activity.

4.4.1 Mechanisms of Antibacterial Action

The unique chemical and structural properties of ZnO-T enhance their S/V ratio, while their tetrapodal shape provides additional active interaction sites. Their sharp,

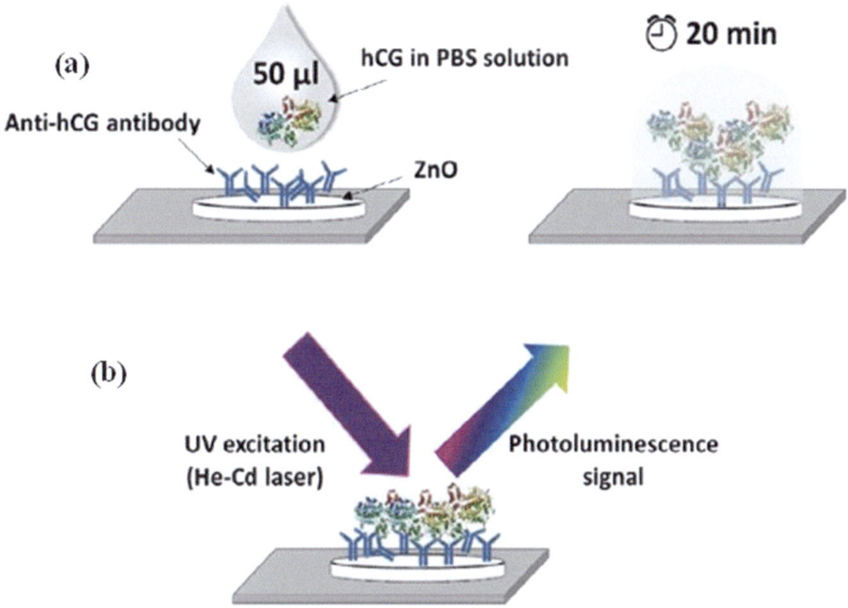

Fig. 13 Schematic illustration of **(a)** hCG incubation on functionalized ZnO substrates and **(b)** the detection of the transduction signal upon UV excitation [74]

needle-like points improve mechanical puncturing efficiency, facilitating bacterial membrane rupture. Another key antibacterial mechanism of ZnO is the generation of reactive oxygen species (ROS) upon exposure to UV or visible light. Additionally, ZnO exhibits low toxicity to mammalian cells, making it well-suited for medical applications.

Generation of Reactive Oxygen Species

ZnO-T generate ROS when exposed to UV or visible light, owing to their semiconducting and photocatalytic properties [23]. Their photoinduced antibacterial action primarily relies on the photocatalytic production of highly ROS including hydroxyl radicals (•OH), superoxide anions (O_2^-), and hydrogen peroxide (H_2O_2) generated through light-driven redox reactions. These transient species exhibit strong biocidal properties, effectively disrupting microbial cells (Fig. 14) [79].

Photoactivation of ZnO: ZnO, with its wide bandgap around 3.37 eV, effectively absorbs ultraviolet light. Upon exposure to photons with energies equal to or exceeding this bandgap, electrons from the VB are excited to the conduction band (CB), leaving behind positively charged holes (h^+) in the valence band, thereby initiating charge carrier generation.

$$ZnO + h\nu \rightarrow e^- + h \quad (2)$$

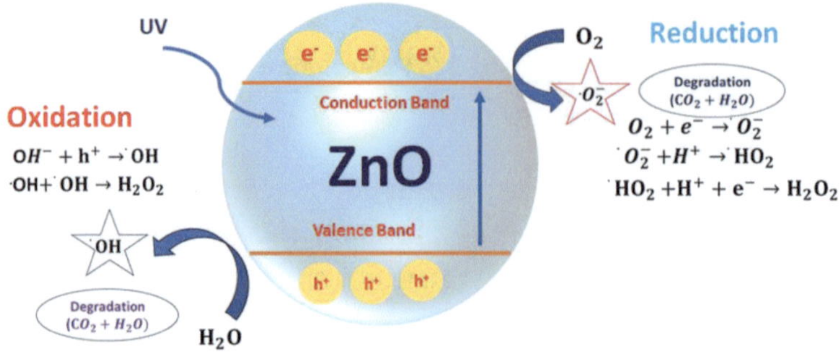

Fig. 14 Mechanism of ZnO nano-photocatalysts towards degradation of chemical species

Superoxide anion radical (O_2^-): The interaction of photogenerated electrons and holes with surrounding molecules results in the production of ROS. The conduction band electrons facilitate the reduction of molecular oxygen (O_2) to superoxide anions.

$$e_{CB}^- + +O_2 \rightarrow O_2^- \tag{3}$$

Hydroxyl radical (OH^-): The valence band holes interact with water (H_2O) or hydroxide ions (OH^-) to generate highly reactive hydroxyl radicals:

$$h_{VB}^+ + H_2O \rightarrow OH^{\cdot} + H^+ \tag{4}$$

$$h_{VB}^+ + OH- \rightarrow OH^{\cdot} \tag{5}$$

Hydrogen peroxide (H_2O_2): Superoxide anions may undergo additional reactions with protons or dismutate, resulting in the formation of hydrogen peroxide [80].

$$O_2^- + 2H^+ \rightarrow H_2O_2 \tag{6}$$

Singlet oxygen ($^1O_2$): Singlet oxygen is generated via energy transfer from excited ZnO to molecular oxygen or through the decomposition of ROS intermediates. Multiple parameters affect the efficacy of ROS production by ZnO-T. Light intensity and wavelength: UV light is more efficacious in stimulating ZnO owing to its appropriate energy aligned with ZnO's bandgap. Doping ZnO with other elements can alter its absorption to the visible spectrum. The high aspect ratio and sharp tips of ZnO-T enhance photocatalytic efficiency by increasing reactive sites and improving light absorption.

The generation of specific ROS is influenced by environmental pH, which impacts ZnO's interactions with the surrounding molecules. Doping ZnO with metals like Ag and Cu or non-metals such as N and S enhances ROS production under visible light, thereby increasing its antimicrobial effectiveness (Fig. 15) [81]. ROS can also increase by surface modification via sensitizing or heterojunction formation.

The unique morphology of ZnO-T makes them well-suited for direct cell contact, as their arms are unable to penetrate or damage eukaryotic cells [78]. The broad surface area of ZnO-T facilitates interactions with bacteria or antibodies, enhancing their potential for antimicrobial applications [72]. Additionally, the nanostructured surfaces of ZnO particles offer dynamic, customizable interfaces that promote effective bacterial attachment and subsequent elimination. The use of solid-state chemistry, involving direct interactions between the active agent and bacteria, presents an alternative mechanism for antibacterial action. Unlike soluble antibiotics, this approach relies on localized contact between ZnO-T or non-soluble ZnO particles and micro-organisms on tissue or material surfaces, expanding the spectrum of antibacterial strategies.

In bacteria, the cell membrane acquires a negative surface charge due to the presence of teichoic acids in Gram-positive strains and lipopolysaccharides in Gram-negative strains. The surface charges and pH of the surrounding environment determine whether ZnO nanoparticles acquire a positive or neutral charge. This charge can lead to various antibacterial effects such as zinc ion release, reactive oxygen species production, or electrostatic interaction with the tetrapod arms (Fig. 15) [68, 82]. Building on these interactions, there are several techniques ZnO-T, for instance, when ZnO exposed to UV light and methylene blue, ZnO-T undergo a photocatalytic reaction with the dye, expanding their reactive surface [83]. Furthermore, the ZnO-T surface may be modified to affix active components for regulated release [21].

In photothermal effects, which occur when light exposure heats up the nanoparticles and raises the local temperature to a point where bacteria are killed. A second consequence of light is the selective breakdown of metal nanomaterial complexes and the subsequent cell injury brought on by one of the many metal-bacteria interactions. Thirdly, it has been shown that the nanoparticles produce ROS by photochemical means. Here, metal oxide nanomaterials semiconducting qualities were used, where the production of high-energy electron–hole pairs can cause medium or the particle itself to oxidize and reduce and activate ROS [84]. Figure 16 illustrates

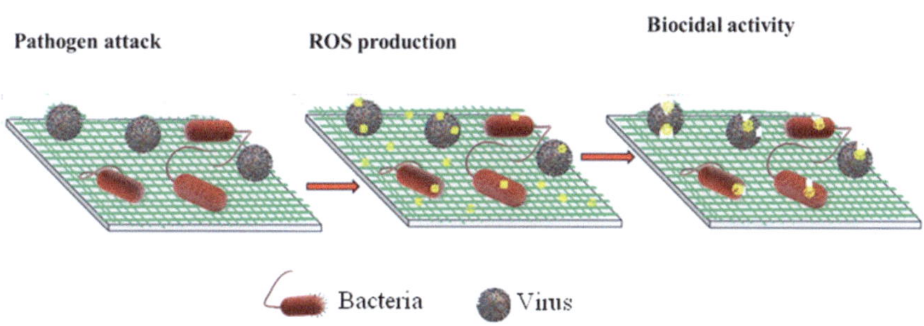

Fig. 15 Illustration of photocatalytic process involved in the disinfection of bacterial and viral pathogens [79]

the formation mechanism of different ROS under various conditions. These potent biocidal ROS effectively neutralize bacteria and viruses by causing damage to key cellular components, including DNA, RNA, proteins, and lipids. The smaller ZnO nanoparticles exhibited superior bactericidal activity [85]. However, evaluating the effectiveness of nanoparticles also requires consideration of the structural variations among different types of pathogens, as these differences can significantly influence their susceptibility [86]. Doped ZnO nanostructures exhibit improved properties, including increased surface area and enhanced charge transfer, which contribute positively to disinfection activities (Fig. 16) [87, 88].

During the inactivation process, the generated ROS primarily target and disrupt the cytoplasmic membrane and cell wall of bacterial and viral cells [89, 90]. However, the efficiency of photo inactivation or disinfection is influenced by several factors,

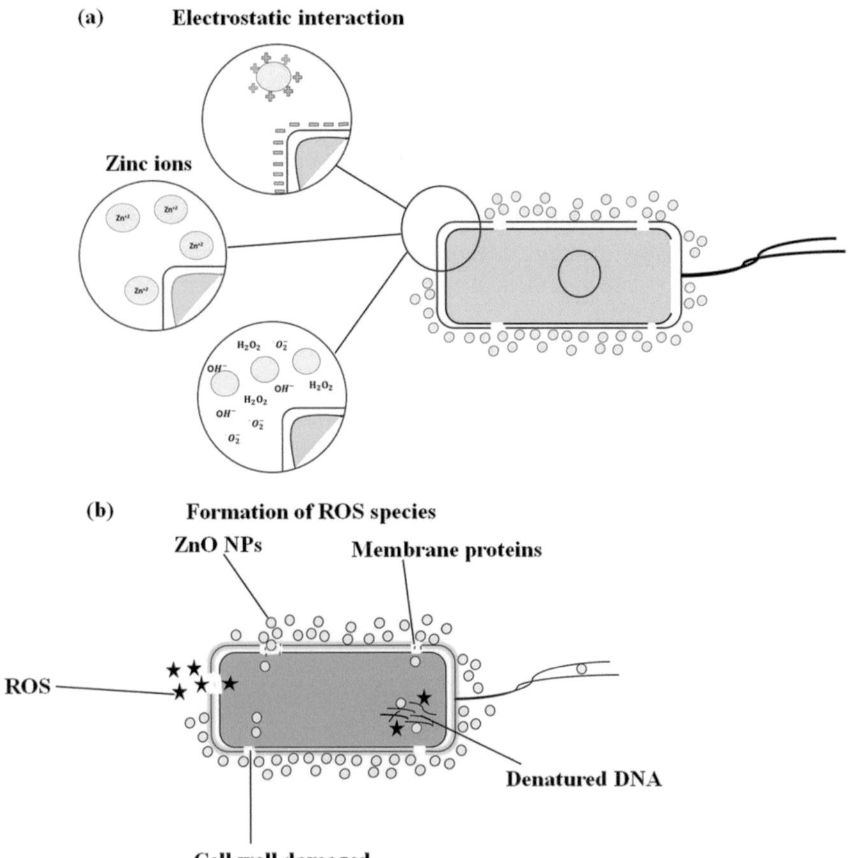

Fig. 16 **(a)** Illustration of the primary bactericidal mechanism of ZnO NPs, targeting the cytoplasmic membrane and cell wall. **(b)** Cell model depicting the key mechanism of bactericidal action of ZnO NPs

including the type of photocatalyst employed, the quantity of ROS generated under specific light wavelengths, and the structural characteristics of the target pathogens. Since different bacteria and viruses possess distinct internal and external cell architectures often comprising complex layers and various forms of RNA, DNA, proteins, and enzymes their susceptibility to photocatalytic disinfection can vary significantly [91]. ZnO exhibits potential disinfection properties because it produces zinc ions and ROS, which cause cell membrane lysis, disintegration, and membrane protein degradation, ultimately leading to cell death [90]. The generation of ROS is a very effective antibacterial mechanism that directly contributes to cell death by destroying essential cellular components including proteins and DNA. In the recent work [92] about the interaction of tailored ZnO nanostructures with glioblastoma cancer cells highlights their promising potential in biomedical engineering. Recent research indicates that ZnO-T can be electrospun with biocompatible polymers to form rose spike-like fibers, exhibiting improved antibacterial properties and enhanced biocompatibility. These advancements pave the way for a wide range of innovative applications in the biomedical field [76, 93].

4.4.2 ZnO Tetrapods as Antiviral Agents

ZnO-T have demonstrated a wide range of applications, including antimicrobial, antifungal, anti-aging treatments, food preservation, and use in cosmetics. Zinc itself is a vital trace element for the human body, as it activates over 300 enzymes involved in numerous metabolic processes, playing a key role in glucose metabolism and the synthesis of insulin [12]. Zinc deficiency in humans can lead to serious health complications, and it has become a significant global concern. ZnO-T have attracted considerable interest as potential antiviral agents due to their unique physicochemical properties, such as high biocompatibility, large surface area, and their ability to generate ROS [94–97].

Virus Neutralization Mechanisms:

In particular, ZnO has promise disinfection properties as a possible antibacterial and antiviral agent because due to its ability to generate reactive oxygen species and release zinc ions. These mechanisms contribute to the disintegration of cell membranes, cell lysis, and damage to membrane proteins, ultimately leading to cell death [90, 98] applications of ZnO nanoparticles. The generation of ROS is a very effective antibacterial mechanism that damages DNA and proteins, two essential biological components, and is directly linked to cell death. ZnO-based nanostructures have been modified through various approaches, including size tailoring, surface modifications via doping, annealing, and the formation of heterojunctions [81, 86, 99–101] to create efficient antimicrobial and antiviral nanomaterials. Zinc exhibits antiviral effects through several mechanisms, including blocking viral entry, interfering with polyprotein processing, and suppressing the activity of viral RNA-dependent RNA polymerase [94–97]. If we compare with ZnO-nanoparticle and ZnO-T. The lower toxicity of the ZnO-T [78] while retaining potential support to the antibacterial activity and theraptic support. This therapy offers improved effectiveness, minimized side effects, and more precise targeting of disease sites. The distinctive three-dimensional architecture and surface properties of the nanoparticles are key factors influencing both their therapeutic performance and potential toxicity. Gupta et-al. investigated that ZnO-T is more effective as an antibacterial particle in antiviral activity compared to ZnO nanoparticles [102]. It prevents Herpes Simplex Virus (HSV) infection by attaching itself to the virus and trapping it, which stops it from entering the host cell. Zinc ion (Zn^{+2}) will attach to viruses (Negative charge due to presence of the glycoprotein on the surface) [103–105] through electrostatic interactions, reducing their ability to replicate [106].

Moreover, viruses exist with certain kind of glycoprotein at the envelope. A unique mechanism for capturing different viruses with oxygen vacancy on our ZnO-T is targeted [107]. The tetrapodal ZnO can be equipped with oxygen vacancy to create polar surface with a positive charge, which we intend to trap the virus with its functional glycoprotein group on the surface like a fly with fly paper (Fig. 17).

The immobilization process was successful, and the binding mechanism was confirmed by illuminating ZnO-T crystals with UV light in ambient air, which facilitated the creation of additional oxygen vacancies through the well-known auto photocatalytic process [107–109]. Subsequently, the trapping efficiency was significantly enhanced [108]. To further validate these findings, in-vivo animal studies were conducted, and Shukla et al. observed that the immune system was able to capture viruses from ZnO-T, leading to a trifunctional vaccination response [110].

Additionally, Pormohammad et al. investigated the interaction of $Zn2^{+}$ ions with two critical proteins involved in the CoV lifecycle, particularly the RNA-dependent

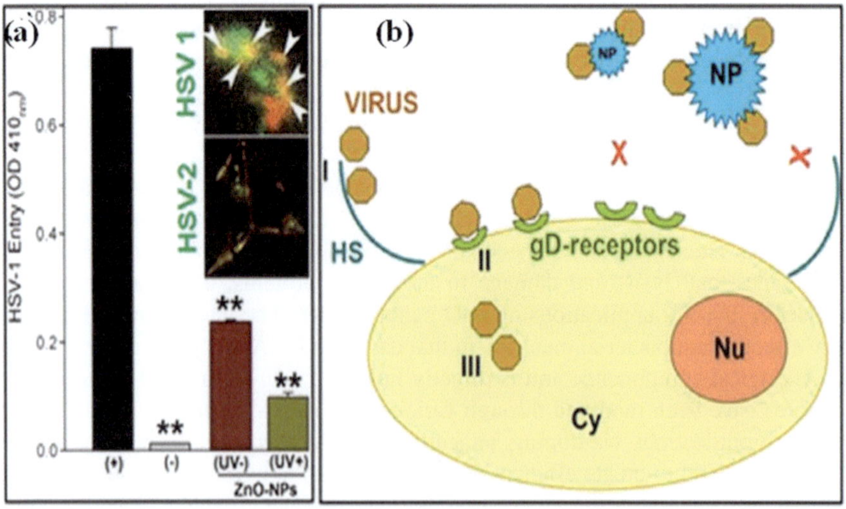

Fig. 17 **a**, **b** Schematics of the capture of HSV-1 viruses by ZnO tetrapods and the associated virus-binding process [107]

RNA polymerase (RdRp), which plays a crucial role in viral replication and whose inhibition is essential for preventing effective replication [111]. Moreover, viruses contain specific glycoproteins on their envelopes that facilitate binding with surface defects on ZnO-T. This enhanced binding capacity suggests that ZnO-T may offer excellent binding possibilities, even with more challenging viruses such as HIV, providing a broad therapeutic window for treating severe viral diseases. Importantly, this antiviral activity is not limited to viruses alone, extending the potential application of ZnO-T in a wide range of therapeutic areas.

4.4.3 Applications in Preventive Therapies

ZnO-T, resembling the filopodia-like structures that viruses use for attachment and entry, offer a promising strategy as viral entry inhibitors by binding to viral glycoproteins. Their virostatic, prophylactic, therapeutic, and neutralizing effects against genital and ocular herpes have been well-documented [107]. Antiviral studies have demonstrated that the application of filopodia-like ZnO-T in vitro can inhibit viral entry through electrostatic interactions between the negatively charged ZnO and the positively charged virus [107, 110]. This interaction is comparable, and possibly stronger, than the ionic binding between positively charged viral glycoproteins and negatively charged heparan sulfate on the cell surface. Furthermore, it was found that the antiviral efficacy of ZnO-T significantly increases when pre-exposed to UV light. As mentioned previously, the UV-induced formation of oxygen vacancies on the ZnO surface enhances the number of active sites available for viral attachment, thereby improving viral inhibition. The majority of these studies have focused on HSV-1, the virus responsible for oral blisters and herpetic keratitis. In the past, zinc oxide-based oral creams have been suggested to help alleviate symptoms and potentially shorten the duration of cold sores [112, 113]. ZnO-T, with their high surface area and minimal agglomeration, can effectively capture a larger number of viral particles compared to bulk ZnO. This property has also been highlighted in studies demonstrating their potential to inhibit HSV-2 genital infections [110, 114]. Mcpherson et al., investigated ZnO as supplementary for prevention of Covid-19 [115]. ZnO-T have been shown to not only reduce viral entry into host cells but also exhibit therapeutic potential by limiting the transmission of infectious particles to adjacent cells. This effect is likely attributed to their enhanced ability to bind with HSV-2 particles (Fig. 17) [108].

4.5 Wound Healing

Wound healing is a cumbersome biological process that restores the integrity of skin after an injury. cells, chemical signals, extracellular matrix (ECM) proteins, and microenvironments all interact intricately during wound healing [116]. Fibroblast migration, proliferation, and myofibroblast trans differentiation are essential phases in the healing of wounds [117]. Following an injury, myofibroblasts give the tissue mechanical support and integrity [118]. When the tissue integrity is sufficiently recovered to be mechanically coherent, they are eliminated by apoptosis under normal circumstances [119]. Myofibroblasts remain in the tissue during pathological wound healing processes, and they cause fibrosis by increasing matrix production and tissue contraction [120]. Consequently, altering fibroblast activities might enhance the available treatments for fibrosis-related illness [121].

Studies have shown that ZnO-T can accelerate wound-healing processes in animal models [110, 122]. For instance, Sonntag and colleagues [122] investigated the effects of ZnO-T on wound healing by assessing its impact on fibroblast proliferation, migration, trans differentiation, and cytokine release in the context of glaucoma surgery. In this study, human tenon's fibroblasts (HTFs) were exposed to ZnO-T, and their proliferation was evaluated using the MTT assay to assess cytotoxicity. The expression of proliferation markers (Ki67), the differentiation marker alpha-smooth muscle actin (α-SMA), and p-SMAD were analysed to evaluate the effects of ZnO-T on fibroblast function. Additionally, cytokine levels, including platelet-derived growth factor (PDGF) and hepatocyte growth factor (HGF), were measured using multiplex ELISA.

The results revealed that exposure to ZnO-T at concentrations of 10–20 μg/mL caused significant toxicity to HTFs. Moreover, there was a marked reduction in the number of Ki67-, α-SMA-, and p-Smad-positive cells, indicating a suppression of fibroblast proliferation, migration, and trans differentiation. Furthermore, ZnO-T treatment led to a significant decrease in IL-6 and HGF levels in the culture supernatants. These findings suggest that ZnO-T exhibits both anti-inflammatory and antiproliferative properties, which could be beneficial for reducing fibrosis, particularly in the context of postoperative wound healing following glaucoma filtering surgery.

Further, the wound healing was much faster in animals treated with ZnO-T [110]. The use of ZnO-T was found to reduce the progression of chronic infection, leading to improved survival rates in treated animals. Additionally, viral particles bound to ZnO-T were more readily internalized by dendritic cells. The application of ZnO-T reduced the chronic infection, as the dendritic cell absorbed the ZnO-T-bound viral particle [110]. After these successful animal studies, initial observations in human patients also corroborated the findings, validating the results obtained in the animal models. Given these promising results, it is crucial to conduct further systematic studies across various animal species to broaden the understanding of ZnO-T therapeutic potential.

4.5.1 ZnO-Infused Hydrogels and Bandages

In biomedical applications, ZnO nanoparticles have drawn a lot of interest because of their antibacterial [78], anti-inflammatory [122], and wound-healing capabilities [123]. A possible method to improve wound care is to include ZnO into hydrogels and bandages, which combine the therapeutic advantages of ZnO with the protective and moisture-retentive properties of these materials.

ZnO-infused hydrogels and bandages are prepared by using polyvinyl alcohol (PVA), polyacrylamide (PAM), To prepare the hydrogel-coated bandages, dissolve another polymer is dissolved distilled water under continuous stirring until a homogeneous solution is obtained [123, 124]. Employ ultrasonication to ensure uniform distribution of ZnO nanoparticles within the polymer solution. Form the hydrogel using chemical or physical crosslinking techniques, and pour the resulting hydrogel solution into molds, allowing it to set. Subsequently, apply a hydrophilic polymer coating or treat the bandage substrate with plasma to enhance adhesion. Evenly spread the ZnO-infused hydrogel onto the prepared bandage substrate, and dry the coated bandages either at ambient temperature or in a humidity-controlled environment [123, 124].

Siebert et al. developed 3D-printed hydrogel patches combining vascular endothelial growth factor (VEGF) and light-activated antibacterial ZnO-T microparticles [21]. These patches support wound healing by promoting blood vessel growth, reducing infections, and releasing VEGF in a controlled manner when exposed to UV or visible light. The ZnO-T particles are chemically treated to enhance protein adhesion, while the hydrogel's stiffness and degradation rate are tailored to match the healing process. The patches can be customized with specific patterns for targeted growth factor release. Lab and early in vivo tests show promising results, highlighting their potential in advanced wound care (Fig. 18).

For production of bandage using ZnO-T dispersed in polycaprolactone (PCL) polymer for wound healing treatment, Nasajpour et al. [125] developed and demonstrated antiviral fibrous arrays inspired by rose thorns generated through electrospinning of a polycaprolactone (PCL) polymer composite integrated with a dispersion of anisotropic ZnO-T nanoparticles (ZOTeN). The engineered mats produced in this way have the potential to be used in the manufacturing of bandages, providing an innovative solution for wound healing with antiviral properties (Fig. 19) [125].

Exploring the wound healing dressing platform based on ZnO-T mix with hydrogel (alginate), Schadte et al. [126] proposed a multifunctional bioink platform composed of sodium alginate hydrogel and nanostructured ZnO-T. This bioink can be functionalized with active pharmaceutical ingredients (APIs). The study demonstrates the printability of this bioink and evaluates its cell compatibility and antibacterial properties in both open- and closed-cell constructs containing varying concentrations of ZnO-T. This approach showcases the potential of combining ZnO-T with hydrogels for enhanced wound healing applications.

The workflow of alginate-based bio-inks, ZnO-T, synthesized through flame transport synthesis, can be functionalized with specific cytokines to enhance biomedical performance. Depending on the wound's condition, these cytokine-functionalized ZnO-T particles can support various therapeutic roles like modulating immune response, reducing inflammation, or promoting angiogenesis. Once incorporated into a sodium alginate solution, the bioink can be used for bioprinting tailored porous scaffolds. Owing to their unique geometry, the tetrapods often extend from the hydrogel surface, allowing direct interaction with the wound environment. These protruding

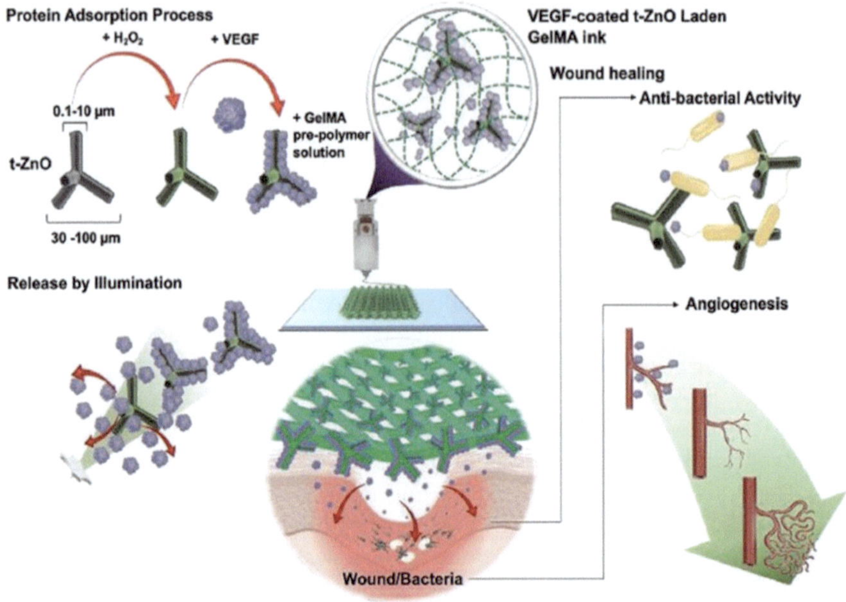

Fig. 18 Schematics of the fabrication process of a smart wound scaffold and its mechanism in promoting wound healing: ZnO-T micro-nano-particles are first prepared and treated with H_2O_2, followed by VEGF adsorption and blending with a GelMA prepolymer solution to form a composite ink. Upon light stimulation, VEGF is released from the H_2O_2-treated ZnO-T particles. The VEGF-coated ZnO-T integrated into the porous printed scaffold enhances angiogenesis and exhibits antibacterial properties at the wounded site [21]

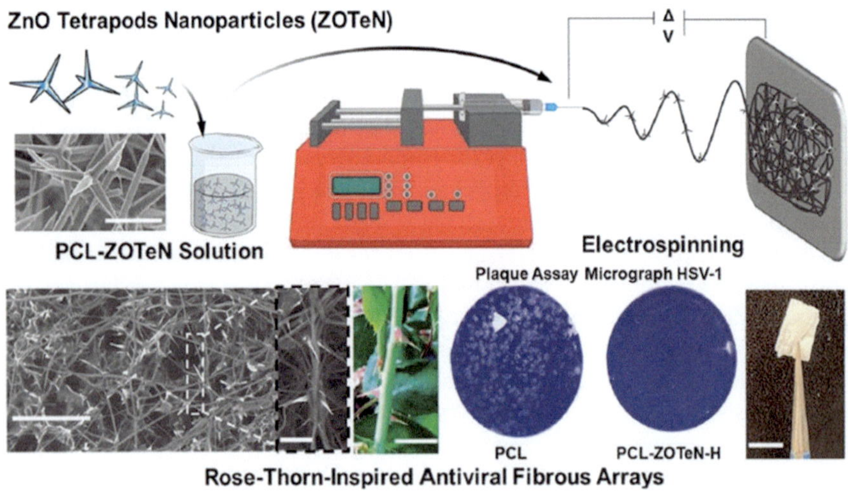

Fig. 19 Nanostructured ZnO-T (ZOTeN) for wound healing application [125]

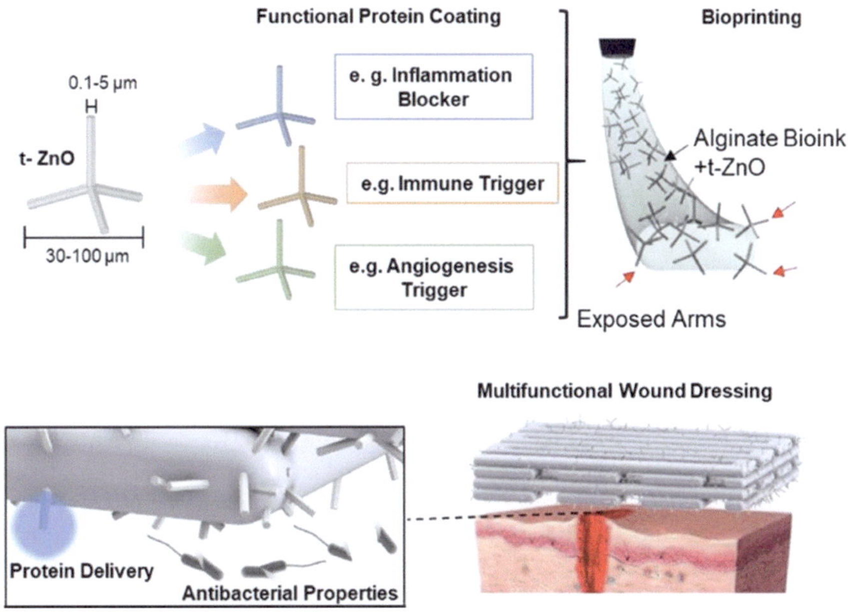

Fig. 20 Mechanism of wound healing using alginate-based bio-ink [126]

arms not only release therapeutic cytokines locally but also contribute to antibacterial protection. Also, the alginate hydrogel acts as a protective barrier, safeguarding the wound from external contaminants while allowing oxygen exchange to support the healing process (Fig. 20).

4.6 Controlling Toxicity

Growing study is being done on the cytotoxicity of ZnO-T, particularly because of their possible uses in biological domains such cancer treatment, biosensing [72, 127], and antimicrobial coatings [21, 128]. ZnO-T have cytotoxic effects via a number of important mechanisms. By producing ROS, ZnO nanostructures can cause oxidative stress in cells, which can harm lipids, proteins, and DNA, among other biological constituent's [129, 130] dissolution and release of Zn ions: ZnO-T can release Zn^{2+} ions when they disintegrate in acidic environments, such as lysosomes. Excessive levels of Zn^{2+} ions can cause cytotoxicity and disturb cellular homeostasis, which can impact enzyme activity. Cell death and the leaking of cellular contents may result from this interaction. Mitochondrial dysfunction: ROS produced by ZnO can disrupt mitochondrial function, resulting in membrane potential loss and intrinsic apoptosis a number of variables determine the degree of cytotoxicity: the shape of ZnO-T is important for their interaction and cellular absorption. Since smaller sized particles are easier for cells to absorb, they are more cytotoxic [131–133]. To mitigate these harmful effects, surface modifications such as coatings made from biocompatible polymers or proteins can be applied. These coatings can reduce ROS production and ion release, thus lessening the cytotoxicity of ZnO-T and enhancing their potential for safe use in biological applications.

Concentration and Exposure Time: Cytotoxic effects are often stronger at higher concentrations and during longer exposure times. Finding the ideal concentration is essential for applications in biomedicine. The effects of ZnO-T on different types of cells are not uniform. Since cancer cells have a faster metabolic rate than normal cells, they can be more vulnerable to damage caused by reactive oxygen species.

Therapeutic Potential Beyond Cancer: ZnO-T are also showing promise in other therapeutic areas, like in the treatment of viral infections. Recent research suggests that ZnO-T could offer excellent binding capabilities, even for deadly viruses like HIV, thereby expanding the therapeutic window for severe diseases [92]. This versatility extends beyond viruses; recent studies have also explored the use of ZnO nanostructures against glioblastoma cancer cells, opening new avenues for their use in cancer treatment.

4.6.1 Biomedical Engineering Applications

ZnO-T have also demonstrated potential in biomedical engineering, especially when combined with biocompatible polymers. These nanostructures can be electrospun into rose- spike-like fibers, which possess enhanced antibacterial properties and other biocompatible features. This combination not only enhances their ability to fight infections but also paves the way for a range of new applications in tissue engineering, wound healing, and drug delivery systems. The recent studies highlight the growing interest in ZnO-T as a versatile material with diverse applications in the medical field [76, 93].

ZnO-T have demonstrated remarkable binding capabilities, potentially extending even to highly challenging viruses like HIV, suggesting a wide therapeutic potential for treating severe diseases. Beyond antiviral applications, recent studies have highlighted the effectiveness of uniquely structured ZnO nanomaterials against glioblastoma cells, further expanding their promise in biomedical engineering. Additionally, ZnO-T can be electrospun with biocompatible polymers to create rose-spike-like fibrous structures, offering improved antibacterial properties and excellent biocompatibility. These advancements have opened exciting new pathways for the use of ZnO-T in diverse biomedical applications. Moreover, research by Büter et al., compared the antibacterial and bactericidal activities of nanostructured ZnO-T with spherical ZnO particles. Compared to ZnO spherical particles, ZnO-T demonstrated significant bactericidal activity against isolates of Klebsiella pneumoniae and Staphylococcus aureus, including multi-resistant strains [78]. The role of ZnO-T for various biomedical application are listed in Table 1.

4.6.2 Strategies to Minimize Toxicity

There have been a number of different ways explored to reduce the cytotoxicity of ZnO-T in order to make it possible for them to be used in biomedical and environmental applications in an efficient manner.

(a) **Surface Coatings and Modifications**

The coatings minimize direct interaction with cells while regulating ion release and ROS production.

Polymer Coating: Coating ZnO-T with biocompatible polymers like polyethylene glycol (PEG), chitosan, or dextran creates a protective barrier that diminishes cellular absorption, restricts $Zn2^{+}$ ion release, and mitigates oxidative stress PEGylated ZnO-T exhibit decreased cytotoxicity across multiple cell lines [135–137].

Lipid or Protein Coating: Encapsulation with lipids or proteins improves biocompatibility by imitating natural cell components, reducing immunological reactions, and decreasing Zn^{2+} dissolution [138, 139].

Silica Coating: Silica (SiO_2) coatings create an inert shell around ZnO-T, reducing ion release and ROS generation while retaining optical and mechanical qualities [140, 141].

Metal and Non-metal Doping: The addition of biocompatible metals like Mg, Ca, or Fe to ZnO-T changes their bandgap energy, lowering UV or visible light ROS generation. Due to oxidative stress, Mg-doped ZnO-T are less hazardous. The addition of non-metal components like nitrogen (N) or sulphur (S) reduces ROS formation and increases stability under physiological control [142–144].

(b) **Optimal Dosage and Safety Assessments**

Prior to their biomedical application, it is essential to determine the ideal dose and safety profiles for ZnO-T. Finding the optimal concentration range for ZnO-T to achieve their therapeutic benefits (antimicrobial, anticancer, etc.) With little harm to healthy cells and tissues is the goal here. The optimal dose is based on the kind of cells or tissues used: the sensitivity of various cell types to ZnO-T varies; for example, cancer cells, fibroblasts, and immune cells [130]. Depending on the intended medical use, lower doses may be enough for antibacterial applications while greater doses may be necessary for anticancer effects. Smaller particle size resulted in a lower dose necessary to trigger cell death, since the cytotoxicity of ZnO NPs found that is significantly dependent on size, in addition to time and dosage [131, 145]. Ratio of in-vitro dosage, the usual doses used in vitro experiments vary from 1 μg/mL to 100 μg/mL [146], contingent upon the kind of cells and the length of time they are exposed. Developing standardized protocols for determining the safe and effective dose of ZnO-T will be critical. Research should also focus on surface modifications and functionalization to enhance the biocompatibility of ZnO-T, thereby expanding their applicability across diverse biomedical fields.

5 Emerging Trends and Future Perspectives

Among different metal nanoparticles, ZnO-T have garnered significant attention from biomedical researchers due to their tunable physicochemical characteristics, such as size, morphology, and surface charge. These attributes make them highly suitable for medicinal applications, as highlighted in studies on their use in label-free optical biosensing and functionalization strategies [57]. Notably, ZnO-T exhibit antiviral properties that can inhibit HSV-1 infection, making them a promising candidate for preventing genital infection caused by herpes simplex virus [107]. Their unique shape facilitates viral attachment, while the optimization of their cytotoxic properties enables effective inhibition of virus-related skin diseases. Modifying the surface of ZnO-T with polymers further enhances their biocompatibility, reduces toxicity, and allows conjugation with biomolecules like drugs, nucleic acids, and proteins. This versatility enables applications in eliminating external bacteria and viruses, wound healing, and drug delivery in cancer therapy.

Despite their promising potential, several challenges hinder the clinical implementation of ZnO-T. A significant obstacle is the limited understanding of their biological properties and toxicity, stemming from gaps in collaboration between biologists and material scientists. Additional challenges include issues with long-term stability, sensitivity to pH changes, impurities, and variations in handling. Addressing these challenges will require interdisciplinary collaboration among material scientists, biologists, and physicians to advance ZnO-T research and facilitate its translation into clinical trials. Moreover, regulatory approval from agencies like the FDA remains a critical step for the commercialization of ZnO-T-based nanomedicines. As the field of ZnO-T research continues to grow, it is reasonable to expect that innovative biosensors and therapeutic applications will receive regulatory approval, enhancing disease diagnosis and treatment.

5.1 Other Potential Applications of ZnO-T

Antibacterial Activity for Cancer Treatment: ZnO-T exhibit strong antibacterial properties, making them potential candidates for cancer therapy by targeting bacterial infections associated with cancer progression [78].

Wound Healing Applications: Due to their lower toxicity compared to ZnO nanoparticles, ZnO-T can be used in Xbandages for wound healing without harming healthy cells [21, 110, 125, 126].

HSV-1 Neutralization: ZnO-T effectively trap and neutralize HSV-1, demonstrating their potential in prophylactic and therapeutic applications, such as treating wounds, ulcers, and microbial infections, as well as delivering drugs in cancer therapy [102].

Antibacterial Drugs: Their robust antibacterial activity positions ZnO-T as potential replacements for antibiotics in the future [78].

Ointments and Creams: ZnO-T can be incorporated into ointments and creams to inhibit the growth of viruses and bacteria on infected areas of the body [78].

Food Industry Applications: ZnO-T can be utilized in biomedical applications, the food industry, and food packaging due to their antibacterial properties [125].

Live Viral Vaccines: ZnO-T exhibit significant antiviral activity against genital herpes infections and could be developed as live vaccines for viral diseases [134].

Electrochemical Immunosensors and PL sensor: ZnO-T enhance the development of electrochemical immunosensors for the rapid detection of pathogens and biomolecule like helicobacter pylori, glucose, providing a foundation for future advancements in biosensing technologies [55, 57, 72].

HSV-2 Trapping Agents: ZnO-T demonstrate potent anti-HSV-2 effects, suggesting their use as prophylactic and therapeutic agents for preventing and treating HSV-2 infections [108].

In short, ZnO-T holds immense promise for various biomedical applications due to their unique characteristics and multifunctional capabilities. Continued research and interdisciplinary collaboration are essential to overcome current challenges and unlock their full potential for clinical and commercial use.

6 Conclusion

To conclude, represent a significant advancement in nanotechnology, offering versatile applications in biomedical research due to their tunable physicochemical properties, such as size, morphology, and surface charge. Their remarkable antibacterial and antiviral activities, particularly against HSV-1 and HSV-2 (Virus), along with their biocompatibility, make them suitable for applications in wound healing, drug delivery, live vaccination, and food safety. Furthermore, their ability to neutralize pathogens and facilitate biosensing highlights their potential in diagnostics and therapeutics.

Challenges such as limited understanding of their biological properties, toxicity concerns, and regulatory hurdles must be addressed through interdisciplinary collaboration. By overcoming these barriers, ZnO-T could achieve widespread clinical and commercial adoption, transforming fields like nanomedicine, biosensing, and antimicrobial therapy. The future of ZnO-T-based applications appears promising, with continued research paving the way for novel solutions in healthcare and beyond.

References

1. Vayssieres, L.: Growth of arrayed nanorods and nanowires of ZnO from aqueous solutions. Adv. Mater. **15**, 464–466 (2003). https://doi.org/10.1002/adma.200390108
2. Pearton, S.J., Norton, D.P., Ip, K., Heo, Y.W., Steiner, T.: Recent advances in processing of ZnO. J. Vac. Sci. & Technol. B: Microelectron. Nanometer Struct. Process., Meas., Phenom. **22**, 932–948 (2004). https://doi.org/10.1116/1.1714985
3. Pearton, S.J., Norton, D.P., Ip, K., Heo, Y.W., Steiner, T.: RETRACTED: recent progress in processing and properties of ZnO. Prog. Mater. Sci. **50**, 293–340 (2005). https://doi.org/10.1016/j.pmatsci.2004.04.001
4. Klingshirn, C.: ZnO: material, physics and applications. ChemPhysChem **8**, 782–803 (2007). https://doi.org/10.1002/cphc.200700002
5. Schmidt-Mende, L., MacManus-Driscoll, J.L.: ZnO – nanostructures, defects, and devices. Mater. Today **10**, 40–48 (2007). https://doi.org/10.1016/s1369-7021(07)70078-0
6. Wang, Z.L.: ZnO nanowire and nanobelt platform for nanotechnology. Mater. Sci. Eng. R. Rep. **64**, 33–71 (2009). https://doi.org/10.1016/j.mser.2009.02.001
7. Zhang, Q., Dandeneau, C.S., Zhou, X., Cao, G.: ZnO Nanostructures for dye-sensitized solar cells. Adv. Mater. **21**, 4087–4108 (2009). https://doi.org/10.1002/adma.200803827
8. Janotti, A., Van de Walle, C.G.: Fundamentals of zinc oxide as a semiconductor. Rep. Prog. Phys. **72**, 126501 (2009). https://doi.org/10.1088/0034-4885/72/12/126501
9. Wu, J., Xue, D.: Progress of science and technology of ZnO as advanced material. Sci Adv Mat **3**, 127–149 (2011). https://doi.org/10.1166/sam.2011.1144
10. Raha, S., Ahmaruzzaman, Md.: ZnO nanostructured materials and their potential applications: progress, challenges and perspectives. Nanoscale Adv. **4**, 1868–1925 (2022). https://doi.org/10.1039/d1na00880c
11. Clarke, B., Ghandi, K.: the interplay of growth mechanism and properties of ZnO nanostructures for different applications. Small **19** (2023). https://doi.org/10.1002/smll.202302864
12. Haase, H., Ober-Blöbaum, J.L., Engelhardt, G., Hebel, S., Heit, A., Heine, H., Rink, L.: Zinc signals are essential for lipopolysaccharide-induced signal transduction in monocytes. J. Immunol. **181**, 6491–6502 (2008). https://doi.org/10.4049/jimmunol.181.9.6491
13. Akhtar S.: Zinc status in south Asian populations—an update. J Health Popul Nutr **31** (2013). https://doi.org/10.3329/jhpn.v31i2.16378
14. Mishra, Y.K., Kaps, S., Schuchardt, A., Paulowicz, I., Jin, X., et al.: Fabrication of macroscopically flexible and highly porous 3d semiconductor networks from interpenetrating nanostructures by a simple flame transport approach, Part; Part Syst Charact **30**, 775–783 (2013). https://doi.org/10.1002/ppsc.201300197
15. Mishra, Y.K., Modi, G., Cretu, V., Postica, V., Lupan, O., et al.: Direct Growth of freestanding ZnO tetrapod networks for multifunctional applications in photocatalysis, UV photodetection, and gas sensing. ACS Appl. Mater. Interfaces **7**, 14303–14316 (2015). https://doi.org/10.1021/acsami.5b02816
16. Wen, L., Liu, N., Wang, S., Zhang, H., Zhao, W., et al.: Enhancing light emission in flexible AC electroluminescent devices by tetrapod-like zinc oxide whiskers. Opt. Express **24**, 23419 (2016). https://doi.org/10.1364/oe.24.023419
17. Hsieh, G.-W., Ling, S.-R., Hung, F.-T., Kao, P.-H., Liu, J.-B.: Enhanced piezocapacitive response in zinc oxide tetrapod–poly (dimethylsiloxane) composite dielectric layer for flexible and ultrasensitive pressure sensor. Nanoscale **13**, 6076–6086 (2021). https://doi.org/10.1039/d0nr06743a
18. Karthikeyan, L., Suma, D.D., Nair, S.G., Ratnam, P., Mathew, D., Devasia, R.: Zinc oxide tetrapod-based thermally conducting epoxy systems for aerospace applications. Trans Indian Natl. Acad. Eng. **6**, 71–77 (2020). https://doi.org/10.1007/s41403-020-00179-7
19. Qiu, H., Hölken, I., Gapeeva, A., Filiz, V., Adelung, R., Baum, M.: Development and characterization of mechanically durable silicone-polythiourethane composites modified with Tetrapodal shaped ZnO particles for the potential application as fouling-release coating in the marine sector. Materials **11**, 2413 (2018). https://doi.org/10.3390/ma11122413

20. Dai, Q., Pan, P., Zhang, R., Liu, J., Yang, Z., Wei, J., Yuan, Q.: A screen-printed piezoelectric energy harvester using ZnO tetrapod arrays. Mater. Tehnol. **52**, 231–234 (2018). https://doi.org/10.17222/mit.2017.157
21. Siebert, L., Luna-Cerón, E., García-Rivera, L.E., Oh, J., Jang, J., et al.: Light-controlled growth factors release on tetrapodal ZnO-incorporated 3D-printed hydrogels for developing smart wound scaffold. Adv Funct Materials **31** (2021). https://doi.org/10.1002/adfm.202007555
22. Lei, Y., Luo, N., Yan, X., Zhao, Y., Zhang, G., Zhang, Y.: A highly sensitive electrochemical biosensor based on zinc oxide nanotetrapods for l-lactic acid detection. Nanoscale **4**, 3438 (2012). https://doi.org/10.1039/c2nr30334e
23. Mishra, Y.K., Adelung, R.: ZnO tetrapod materials for functional applications. Mater. Today **21**, 631–651 (2018). https://doi.org/10.1016/j.mattod.2017.11.003
24. Yi, Z., Wang, J., Jiang, T., Tang, Q., Cheng, Y.: Photocatalytic degradation of sulfamethazine in aqueous solution using ZnO with different morphologies. R. Soc. Open Sci. **5**, 171457 (2018). https://doi.org/10.1098/rsos.171457
25. Dejneka, A., Aulika, I., Makarova, M.V., Hubicka, Z., Churpita, A., Chvostova, D., et al.: Optical spectra and direct optical transitions in amorphous and crystalline ZnO thin films and powders. J. Electrochem. Soc. **157**, G67 (2010). https://doi.org/10.1149/1.3267882
26. Kitano, M., Hamabe, T., Maeda, S., Okabe, T.: Growth of large tetrapod-like ZnO crystals. J. Cryst. Growth **102**, 965–973 (1990). https://doi.org/10.1016/0022-0248(90)90867-k
27. Zhang, X.-H., Xie, S.-Y., Jiang, Z.-Y., Xie, Z.-X., Huang, R.-B., et al.: Microwave plasma growth and high spatial resolution cathodoluminescent spectrum of tetrapod ZnO nanostructures. J. Solid State Chem. **173**, 109–113 (2003). https://doi.org/10.1016/s0022-4596(03)00099-9
28. Zhou, Z., Deng, H., Yi, J., Liu, S.: A new method for preparation of zinc oxide whiskers. Mater. Res. Bull. **34**, 1563–1567 (1999). https://doi.org/10.1016/s0025-5408(99)00183-x
29. Delaunay, J.-J., Kakoiyama, N., Yamada, I.: Fabrication of three-dimensional network of ZnO tetratpods and its response to ethanol. Mater. Chem. Phys. **104**, 141–145 (2007). https://doi.org/10.1016/j.matchemphys.2007.02.099
30. Wang, B.-B., Xie, J.-J., Yuan, Q., Zhao, Y.-P.: Growth mechanism and joint structure of ZnO tetrapods. J. Phys. D Appl. Phys. **41**, 102005 (2008). https://doi.org/10.1088/0022-3727/41/10/102005
31. Ronning, C., Shang, N.G., Gerhards, I., Hofsäss, H., Seibt, M.: Nucleation mechanism of the seed of tetrapod ZnO nanostructures. J. Appl. Phys. **98** (2005). https://doi.org/10.1063/1.1997290
32. Lupan, O., Chow, L., Chai, G.: A single ZnO tetrapod-based sensor. Sens. Actuators, B Chem. **141**, 511–517 (2009). https://doi.org/10.1016/j.snb.2009.07.011
33. Newton, M.C., Warburton, P.A.: ZnO tetrapod nanocrystals. Mater. Today **10**, 50–54 (2007). https://doi.org/10.1016/S1369-7021(07)70079-2
34. Paulowicz, I., Postica, V., Lupan, O., Wolff, N., Shree, S., Cojocaru, A., Deng, M., Mishra, Y.K., Tiginyanu, I., Kienle, L., Adelung, R.: Zinc oxide nanotetrapods with four different arm morphologies for versatile nanosensors. Sensors Actuators B Chem. **262**, 425–435 (2018). https://doi.org/10.1016/j.snb.2018.01.206
35. Shiojiri, M., Kaito, C.: Structure and growth of ZnO smoke particles prepared by gas evaporation technique. J. Cryst. Growth **52**, 173–177 (1981). https://doi.org/10.1016/0022-0248(81)90189-5
36. Nishio, K., Isshiki, T., Kitano, M., Shiojiri, M.: Structure and growth mechanism of tetrapod-like ZnO particles. Philos. Mag. A **76**, 889–904 (1997). https://doi.org/10.1080/01418619708214216
37. Fujii, M., Iwanaga, H., Ichihara, M., Takeuchi, S.: Structure of tetrapod-like ZnO crystals. J. Cryst. Growth **128**, 1095–1098 (1993). https://doi.org/10.1016/s0022-0248(07)80104-7
38. Iwanaga, H., Fujii, M., Takeuchi, S.: Growth model of tetrapod zinc oxide particles. J. Cryst. Growth **134**, 275–280 (1993). https://doi.org/10.1016/0022-0248(93)90136-k
39. Dai, Y., Zhang, Y., Wang, Z.L.: The octa-twin tetraleg ZnO nanostructures. Solid State Commun. **126**, 629–633 (2003). https://doi.org/10.1016/s0038-1098(03)00277-1

40. Sirbuly, D.J., Law, M., Yan, H., Yang, P.: Semiconductor nanowires for subwavelength photonics integration. J. Phys. Chem. B **109**, 15190–15213 (2005). https://doi.org/10.1021/jp051813i
41. Huang, M.H., Mao, S., Feick, H., Yan, H., Wu, Y., Kind, H., Weber, E., Russo, R., Yang, P.: Room-temperature ultraviolet nanowire nanolasers. Science **292**, 1897–1899 (2001). https://doi.org/10.1126/science.1060367
42. Fischer, A.M., Srinivasan, S., Garcia, R., Ponce, F.A., Guaño, S.E., et al.: Optical properties of highly luminescent zinc oxide tetrapod powders. Appl. Phys. Lett. **91** (2007). https://doi.org/10.1063/1.2787875
43. Dai, Y., Zhang, Y., Li, Q.K., Nan, C.W.: Synthesis and optical properties of tetrapod-like zinc oxide nanorods. Chem. Phys. Lett. **358**, 83–86 (2002). https://doi.org/10.1016/s0009-2614(02)00582-1
44. Han, J., Zhu, Z., Ray, S., Azad, A.K., Zhang, W., He, M., Li, S., Zhao, Y.: Optical and dielectric properties of ZnO tetrapod structures at terahertz frequencies. Appl. Phys. Lett. **89** (2006). https://doi.org/10.1063/1.2222329
45. Vanheusden, K., Warren, W.L., Seager, C.H., Tallant, D.R., Voigt, J.A., Gnade, B.E.: Mechanisms behind green photoluminescence in ZnO phosphor powders. J. Appl. Phys. **79**, 7983–7990 (1996). https://doi.org/10.1063/1.362349
46. Wang, Y.W., Zhang, L.D., Wang, G.Z., Peng, X.S., Chu, Z.Q., Liang, C.H.: Catalytic growth of semiconducting zinc oxide nanowires and their photoluminescence properties. J. Cryst. Growth **234**, 171–175 (2002). https://doi.org/10.1016/s0022-0248(01)01661-x
47. Newton, M.C., Firth, S., Matsuura, T., Warburton, P.A.: Synthesis and characterisation of zinc oxide tetrapod nanocrystals. J. Phys. Conf. Ser. **26**, 251–255 (2006). https://doi.org/10.1088/1742-6596/26/1/060
48. Newton, M.C., Firth, S., Warburton, P.A.: ZnO tetrapod Schottky photodiodes. Appl. Phys. Lett. **89** (2006). https://doi.org/10.1063/1.2335949
49. Kwok, W.M., Djurišić, A.B., Leung, Y.H., Chan, W.K., Phillips, D.L.: Time-resolved photoluminescence from ZnO nanostructures. Appl. Phys. Lett. **87** (2005). https://doi.org/10.1063/1.2137456
50. Djurišić, A.B., Kwok, W.M., Leung, Y.H., Chan, W.K., Phillips, D.L., et al.: Ultrafast spectroscopy of stimulated emission in single ZnO tetrapod nanowires. Nanotechnology **17**, 244–249 (2005). https://doi.org/10.1088/0957-4484/17/1/041
51. Leung, Y.H., Kwok, W.M., Djurišić, A.B., Phillips, D.L., Chan, W.K.: Time-resolved study of stimulated emission in ZnO tetrapod nanowires. Nanotechnology **16**, 579–582 (2005). https://doi.org/10.1088/0957-4484/16/4/040
52. Roduner, E.: Size matters: why nanomaterials are different. Chem. Soc. Rev. **35**, 583 (2006). https://doi.org/10.1039/b502142c
53. Heo, Y.W., Tien, L.C., Norton, D.P., Pearton, S.J., Kang, B.S., et al.: Pt / ZnO nanowire Schottky diodes. Appl. Phys. Lett. **85**, 3107–3109 (2004). https://doi.org/10.1063/1.1802372
54. Wan, Q., Yu, K., Wang, T.H., Lin, C.L.: Low-field electron emission from tetrapod-like ZnO nanostructures synthesized by rapid evaporation. Appl. Phys. Lett. **83**, 2253–2255 (2003). https://doi.org/10.1063/1.1612899
55. Myndrul, V., Coy, E., Babayevska, N., Zahorodna, V., Balitskyi, V., et al.: MXene nanoflakes decorating ZnO tetrapods for enhanced performance of skin-attachable stretchable enzymatic electrochemical glucose sensor. Biosens. Bioelectron. **207**, 114141 (2022). https://doi.org/10.1016/j.bios.2022.114141
56. Nie, L., Gao, L., Yan, X., Wang, T.: Functionalized tetrapod-like ZnO nanostructures for plasmid DNA purification, polymerase chain reaction and delivery. Nanotechnology **18**, 015101 (2006). https://doi.org/10.1088/0957-4484/18/1/015101
57. Terracciano, M., Račkauskas, S., Falanga, A.P., Martino, S., Chianese, G., et al.: ZnO Tetrapods for label-free optical biosensing: physicochemical characterization and functionalization strategies. IJMS **24**, 4449 (2023). https://doi.org/10.3390/ijms24054449
58. Lei, Y., Liu, X., Yan, X., Song, Y., Kang, Z., et al.: Multicenter uric acid biosensor based on tetrapod-shaped ZnO nanostructures. J. Nanosci. Nanotech. **12**, 513–518 (2012). https://doi.org/10.1166/jnn.2012.5336

59. Kim, J.Y., Jo, S.-Y., Sun, G.-J., Katoch, A., Choi, S.-W., Kim, S.S.: Tailoring the surface area of ZnO nanorods for improved performance in glucose sensors. Sens. Actuators, B Chem. **192**, 216–220 (2014). https://doi.org/10.1016/j.snb.2013.10.113
60. Yang, Y., Yang, H., Yang, M., liu, Y., Shen, G., Yu, R.: Amperometric glucose biosensor based on a surface treated nanoporous ZrO_2/Chitosan composite film as immobilization matrix. Anal. Chim. Acta **525** 213–220 (2004). https://doi.org/10.1016/j.aca.2004.07.071
61. Lei, Y., Yan, X., Luo, N., Song, Y., Zhang, Y.: ZnO nanotetrapod network as the adsorption layer for the improvement of glucose detection via multiterminal electron-exchange. Colloids Surf. Physicochem. Eng. Asp. **361**, 169–173 (2010). https://doi.org/10.1016/j.colsurfa.2010.03.029
62. Myndrul, V., Iatsunskyi, I., Babayevska, N., Jarek, M., Jesionowski, T.: Effect of electrode modification with chitosan and nafion on the efficiency of real-time enzyme glucose biosensors based on ZnO tetrapods. Materials **15**, 4672 (2022). https://doi.org/10.3390/ma15134672
63. Zhao, W., Zhao, Y., Karlsson, M., Wang, Q., Toprak, M.S.: Evaluation of zinc oxide nano-microtetrapods for biomolecule sensing applications. SPIE Proceedings **9668**, 966833 (2015). https://doi.org/10.1117/12.2202529
64. Terracciano, M., Rea, I., Borbone, N., Moretta, R., Oliviero, G., Piccialli, G., De Stefano, L.: Porous silicon-based aptasensors: the next generation of label-free devices for health monitoring. Molecules **24**, 2216 (2019). https://doi.org/10.3390/molecules24122216
65. Chianese, G., Terracciano, M., Moretta, R., Cappiello, P., Vitiello, G., et al.: Synthesis and surface modification of nanostructured F-doped ZnO: toward a transducer for label-free optical biosensing. Appl. Sci. **9**, 3380 (2019). https://doi.org/10.3390/app9163380
66. Terracciano, M., Galstyan, V., Rea, I., Casalino, M., De Stefano, L., Sbervegleri, G.: Chemical modification of TiO_2 nanotube arrays for label-free optical biosensing applications. Appl. Surf. Sci. **419**, 235–240 (2017). https://doi.org/10.1016/j.apsusc.2017.05.029
67. Moretta, R., De Stefano, L., Terracciano, M., Rea, I.: Porous silicon optical devices: recent advances in biosensing applications. Sensors **21**, 1336 (2021). https://doi.org/10.3390/s21041336
68. Abebe, B., Zereffa, E.A., Tadesse, A., Murthy, H.C.A.: A review on enhancing the antibacterial activity of ZnO: mechanisms and microscopic investigation. Nanoscale Res. Lett. **15** (2020). https://doi.org/10.1186/s11671-020-03418-6
69. Mardosaitė, R., Jurkevičiūtė, A., Račkauskas, S.: Superhydrophobic ZnO nanowires: wettability mechanisms and functional applications. Cryst. Growth Des. **21** 4765–4779 (2021). https://doi.org/10.1021/acs.cgd.1c00449
70. Rackauskas, S., Barbero, N., Barolo, C., Viscardi, G.: ZnO nanowire application in chemoresistive sensing: a review. Nanomaterials **7**, 381 (2017). https://doi.org/10.3390/nano7110381
71. Viter, R., Tepliakova, I., Drobysh, M., Zbolotnii, V., Rackauskas, S., et al.: Photoluminescence-based biosensor for the detection of antibodies against SARS-CoV-2 virus proteins by ZnO tetrapod structure integrated within microfluidic system. Sci. Total. Environ. **939**, 173333 (2024). https://doi.org/10.1016/j.scitotenv.2024.173333
72. Chauhan, N., Gupta, S., Avasthi, D.K., Adelung, R., Mishra, Y.K., Jain, U.: Zinc oxide tetrapods based biohybrid interface for voltammetric sensing ofhelicobacter pylori. ACS Appl. Mater. Interfaces **10**, 30631–30639 (2018). https://doi.org/10.1021/acsami.8b08901
73. Brás, M., Zanoni, J., Falcão, B.P., Leitão, J.P., Costa, F.M., et al.: Label-free nanoscale ZnO tetrapod-based transducers for tetracycline detection. ACS Appl. Nano Mater. **5**, 1232–1243 (2022). https://doi.org/10.1021/acsanm.1c03838
74. Rodrigues, J., Pereira, S.O., Santos, N.F., Rodrigues, C., Costa, F.M., Monteiro, T.: Insights on luminescence quenching of ZnO tetrapods in the detection of hCG. Appl. Surf. Sci. **527**, 146813 (2020). https://doi.org/10.1016/j.apsusc.2020.146813
75. Augustine, R., Malik, H.N., Singhal, D.K., Mukherjee, A., Malakar, D., Kalarikkal, N., Thomas, S.: Electrospun polycaprolactone/ZnO nanocomposite membranes as biomaterials with antibacterial and cell adhesion properties. J Polym Res **21** (2014). https://doi.org/10.1007/s10965-013-0347-6

76. Nasajpour, A., Mandla, S., Shree, S., Mostafavi, E., Sharifi, R., et al.: Nanostructured fibrous membranes with rose spike-like architecture. Nano Lett. **17**, 6235–6240 (2017). https://doi.org/10.1021/acs.nanolett.7b02929
77. Rackauskas, S., Klimova, O., Jiang, H., Nikitenko, A., Chernenko, K.A., et al.: A novel method for continuous synthesis of ZnO tetrapods. J. Phys. Chem. C **119**, 16366–16373 (2015). https://doi.org/10.1021/acs.jpcc.5b03702
78. Büter, A., Maschkowitz, G., Baum, M., Mishra, Y.K., Siebert, L., Adelung, R., Fickenscher, H.: Antibacterial activity of nanostructured zinc oxide tetrapods. IJMS **24**, 3444 (2023). https://doi.org/10.3390/ijms24043444
79. Si, Y., Zhang, Z., Wu, W., Fu, Q., Huang, K., Nitin, N., Ding, B., Sun, G.: Daylight-driven rechargeable antibacterial and antiviral nanofibrous membranes for bioprotective applications. Sci. Adv. **4** (2018). https://doi.org/10.1126/sciadv.aar5931
80. Hayyan, M., Hashim, M.A., Alnashef, I.M.: Superoxide ion: generation and chemical implications. Chem. Rev. **116**, 3029–3085 (2016). https://doi.org/10.1021/acs.chemrev.5b00407
81. Kumar, V., Prakash, J., Singh, J.P., Chae, K.H., Swart, C., Ntwaeaborwa, O.M., Swart, H.C., Dutta, V.: Role of silver doping on the defects related photoluminescence and antibacterial behaviour of zinc oxide nanoparticles. Colloids Surf. B Biointerfaces **159**, 191–199 (2017). https://doi.org/10.1016/j.colsurfb.2017.07.071
82. Mendes, C.R., Dilarri, G., Forsan, C.F., Sapata, V.D.M.R., Lopes, P.R.M., et al.: Antibacterial action and target mechanisms of zinc oxide nanoparticles against bacterial pathogens. Sci Rep **12** (2022). https://doi.org/10.1038/s41598-022-06657-y
83. Olejnik, M., Kersting, M., Rosenkranz, N., Loza, K., Breisch, M., et al.: Cell-biological effects of zinc oxide spheres and rods from the nano-to the microscale at sub-toxic levels. Cell Biol. Toxicol. **37**, 573–593 (2020). https://doi.org/10.1007/s10565-020-09571-z
84. Yougbaré, S., Mutalik, C., Okoro, G., Lin, I.-H., Krisnawati, D.I., Jazidie, A., Nuh, M., Chang, C.-C., Kuo, T.-R.: Emerging trends in nanomaterials for antibacterial applications. IJN **16**, 5831–5867 (2021). https://doi.org/10.2147/ijn.s328767
85. Prakash, J., Krishna, S.B.N., Kumar, P., Kumar, V., Ghosh, K.S., et al.: Recent advances on metal oxide based nano-photocatalysts as potential antibacterial and antiviral agents. Catalysts **12**, 1047 (2022). https://doi.org/10.3390/catal12091047
86. Raj, N.B., PavithraGowda, N.T., Pooja, O.S., Purushotham, B., Kumar, M.R.A., Sukrutha, S.K., Ravikumar, C.R., Nagaswarupa, H.P., Murthy, H.C.A., Boppana, S.B.: Harnessing ZnO nanoparticles for antimicrobial and photocatalytic activities. J Photochem Photobiol **6**, 100021 (2021). https://doi.org/10.1016/j.jpap.2021.100021
87. Kumar, R.S., Dananjaya, S.H.S., De Zoysa, M., Yang, M.: Enhanced antifungal activity of Ni-doped ZnO nanostructures under dark conditions. RSC Adv. **6**, 108468–108476 (2016). https://doi.org/10.1039/c6ra18442a
88. Vijayalakshmi, K., Sivaraj, D.: Enhanced antibacterial activity of Cr doped ZnO nanorods synthesized using microwave processing. RSC Adv. **5**, 68461–68469 (2015). https://doi.org/10.1039/c5ra13375k
89. Kaushik, N., Mitra, S., Baek, E.J., Nguyen, L.N., Bhartiya, P., et al.: The inactivation and destruction of viruses by reactive oxygen species generated through physical and cold atmospheric plasma techniques: current status and perspectives. J. Adv. Res. **43**, 59–71 (2023). https://doi.org/10.1016/j.jare.2022.03.002
90. Sharmila, G., Muthukumaran, C., Sandiya, K., Santhiya, S., Pradeep, R.S., Kumar, N.M., Suriyanarayanan, N., Thirumarimurugan, M.: Biosynthesis, characterization, and antibacterial activity of zinc oxide nanoparticles derived from Bauhinia tomentosa leaf extract. J Nanostruct Chem **8**, 293–299 (2018). https://doi.org/10.1007/s40097-018-0271-8
91. Kumar, A., Soni, V., Singh, P., Khan, A.A.P., Nazim, M., et al.: Green aspects of photocatalysts during corona pandemic: a promising role for the deactivation of COVID-19 virus. RSC Adv. **12**, 13609–13627 (2022). https://doi.org/10.1039/d1ra08981a
92. Wahab, R., Kaushik, N., Khan, F., Kaushik, N.K., Choi, E.H., Musarrat, J., Al-Khedhairy, A.A.: Self-styled ZnO nanostructures promotes the cancer cell damage and supresses the epithelial phenotype of glioblastoma. Sci. Rep. **6** (2016). https://doi.org/10.1038/srep19950

93. Nasajpour, A., Ansari, S., Rinoldi, C., Rad, A.S., Aghaloo, T., et al.: A multifunctional polymeric periodontal membrane with osteogenic and antibacterial characteristics. Adv. Funct. Materials **28** (2017). https://doi.org/10.1002/adfm.201703437
94. Haraguchi, Y., Sakurai, H., Hussain, S., Anner, B.M., Hoshino, H.: Inhibition of HIV-1 infection by zinc group metal compounds. Antiviral Res. **43**, 123–133 (1999). https://doi.org/10.1016/s0166-3542(99)00040-6
95. Krenn, B.M., Gaudernak, E., Holzer, B., Lanke, K., Van Kuppeveld, F.J.M., Seipelt, J.: Antiviral activity of the zinc ionophores pyrithione and hinokitiol against picornavirus infections. J. Virol. **83**, 58–64 (2009). https://doi.org/10.1128/jvi.01543-08
96. te Velthuis, A.J.W., van den Worm, S.H.E., Sims, A.C., Baric, R.S., Snijder, E.J., van Hemert, M.J.: Zn^{2+} inhibits coronavirus and arterivirus RNA polymerase activity in vitro and zinc ionophores block the replication of these viruses in cell culture. PLoS Pathog. **6**, e1001176 (2010). https://doi.org/10.1371/journal.ppat.1001176
97. Anang, S., Kaushik, N., Hingane, S., Kumari, A., Gupta, J., Asthana, S., Shalimar, Nayak, B., Ranjith-Kumar, C.T., Surjit, M.: Potent inhibition of hepatitis E Virus release by a cyclic peptide inhibitor of the interaction between viral open reading frame 3 protein and host tumor susceptibility gene 101. J Virol **92** (2018). https://doi.org/10.1128/jvi.00684-18
98. Asif, N., Amir, M., Fatma, T.: Recent advances in the synthesis, characterization and biomedical applications of zinc oxide nanoparticles. Bioprocess Biosyst. Eng. **46**, 1377–1398 (2023). https://doi.org/10.1007/s00449-023-02886-1
99. Singh, N., Prakash, J., Gupta, R.K.: Design and engineering of high-performance photocatalytic systems based on metal oxide–graphene–noble metal nanocomposites. Mol. Syst. Des. Eng. **2**, 422–439 (2017). https://doi.org/10.1039/c7me00038c
100. Lallo da Silva, B., Caetano, B.L., Chiari-Andréo, B.G., Pietro, R.C.L.R., Chiavacci, L.A.: Increased antibacterial activity of ZnO nanoparticles: Influence of size and surface modification. Colloids Surf. B **177**, 440–447 (2019). https://doi.org/10.1016/j.colsurfb.2019.02.013
101. Milionis, A., Tripathy, A., Donati, M., Sharma, C.S., Pan, F., Maniura-Weber, K., Ren, Q., Poulikakos, D.: Water-based scalable methods for self-cleaning antibacterial ZnO-nanostructured surfaces. Ind. Eng. Chem. Res. **59**, 14323–14333 (2020). https://doi.org/10.1021/acs.iecr.0c01998
102. Gupta, J., Irfan, M., Ramgir, N., Muthe, K.P., Debnath, A.K., Ansari, S., Gandhi, J., Ranjith-Kumar, C.T., Surjit, M.: Antiviral activity of zinc oxide nanoparticles and tetrapods against the hepatitis E and hepatitis C viruses. Front. Microbiol. **13** (2022). https://doi.org/10.3389/fmicb.2022.881595
103. Kang, S., Yang, M., Hong, Z., Zhang, L., Huang, Z., et al.: Crystal structure of SARS-CoV-2 nucleocapsid protein RNA binding domain reveals potential unique drug targeting sites. Acta Pharm. Sin. B **10**, 1228–1238 (2020). https://doi.org/10.1016/j.apsb.2020.04.009
104. Cubuk, J., Alston, J.J., Incicco, J.J., Singh, S., Stuchell-Brereton, M.D., et al.: The SARS-CoV-2 nucleocapsid protein is dynamic, disordered, and phase separates with RNA. Nat Commun **12** (2021). https://doi.org/10.1038/s41467-021-21953-3
105. Zeng, W., Liu, G., Ma, H., Zhao, D., Yang, Y., et al.: Biochemical characterization of SARS-CoV-2 nucleocapsid protein. Biochem. Biophys. Res. Commun. **527**, 618–623 (2020). https://doi.org/10.1016/j.bbrc.2020.04.136
106. Sportelli, M.C., Izzi, M., Loconsole, D., Sallustio, A., Picca, R.A., Felici, R., Chironna, M., Cioffi, N.: On the efficacy of ZnO nanostructures against SARS-CoV-2. IJMS **23**, 3040 (2022). https://doi.org/10.3390/ijms23063040
107. Mishra, Y.K., Adelung, R., Röhl, C., Shukla, D., Spors, F., Tiwari, V.: Virostatic potential of micro–nano filopodia-like ZnO structures against herpes simplex virus-1. Antivir. Res. **92**, 305–312 (2011). https://doi.org/10.1016/j.antiviral.2011.08.017
108. Antoine, T.E., Mishra, Y.K., Trigilio, J., Tiwari, V., Adelung, R., Shukla, D.: Prophylactic, therapeutic and neutralizing effects of zinc oxide tetrapod structures against herpes simplex virus type-2 infection. Antivir. Res. **96**, 363–375 (2012). https://doi.org/10.1016/j.antiviral.2012.09.020

109. Duggal, N., Jaishankar, D., Yadavalli, T., Hadigal, S., Mishra, Y.K., Adelung, R., Shukla, D.: Zinc oxide tetrapods inhibit herpes simplex virus infection of cultured corneas. Mol. Vis. **23**, 26–38 (2017)
110. Antoine, T.E., Hadigal, S.R., Yakoub, A.M., Mishra, Y.K., Bhattacharya, P., Haddad, C., Valyi-Nagy, T., Adelung, R., Prabhakar, B.S., Shukla, D.: Intravaginal zinc oxide tetrapod nanoparticles as novel immunoprotective agents against genital herpes. J. Immunol. **196**, 4566–4575 (2016). https://doi.org/10.4049/jimmunol.1502373
111. Pormohammad, A., Monych, N., Turner, R.: Zinc and SARS-CoV-2: a molecular modeling study of Zn interactions with RNA-dependent RNA-polymerase and 3C-like proteinase enzymes. Int. J. Mol. Med. **47**, 326–334 (2020). https://doi.org/10.3892/ijmm.2020.4790
112. Godfrey, H.R., Godfrey, N.J., Godfrey, J.C., Riley, D.: A randomized clinical trial on the treatment of oral herpes with topical zinc oxide/glycine. Altern. Ther. Health Med. **7**, 49–56 (2001). http://www.ncbi.nlm.nih.gov/pubmed/11347285
113. Singh, B.B., Udani, J., Vinjamury, S.P., Der-Martirosian, C., Gandhi, S., Khorsan, R., Nanjegowda, D., Singh, V.: Safety and effectiveness of an L-lysine, zinc, and herbal-based product on the treatment of facial and circumoral herpes. Altern. Med. Rev. **10**, 123–7 (2005). http://www.ncbi.nlm.nih.gov/pubmed/15989381
114. Agelidis, A., Koujah, L., Suryawanshi, R., Yadavalli, T., Mishra, Y.K., Adelung, R., Shukla, D.: An intra-vaginal zinc oxide tetrapod nanoparticles (ZOTeN) and genital herpesvirus cocktail can provide a novel platform for live virus vaccine. Front. Immunol. **10** (2019). https://doi.org/10.3389/fimmu.2019.00500
115. McPherson, S.W., Keunen, J.E., Bird, A.C., Chew, E.Y., van Kuijk, F.J.: Investigate oral zinc as a prophylactic treatment for those at risk for COVID-19. Am. J. Ophthalmol. **216**, A5–A6 (2020). https://doi.org/10.1016/j.ajo.2020.04.028
116. Hong, C.-H., Arosemena, A., Zurakowski, D., Ayyala, R.S.: Glaucoma drainage devices: a systematic literature review and current controversies. Surv. Ophthalmol. **50**, 48–60 (2005). https://doi.org/10.1016/j.survophthal.2004.10.006
117. Tomasek, J.J., Gabbiani, G., Hinz, B., Chaponnier, C., Brown, R.A.: Myofibroblasts and mechano-regulation of connective tissue remodelling. Nat. Rev. Mol. Cell Biol. **3**, 349–363 (2002). https://doi.org/10.1038/nrm809
118. Lemoinne, S., Cadoret, A., El Mourabit, H., Thabut, D., Housset, C.: Origins and functions of liver myofibroblasts. Biochim. Biophys. Acta–Mol. Basis Dis. **2013**, 948–954 (1832). https://doi.org/10.1016/j.bbadis.2013.02.019
119. Desmoulière, A., Redard, M., Darby, I., Gabbiani, G.: Apoptosis mediates the decrease in cellularity during the transition between granulation tissue and scar. Am. J. Pathol. **146** (1995) 56–66. http://www.ncbi.nlm.nih.gov/pubmed/7856739
120. Darby, I.A., Hewitson, T.D.: Fibroblast differentiation in wound healing and fibrosis. 143–179 (2007). https://doi.org/10.1016/S0074-7696(07)57004-X
121. Heckmann, M., Krieg, T.: Biological and pharmacological modulations of fibroblast functions. Skin Pharmacol. Physiol. **2**, 125–137 (1989). https://doi.org/10.1159/000210810
122. Sonntag, S.R., Gniesmer, S., Gapeeva, A., Adelung, R., Cojocaru, A., Mishra, Y.K., Kaps, S., Tura, A., Grisanti, S., Grisanti, S., Nassar, K.: Zinc oxide tetrapods modulate wound healing and cytokine release in vitro—a new antiproliferative substance in glaucoma filtering surgery. Life. **12**, 1691 (2022). https://doi.org/10.3390/life12111691
123. Cleetus, C.M., Alvarez Primo, G., Fregoso, N. L., Raveendran, J.C., Noveron et al.: Alginate hydrogels with embedded ZnO nanoparticles for wound healing therapy. Int. J. Nanomedicine **15** 5097–5111 (2020). https://doi.org/10.2147/IJN.S255937
124. Khorasani, M.T., Joorabloo, A., Adeli, H., Milan, P.B., Amoupour, M.: Enhanced antimicrobial and full-thickness wound healing efficiency of hydrogels loaded with heparinized ZnO nanoparticles: In vitro and in vivo evaluation. Int. J. Biol. Macromol. **166**, 200–212 (2021). https://doi.org/10.1016/j.ijbiomac.2020.10.142
125. Nasajpour, A., Samandari, M., Patil, C.D., Abolhassani, R., Suryawanshi, R.K., et al.: Nanoengineered antiviral fibrous arrays with rose-thorn-inspired architectures. ACS Mater. Lett. **3**, 1566–1571 (2021). https://doi.org/10.1021/acsmaterialslett.1c00419

126. Schadte, P., Rademacher, F., Andresen, G., Hellfritzsch, M., Qiu, H., et al.: 3D-printed wound dressing platform for protein administration based on alginate and zinc oxide tetrapods. Nano Converg. **10**, 53 (2023). https://doi.org/10.1186/s40580-023-00401-6
127. Gautam, A., Singhal, C., Mishra, A., John, A.T., Gautam, U.K., et al.: Detection of prostate cancer DNA using tetrapods based disposable paper ecofriendly biosensor device. Med. Devices Sens. **3** (2020). https://doi.org/10.1002/mds3.10122
128. Krishna, S.B.N., Jakmunee, J., Mishra, Y.K., Prakash, J.: ZnO based 0–3D diverse nano-architectures, films and coatings for biomedical applications. J. Mater. Chem. B. **12**, 2950–2984 (2024). https://doi.org/10.1039/D4TB00184B
129. Suwannasing, C., Suwannasom, N., Soonthornchookiat, P., Srisai, P., Pattaweerakul, C., et al.: The potential of HSA-stabilized zinc oxide nanoparticles as radiosensitizers to enhance the cytotoxic effects and radiosensitivity of cervical cancer cells. Cancer Nanotechnol **15**, 59 (2024). https://doi.org/10.1186/s12645-024-00298-8
130. Papavlassopoulos, H., Mishra, Y.K., Kaps, S., Paulowicz, I., Abdelaziz, R., et al.: Toxicity of functional nano-micro zinc oxide tetrapods: impact of cell culture conditions, cellular age and material properties. PLoS One **9** e84983 (2014). https://doi.org/10.1371/journal.pone.0084983
131. Florece, K.N.P., Bondoc, J.M.R., Espejo, A.A.R., Felix, B.F.P., Liberato, L.M.A., et al.: Cytotoxic activity of zinc oxide nanoparticles against cancer cells: a systematic review. Asian J. Biol. Life Sci. **11** 237–248 (2022). https://doi.org/10.5530/ajbls.2022.11.33
132. Joshi, M., Bhatt, P.: Deciphering the anticancer activity of biocompatible zinc oxide nanoparticles loaded with methotrexate on breast cancer cells. Bull. Mater. Sci. **46**, 192 (2023). https://doi.org/10.1007/s12034-023-03044-9
133. Roy, R., Das, M., Dwivedi, P.D.: Toxicological mode of action of ZnO nanoparticles: impact on immune cells. Mol. Immunol. **63**, 184–192 (2015). https://doi.org/10.1016/j.molimm.2014.08.001
134. Duggal, N., Jaishankar, D., Yadavalli, T., Hadigal, S., Mishra, Y.K., Adelung, R., Shukla, D.: Zinc oxide tetrapods inhibit herpes simplex virus infection of cultured corneas. Mol. Vis. **23** 26–38 (2017). http://www.ncbi.nlm.nih.gov/pubmed/28275313
135. Masud, R.A., Islam, M.S., Haque, P., Khan, M.N.I., Shahruzzaman, M., Khan, M., Takafuji, M., Rahman, M.M.: Preparation of novel chitosan/poly (ethylene glycol)/ZnO bionanocomposite for wound healing application: effect of gentamicin loading. Materialia **12**, 100785 (2020). https://doi.org/10.1016/j.mtla.2020.100785
136. Hsiao, I.-L., Huang, Y.-J.: Titanium oxide shell coatings decrease the cytotoxicity of ZnO nanoparticles. Chem. Res. Toxicol. **24**, 303–313 (2011). https://doi.org/10.1021/tx1001892
137. Bui, V., Park, D., Lee, Y.-C.: Chitosan combined with ZnO, TiO_2 and Ag nanoparticles for antimicrobial wound healing applications: a mini review of the research trends. Polymers (Basel) **9**, 21 (2017). https://doi.org/10.3390/polym9010021
138. Cao, D., Shu, X., Zhu, D., Liang, S., Hasan, M., Gong, S.: Lipid-coated ZnO nanoparticles synthesis, characterization and cytotoxicity studies in cancer cell. Nano Converg. **7**, 14 (2020). https://doi.org/10.1186/s40580-020-00224-9
139. Ancona, A., Dumontel, B., Garino, N., Demarco, B., Chatzitheodoridou, D., Fazzini, W., Engelke, H., Cauda, V.: Lipid-coated zinc oxide nanoparticles as innovative ROS-generators for photodynamic therapy in cancer cells. Nanomaterials **8**, 143 (2018). https://doi.org/10.3390/nano8030143
140. Chia, S.L., Leong, D.T.: Reducing ZnO nanoparticles toxicity through silica coating. Heliyon. **2**, e00177 (2016). https://doi.org/10.1016/j.heliyon.2016.e00177
141. Sotiriou, G.A., Watson, C., Murdaugh, K.M., Darrah, T.H., Pyrgiotakis, G., Elder, A., Brain, J.D., Demokritou, P.: Engineering safer-by-design silica-coated ZnO nanorods with reduced DNA damage potential. Environ. Sci. Nano **1**, 144 (2014). https://doi.org/10.1039/c3en00062a
142. Singh, J., Gill, P.S., Vashihth, A., Verma, N.K.: Room temperature ferromagnetism in Mg-doped ZnO nanoparticles. p. 050192. (2015). https://doi.org/10.1063/1.4917833

143. Xia, T., Zhao, Y., Sager, T., George, S., Pokhrel, S., et al.: Decreased dissolution of ZnO by iron doping yields nanoparticles with reduced toxicity in the rodent lung and zebrafish embryos. ACS Nano **5**, 1223–1235 (2011). https://doi.org/10.1021/nn1028482
144. Guo, B.-L., Han, P., Guo, L.-C., Cao, Y.-Q., Li, A.-D., Kong, J.-Z., Zhai, H.-F., Wu, D.: The antibacterial activity of Ta-doped ZnO nanoparticles. Nanoscale Res. Lett. **10**, 336 (2015). https://doi.org/10.1186/s11671-015-1047-4
145. Arooj, S., Nazir, S., Nadhman, A., Ahmad, N., Muhammad, B., Ahmad, I., Mazhar, K., Abbasi, R.: Novel ZnO: Ag nanocomposites induce significant oxidative stress in human fibroblast malignant melanoma (Ht144) cells. Beilstein J. Nanotechnol. **6**, 570–582 (2015). https://doi.org/10.3762/bjnano.6.59
146. Li, Z., Yang, R., Yu, M., Bai, F., Li, C., Wang, Z.L.: Cellular level biocompatibility and biosafety of ZnO nanowires. J. Phys. Chem. C **112**, 20114–20117 (2008). https://doi.org/10.1021/jp808878p

Jayalakshmi Vallamkondu , is currently an Associate Professor in the Dept. of Physics, NIT-Warangal. She had postdoc experience from Queen's Univ., Canada and Georgia Institute of Tech., USA. She completed Ph.D. (2010) at the Centre for Nano and Soft Matter Sciences, Bangalore following her undergraduate and postgraduate at SV Univ., India with first-class distinctions. Her research career is distinguished by contributions to the field of materials science, including over 50 international journal papers and a significant presence in conference proceedings. Her research is supported by high-profile grants from SERB, DBT, and SPARC, Govt. of India, focusing on topics like synthesis of nematic and chiral nematic emulsions, development of technologies for detecting Alzheimer's biomarkers using liquid crystals and ZnO tetrapods for novel applications. Alongside her research, she is actively involved in academic and administrative roles, significantly contributing to curriculum development, student guidance, member of academic committees, organizing workshops and educational programs, enhancing the learning environment and research capabilities at NIT Warangal.

Statistical Physics View on the Dynamics and Self-Organization of Tetrapods

Vrinda Garg and Surya K. Ghosh

1 Introduction

The modeling and synthesis of tetrapods with precise, well-defined shapes has sparked significant interest due to their potential for forming self-assembled structures. The exceptional stability and porosity of tetrapods, like ZnO and DNA hydrogels, facilitate studying molecular and self-assembly processes through tetrapod-based model systems. Surface functionalization [1, 2], either on the entire surface or tip-specific, has expanded its applications in biosensing, controlled drug delivery, bioprinting, tissue engineering, electronics, lasers, and 3D cell culture [3, 4].

ZnO, a direct wide-bandgap metal oxide semiconductor, is a remarkable material for nano- and micro-structuring applications. Its exceptional physical robustness, characterized by a high Young's modulus (~150 GPa) and stability at high temperatures (~1800 °C), makes it ideal for modeling tetrapods [5]. These properties enable ZnO-Tetrapods (ZnO-T) to maintain structural integrity under diverse environmental conditions, providing a reliable platform for studying self-assembly and structural dynamics in 2D and 3D systems.

The incredible structure of tetrapods can be utilized to study the polymer network-like structure mimicking DNA hydrogels. DNA nanotechnology originated in the 1980s when Ned Seeman at NYU designed a stationary four-way DNA junction resembling a tetrapod. This innovation laid the groundwork for structural DNA nanotechnology, enabling the construction of synthetic 2D and 3D DNA nanostructures. The advent of the DNA origami technique revolutionized the field, allowing the fabrication of larger and more intricate designs for studying self-assembled structures [6]. Since then, DNA nanotechnology has rapidly evolved, leading to a broad spectrum of applications in nanomedicine, nanoelectronics, synthetic biology, and nanorobotics [2, 7, 8]. Building on these advancements, this chapter explores

V. Garg · S. K. Ghosh (✉)
Department of Physics, National Institute of Technology, Warangal, Telangana, India
e-mail: skghosh@nitw.ac.in

K. Thangaraju et al. (eds.), *ZnO Tetrapods*, Engineering Materials,
https://doi.org/10.1007/978-981-95-1614-8_5

various statistical models of tetrapods, ranging from ZnO to DNA hydrogels. Tetrapods exhibit unique stability due to their symmetrical geometry, fully accessible surface area on each arm, high porosity arising from self-organization, and uniform force distribution. To elucidate the dynamics of these systems, we discuss several modeling approaches and coarse-grained models of tetrahedral particles, with four arms extending from a central core, in both 2D and 3D environments in Sect. 4. The dynamics of tetrapod models are examined under inert and non-inert conditions, both with and without gravitational influence. Furthermore, we investigate the impact of physical properties such as shape, size, morphology, arm length, and hydrophobic ratio in Sect. 5. The integration of machine learning and artificial intelligence in modeling tetrapod structures is addressed in Sect. 6. Finally, Sect. 7 presents perspectives on future research directions.

2 Tetrapods

2.1 Material Properties of ZnO Tetrapods

Zinc oxide (ZnO) is an adaptable semiconductor material with extensive applications due to its unique physical and chemical properties. These properties include a 3.37 eV direct bandgap, 60 meV exciton binding energy, strong luminescence, and high piezoelectric constants. These features make ZnO a critical material for innovations in electronics, optoelectronics, spintronics, and piezoelectric transducers. Its ease of synthesis into diverse morphologies, including nanorods, nanowires, nanobelts, and particularly tetrapods, further enhances its functional diversity. Among these morphologies, ZnO-T are distinguished by their distinctive geometry: a central core with four arms branching out. This configuration imparts exceptional properties, including enhanced electron extraction and transport capabilities, making them ideal for photovoltaic devices. The unique structure of ZnO-T forms porous networks that provide superior electron pathways and mechanical strength, which are essential for energy conversion systems and hybrid solar cells. Furthermore, these porous networks can create gel-type structures with various biomedical applications.

ZnO-T demonstrates multifunctionality across several domains. Nanoelectronics and optoelectronics' excellent optical and electronic properties enable applications in UV detectors, light-emitting diodes, and transistors [9]. Its high surface area and structural stability make sensing devices effective for gas and biosensors. In biomedical engineering, it serves as a reinforcing material in composites, offering the potential for advanced linking technologies and biocompatible materials. Additionally, its unique morphology facilitates efficient charge collection and transport in energy harvesting, thereby enhancing solar cell performance [10, 11].

The tetrapodal morphology arises from a combination of hexagonal wurtzite arms and a central zinc-blende core, which form through controlled nucleation and growth processes. These structural traits ensure better substrate orientation and lower defect densities than other nanostructures. Optimizing arm length-to-diameter ratios is crucial for enhancing packing density and device interface charge separation.

In conclusion, ZnO-T exemplify a class of materials with unparalleled versatility and potential, offering solutions for current and emerging technological challenges. Their unique structural and functional attributes continue to drive research in nanotechnology, energy systems, and biomedical applications.

2.2 *Physical Properties of DNA Tetrapods*

Amphiphilic molecules and DNA nanotechnology provide significant insights into self-assembly phenomena, facilitating the formation of ordered structures with diverse functionalities. DNA, with its programmable base-pairing specificity, has been exploited to create nanostructures and hydrogels with remarkable applications in biomedicine, synthetic biology, and material science. DNA hydrogels, composed of self-assembled nanostars or Y-shaped motifs, exhibit tunable mechanical properties, bio-compatibility, and viscoelasticity, enabling their use in drug delivery, tissue engineering, and cell culture.

Similarly, due to their unique branched geometry, tetrapod nanocrystals, such as ZnO-based structures, show great promise in optoelectronics and self-assembly applications. Computational modeling, using coarse-grained techniques like bead-spring models, enhances our understanding of these systems' self-assembly mechanisms and gelation behaviors. By integrating DNA hydrogel self-assembly principles with tetrapod nanostructure design, it is possible to explore hybrid systems that mimic extracellular matrices or investigate gelation processes. Such models unlock potential applications in multifunctional materials, from bio-inspired scaffolds to advanced optoelectronic devices. This cross-disciplinary approach enables the study of tetrapods as analogs to DNA hydrogels, emphasizing their shared self-assembly mechanisms and expanding their utility in innovative scientific and engineering applications.

With their unique branched geometry and exceptional optoelectronic properties, ZnO-T are versatile building blocks in multifunctional devices. These nanostructures are utilized in applications ranging from photodetectors and gas sensors to biocompatible platforms for biomedical research. The hierarchical assembly of tetrapods facilitates the design of materials with tunable properties, bridging the gap between nanoscale functionality and macroscopic performance.

In parallel, DNA hydrogels demonstrate a tendency for self-assembly, forming networks through predictable base-pairing interactions. These materials, characterized by high bio-compatibility and programmable mechanical properties, are pivotal in biomedical applications such as drug delivery and tissue scaffolding. Computational modeling of such systems using coarse-grained methods provides insights into the dynamics of self-assembly and gelation. One can explore hybrid systems that integrate biological and inorganic components by drawing analogies between the self-assembly of DNA hydrogels and the organization of ZnO-T. Such interdisciplinary approaches enable the design of novel materials combining the optoelectronic functionality of tetrapods with the dynamic and adaptive nature of DNA-based networks. These hybrid materials hold great potential for creating innovative scaffolds, stimuli-responsive devices, and bio-inspired multifunctional platforms, pushing the frontiers of material science and nanotechnology.

3 Modeling Techniques

3.1 Langevin Dynamics

In Langevin dynamics, random forces arise from continuous collisions between solvent molecules and larger particles, such as colloids. The time and length scales of these interactions differ significantly, with colloids and surrounding solvent molecules separated by many orders of magnitude. The stochastic forces exerted on colloids are modeled by extending Newton's equation of motion to include a frictional term and a random, stochastic term that accounts for collisions with the surrounding fluid. This results in the generalized Langevin equation, which describes the dynamics of colloidal systems under these conditions.

Langevin dynamics has been used to perform coarse-grained molecular dynamics simulations [12, 13]. Each bead trajectories were calculated using the following equation:

$$mx'' = -\nabla U_h(x) - \xi x^{'(t)} + F^R(t) \tag{1}$$

where m is bead mass, $U_h(x)$ is the harmonic potential governing the bead interactions and the damping parameter is given by ξ. The noise term $F^R(t)$ interacts with a stochastic heat bath. Simulations equilibrated at a fixed temperature, T, with a time step τ.

3.2 Dissipative Particle Dynamics

Dissipative particle dynamics (DPD) is a mesoscopic simulation method that models clusters of molecules moving collectively in a Lagrangian framework. It is particularly well-suited for studying fluids' dynamic and rheological properties [14]. DPD incorporates features of molecular dynamics (MD) and lattice-gas automata (LGA), offering higher computational efficiency than MD while maintaining greater flexibility compared to LGA. The theoretical formulation of DPD is discussed in a subsequent section.

4 Simulation Models in 2D and 3D

Building on the unique structural and functional attributes of ZnO-T, this section explores their modeling through coarse-grained simulations. This approach, mainly using tools like LAMMPS, simplifies complex atomic systems while retaining critical physical characteristics. Coarse-grained modeling provides a robust framework for investigating large-scale behaviors, including forming porous networks, mechanical properties, and charge transport dynamics. By focusing on mesoscale interactions, this method bridges the gap between atomistic details and macroscopic applications, offering profound insights into the design and optimization of tetrapods for diverse technological advancements.

4.1 Non-Inert Model of Tetrapod (Y-Shaped Tetrapod in 2D)

4.1.1 Modeling Details

This model employs a bead-spring approach to simulate the interaction and self-assembly of Y-shaped DNA units. The geometric structure consists of ten beads per Y-shaped unit: one bead at the core, followed by six structural beads leading to the formation of three distinct arms and three patch beads at the terminal ends of each arm. These patch beads are designed to mimic complementary sticky ends (A-B pairs) and provide the basis for interaction between distinct DNA sequences (Fig. 1).

To maintain the structural rigidity in each arm, beads on the neighboring arm are connected through harmonic springs. At the same time, an angular potential ensures the appropriate geometric distribution, with a 120° separation between arms around the central bead. The attractive interaction between complementary patch types (A

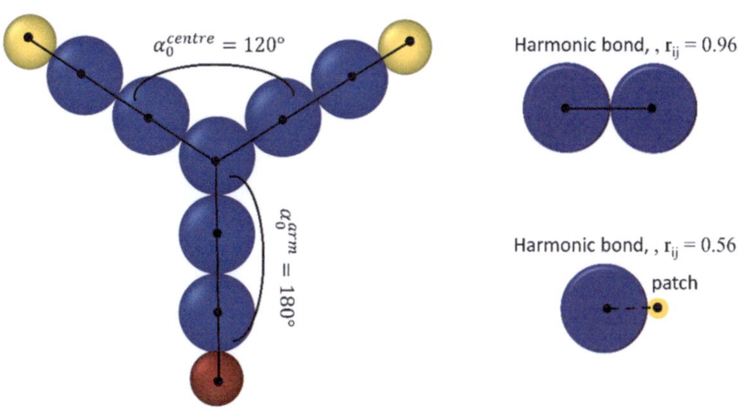

Fig. 1 Schematic representation of the bead spring Y-shaped 2D inert tetrapod model

and B) is modeled using a truncated Lennard–Jones (LJ) potential. In contrast, non-complementary patches are repelled using a Weeks-Chandler-Andersen (WCA) potential. This combination of repulsive and short-range attractive forces prevents multiple patch interactions, ensuring a strict valency of 1 per arm.

The model is implemented using LAMMPS, using reduced units to simulate interactions between beads, bonds, and patches. Parameters are calibrated to reflect experimental hydrogel setups. Unit lengths and bond energies are in the nanometer and picojoule ranges, respectively, ensuring a realistic representation of DNA assembly at the sub-micrometer scale. Additionally, the coarse-grained modeling of Y-shaped DNA units allows for the simulation of dynamic interactions, such as assembling micrometer-scale structures through controlled sticky-end interactions.

4.1.2 Simulation Details

Self-avoiding Y-shaped molecules are used as building blocks to generate a random arrangement inside a cubic simulation box in 3D with periodic boundary conditions [15]. Initially, a central Y-shaped unit is seeded, and additional Y-shaped units are introduced by translating and rotating in random directions. These Y-shaped units are tagged as type A or B with equal probability using a Monte Carlo algorithm to ensure randomness and prevent overlap. The process continues until the desired number density of Y-shapes is achieved. The final arrangement serves as the input for further molecular dynamics simulations.

The simulation box has a length of $30\sigma_{LJ}$, where σ_{LJ} represents the Lennard–Jones (LJ) unit of length. The number density n_D is adjusted by varying the number of Y-shaped units, and the volume of each Y-shape is calculated as $V = \left(\pi r_Y^2 \times l_Y\right) \times 3 \approx 7.3\sigma LJ^3$, with $r_Y = 0.56\sigma_{LJ}$ and $l_Y = 2.48\sigma_{LJ}$ defining the radius and length of each arm, respectively. The concentration is characterized by the volume fraction ϕ_V and number density ρ.

Langevin dynamics is employed to model the system's behavior, where the generalized Langevin equation governs the trajectory of each bead and is given by Eq. 1. In this equation, m is the bead mass, $U_h(x)$ is the harmonic potential governing bead interactions, ξ represents the damping coefficient, and $F^R(t)$ is the random force accounting for thermal fluctuations. The harmonic potential ensures structural integrity between beads in the arms of the Y-shape, while the damping term models the frictional forces exerted by the surrounding solvent. The random force term represents the stochastic interactions with the heat bath. The system is equilibrated at a fixed temperature T ranging from $0.05\varepsilon_{LJ}/k_B$ to $0.7\varepsilon_{LJ}/k_B$, and the simulation runs with a time step of $0.005\tau_{LJ}$ over $50{,}000\tau_{LJ}$ to reach equilibrium. This simulation setup explores self-assembly behaviors of Y-shaped DNA units, focusing on interactions between complementary and non-complementary patches modeled by truncated Lennard–Jones and Weeks-Chandler-Andersen potentials, respectively. This method allows for the study of dynamic processes, such as the assembly of micrometer-scale structures via controlled sticky-end interactions, providing insight into the behavior of DNA-based hydrogels and their potential applications in fields like tissue engineering and drug delivery.

4.2 Non-Inert Model of Tetrapod (X, Star-Shaped, 5-Armed Tetrapods in 2D)

4.2.1 Modeling Details

The model utilizes a coarse-grained (CG) bead-spring approach to simulate the structure of DNA nanostars, initially established by Eiser et al. [16]. The force field parameters, including energy ($\in_{LJ}$), length (σ_{LJ}), and mass (m_{LJ}), are represented in Lennard–Jones (LJ) reduced units. The CG representation consists of two distinct bead types: large beads representing the double-stranded nature of DNA and small patch beads modeling the sticky ends (Eiser et al. [14]). Two patch types, designated patch-I, and patch-II, form complementary interactions, with a repulsive potential between identical patches to prevent unwanted three-body interactions. Non-bonded interactions are modeled using the Weeks-Chandler-Andersen (WCA) potential, ensuring repulsion at a cutoff distance of 1.12 σ_{LJ}.

For complementary patches, an attractive Lennard–Jones (LJ) potential is applied with a cutoff distance of 5 σ_{LJ}. Neighboring beads are connected via a harmonic potential with high stiffness, ensuring minimal fluctuation around equilibrium distances. The harmonic bond potential maintains structural integrity, while angular bending potentials enforce the correct geometrical configuration of the nanostars. The equilibrium angles are set at 120°, 90°, and 180° for X-shaped, Y-shaped, and 5-way junction nanostars, respectively (Eiser et al. [14]). This model effectively captures the complex structural dynamics of DNA nanostars.

4.2.2 Simulation Details

This simulation work has been done by Supriyo Naskar et al. [16], who employed a coarse-grained approach to model X-shaped, Y-shaped, and five-way junction (5WJ) DNA nanostars (Fig. 2). Three distinct systems containing 50, 150, and 250 nanostars, were constructed and randomly placed within a periodic simulation box of dimensions $30 \times 30 \times 30\sigma_{LJ}^3$. The volume fraction was calculated based on the individual volume of each nanostar, with approximate volumes of $7.33\sigma_{LJ}^3$ for Y-shaped, $9.77\sigma_{LJ}^3$ for X-shaped, and $12.22\sigma_{LJ}^3$ for 5WJ nanostars.

Simulations were performed using the LAMMPS package, employing Langevin dynamics to evolve the positions of beads, as described by Eq. 1. The modeling framework incorporated WCA repulsion, harmonic bonds, and angular potentials, with an overdamped approximation for the gel phase achieved by setting a high

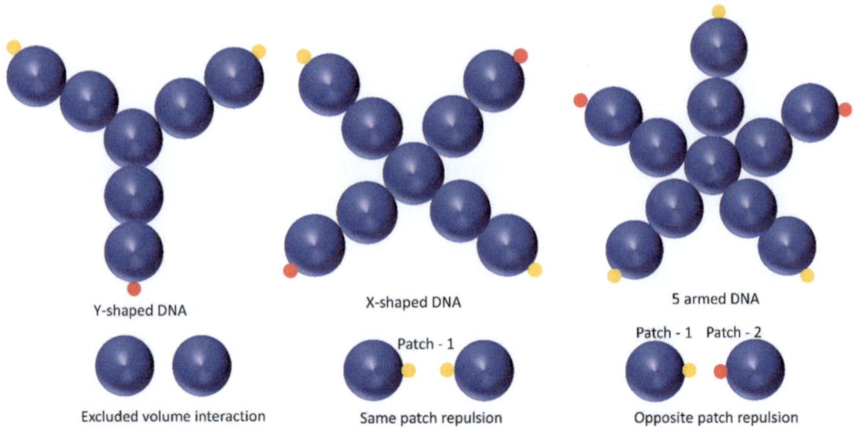

Fig. 2 Schematic representation of the coarse-grained model of Y-shaped, X-shaped, 5-armed-shaped 2D inert tetrapod model mimicking DNA nanostars

damping factor (ξ). The stochastic noise was modeled as a Gaussian process, $F^R(t)$, with a zero-mean, delta-correlated structure.

Simulations were carried out at temperatures ranging from 0.075 $\in_{LJ}/k_B$ to 0.775 $\in_{LJ}/k_B$, with step intervals of 0.025 $\in_{LJ}/k_B$. A time step of 0.005 τ_{LJ} was used for each simulation, and the system was integrated for up to 10^7 steps per temperature to ensure equilibrium, with energy and bond formation reaching steady-state values. This systematic approach guarantees the reliability of the simulation results across a range of temperatures and system sizes.

4.3 Hydro Mesh and Amphi Branch Tetrapod (3D)

4.3.1 Modeling Details

Dissipative Particle Dynamics (DPD) is used in this simulation to explore the morphologies and self-assembly of tetrapod nanoparticles in solution [17]. DPD, a coarse-grained simulation technique, enables the study of mesoscale systems over millisecond timescales and micrometer length scales by modeling the motion of particles representing groups of atoms or molecules. In the simulation, each tetrapod nanoparticle was considered as a rigid body, ensuring that all DPD beads within the same nanoparticle moved with uniform translational velocity. The dynamics of these particles were governed by Newton's equation of motion, excluding inter-nanoparticle forces, and is expressed as:

$$m_i \frac{dv_i}{dt} = f_i = \sum_{j \neq i} \left(F_{ij}^{cons} + F_{ij}^{random} + F_{ij}^{diss} \right) \tag{2}$$

where F_{ij}^{cons} is the conservative force, F_{ij}^{random} is the pairwise random force, and F_{ij}^{diss} is the dissipative force. The conservative force, derived from a prescribed potential between particles ii and jj, acts as a weakly repulsive force that ensures an even spatial distribution of particles and can be interpreted as a pressure force. It is defined as:

$$F_{ij}^{cons} = \begin{cases} -\alpha_{\mathrm{i}}\left(1 - \frac{|l_{ij}|}{l_{cut}}\right)n_{ij}, & |l_{ij}| \le l_{cut} \\ 0, & |l_{ij}| > l_{cut}, \end{cases} \tag{3}$$

where α_{ij} is the parameter responsible for repulsive interaction between i and j beads, the cutoff length is given by l_{cut}, $l_{ij} = l_j - l_i$ and $n_{ij} = l_{ij}/|l_{ij}|$. The dissipative force is proportional to the velocity difference between two particles and opposes their relative motion, dissipating kinetic energy from the system. To maintain a constant temperature, this lost energy must be replenished. The random force F_{ij}^{random} replenishes kinetic energy in the system, complementing the dissipative force F_{ij}^{diss}. Together, these forces act as a thermostat to maintain thermal equilibrium, with their magnitudes and forms governed by the fluctuation–dissipation theorem. They can be written as

$$F_{ij}^{diss} = \begin{cases} -\gamma w^{diss}(|l_{ij}|)(n_{ij} \cdot v_{ij})n_{ij}, & |l_{ij}| \le l_{cut} \\ 0 & |l_{ij}| > l_{cut} \end{cases} \tag{4}$$

$$F_{ij}^{diss} = \begin{cases} -\gamma w^{diss}(|l_{ij}|)(n_{ij} \cdot v_{ij})n_{ij}, & |l_{ij}| \le l_{cut} \\ 0 & |l_{ij}| > l_{cut}, \end{cases} \tag{5}$$

$$F_{ij}^{random} = \begin{cases} -\beta w^{random}(|l_{ij}|)\xi_{ij}\Delta t^{-\frac{1}{2}}n_{ij}, & |l_{ij}| \le l_{cut} \\ 0 & |l_{ij}| > l_{cut}, \end{cases} \tag{6}$$

where γ is the friction coefficient, ξ_{ij} is the random number as per the Gaussian distribution, β is the noise amplitude, Δt is time step, w^{diss} and w^{random} are the length-dependent weight functions.

$$w^{diss}(l) = \left[w^{random}(l)\right]^2 = \begin{cases} \left(1 - \frac{|l_{ij}|}{l_{cut}}\right)^2, & |l_{ij}| \le l_{cut} \\ 0 & |l_{ij}| > l_{cut}, \end{cases} \tag{7}$$

The fluctuation–dissipation theorem ensures temperature control, relating $\beta = 2\gamma k_B T$ where the Boltzmann constant is k_B and temperature T. Length and time scaling followed the method of Groot and Rabone. This DPD-based model provides a robust framework for exploring the self-assembly behavior of tetrapod nanoparticles and their potential applications in nanomaterials.

4.3.2 Simulation Details

The simulation study investigates the self-assembly of tetrapod nanoparticles with amphiphilic arms. It can be done via dissipative particle dynamics [17]. Each tetrapod consisted of hydrophilic segments along the arms and hydrophobic tips, arranged in a diamond lattice structure (Fig. 3). The Groot–Warren theory was used to determine the interaction parameters, with the hydrophobic–hydrophilic repulsion parameter set to 25 k_BT to ensure phase separation and drive self-assembly.

A periodic cubic simulation box of size 42 × 42 × 42 was used, containing 125 tetrapod nanoparticles, and the system was maintained at a temperature of 1.0 k_BT. The tetrapod arms were modeled as spheroids, scaled to one-tenth of their actual size, ensuring computational feasibility while preserving essential assembly behavior. The simulation was run over 10^6 DPD time steps, allowing the system to reach equilibrium.

The self-assembly process was analyzed by quantifying tip-connecting clusters, defined by the average number of mutually connected hydrophobic tips $\langle N_{tip} \rangle$. A threshold cutoff distance of 1.5σ was used to identify hydrophobic tip interactions. The time evolution of cluster formation was monitored to track the emergence of stable aggregates, providing insight into the structural organization of tetrapod nanoparticles in solution.

4.4 Meta-DNA Structure (2D)

Tetrapods offer a versatile platform for DNA hydrogels, utilizing DNA origami to construct complex nanostructures [18]. DNA origami has emerged as a powerful

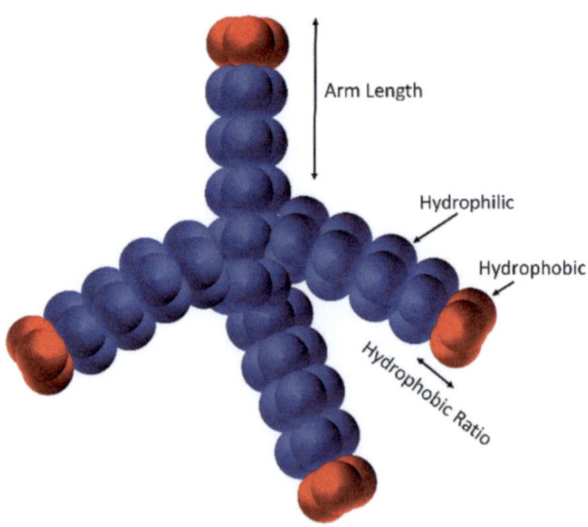

Fig. 3 Schematic representation of the coarse-grained model of an Amphi Branch tetrapod model. The four arms extend from a central core, with hydrophilic (blue) and hydrophobic (red) particles

method for creating objects at the 10–100 nm scale, with applications ranging from metamaterials to surface-based biophysical assays. Research has shown that meta-DNA, a six-helix bundle structure at the submicrometre scale, facilitates the assembly of diverse 2D and 3D architectures. These structures mimic the molecular behavior of DNA strands, employing sticky-end cohesion or blunt-end stacking to facilitate self-assembly into micrometer-sized hydrogels.

Meta-DNA units are characterized by their rigid yet programmable flexibility, allowing for various shapes and geometric configurations. Through tailored interactions between M-DNA units, coarse-grained computational models simulate self-assembly, enabling the design of left-handed and right-handed structures. This approach facilitates the construction of tetrahedrons, octahedrons, prisms, and closely packed lattices, which apply to various scientific and biomedical fields. The dynamic strand-displacement reactions observed in meta-DNA architectures also create static and dynamic DNA hydrogels, extending their potential beyond the nanoscale into macroscopic systems. This methodology opens new possibilities for coarse-grained modeling of tetrapod structures using sticky ends, enhancing the understanding of complex DNA hydrogel systems at various scales.

4.5 3-Armed DNA Hydrogel

This coarse-grained model, developed by Yair Augusto Gutierrez Fosado [19], employs the oxDNA2 framework to simulate DNA nanostructures, specifically focusing on tetrapod geometries and DNA hydrogels (Fig. 4). The oxDNA2 model represents single-stranded DNA (ssDNA) as chains of rigid nucleotide beads with well-defined interaction sites, capturing the essential thermodynamics and structural properties of DNA with high accuracy.

In this model, each nucleotide is considered as a rigid body, connected by FENE (finitely extensible nonlinear elastic) springs, which are used to represent the backbone connectivity between nucleotides. This allows for the modeling of DNA duplex structures, which is essential for simulating self-assembly processes within tetrapod and hydrogel networks. The model accounts for excluded-volume effects and specific interactions such as stacking, cross-stacking, and coaxial stacking, essential for accurately modeling DNA's 3D structure. To model these interactions, a combination of Morse and harmonic potentials is used for the stacking and angle interactions, respectively, ensuring that DNA strands form correct structural configurations, including double helices and multi-stranded structures.

The oxDNA2 model also includes crucial details like hydrogen bonding between complementary base pairs, which governs the self-assembly behavior of DNA strands. These features enable the model to capture the fundamental behavior of DNA-based materials, including their assembly into complex nanostructures like tetrapods.

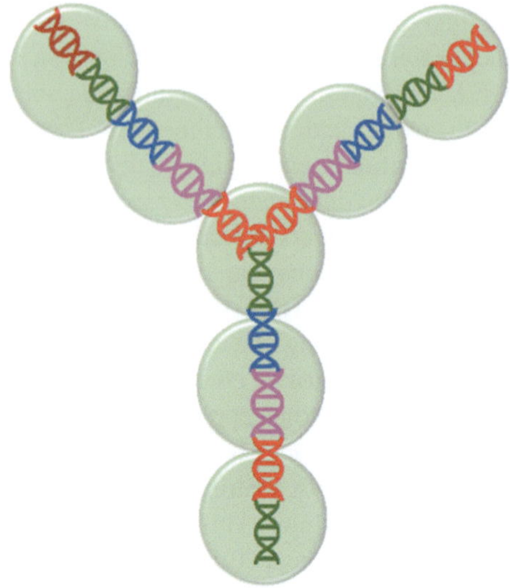

Fig. 4 Schematics of a 3-armed DNA structure

Its parameters are specifically tuned to match experimental values such as DNA strands' persistence lengths and melting temperatures to ensure the model is thermodynamically accurate. The model also incorporates implicit ion effects via Debye-Hückel electrostatics, improving the accuracy in simulating real-world conditions where salt and ionic interactions affect the behavior of DNA molecules. Moreover, the oxDNA2 model includes the major and minor grooves of DNA, essential for understanding the physical interactions between DNA strands and the structural characteristics of DNA hydrogels.

These features make the oxDNA2 model highly suitable for simulating DNA nanostructures. It offers a robust framework for studying self-assembly processes, network formation, and DNA hydrogel behavior, particularly for tetrapod-shaped assemblies. This approach allows one to explore various configurations of DNA nanostructures and their future utility in fields like nanotechnology and biomedical research.

4.6 DNA Hydrogel Model

The oxDNA model has been adapted to simulate tetrapod-like structures, focusing on key features of DNA nanostars with three arms [20]. The DNA hydrogel system in this model consists of Y-shaped DNA nanostars, denoted as S-DNA and S'-DNA, each featuring complementary sticky ends that enable binding through DNA hybridization. These nanostars are composed of three single-stranded (ss) oligonucleotides: S1, S2, S3 for S-DNA, and S'1, S'2, S'3 for S'-DNA. Each oligonucleotide is 46 bases in length and incorporates three functional components: (i) a 30-base core that forms double-stranded arms (15 bases per arm); (ii) 12-base sticky ends that facilitate crosslinking between nanostars; and (iii) a 4-base flexible joint linking the core to the sticky ends, providing structural flexibility to the system.

At low temperatures, the core sections of these nanostars hybridize to form double-stranded DNA. The duplex formed by S-DNA and S'-DNA follows canonical Watson–Crick base pairing (A-T and C-G), resulting in tightly twisted double helices within the nanostar cores. The oxDNA model employed in these simulations supports only Watson–Crick base pairs, which ensures the accurate representation of the nanostar structure but excludes non-canonical base pairs or Hoogsteen interactions. It is important to note that, due to these interactions, the junction where the three arms meet can experience strain, especially at higher temperatures or when the system undergoes significant deformation.

Modifications were introduced to mitigate the structural strain at the junction of the nanostars by adding inert nucleotides to the core regions. These modified molecules, denoted as Sn1S1n2S2n3S3 for S-DNA (and similarly for S'-DNA), incorporate additional nucleotides that do not participate in hydrogen bonding, providing enhanced flexibility at the junction. This modification allows the system to accommodate better the mechanical stress at the nanostar junctions, which is critical for forming stable, flexible nanostar assemblies.

This model, leveraging the oxDNA framework, offers a powerful tool for simulating the self-assembly and structural dynamics of DNA nanostars in hydrogels. It also provides a solid foundation for investigating their potential applications in various nanotechnological and biomedical fields.

4.7 Linear Linker Model

This model, proposed by Łukasz Baran [21], employs a linear linker molecule to achieve a tetrapodal geometry. The structure consists of a central core to which three arms are attached, forming a tetrahedral shape (Fig. 5a). In this study, each green segment represents a functional group, while yellow entities serve as active interaction centers. The segment length of the linkers was set to $\sigma_{\text{linker}} = \sigma$, and the active sites were smaller, with $\sigma_{\text{active}} = 0.2\sigma_{\text{linker}}$. The linker segments are tangentially bonded with a distance of σ_{linker}, and the active sites are embedded at the terminal units with a bonding distance of $d_{\text{bonding}} = 0.36\sigma_{\text{linker}}$. Explicit metal atoms (Fig. 5b) were incorporated, ranging in size from $\sigma_{\text{metal}} = 0.5\sigma_{\text{linker}}$ to $\sigma_{\text{metal}} = 1.0\sigma_{\text{linker}}$.

Molecular dynamics simulations treated all entities as flat, rigid objects, maintaining bonds via harmonic potentials. The potential between linker-linker is written as:

$$U_{lin\,\ker-lin\,\ker} = -k_{lin\,\ker-lin\,\ker}(r - \sigma_{lin\,\ker-lin\,\ker})^2 \tag{8}$$

and the harmonic potential between the active site and the linker is given as

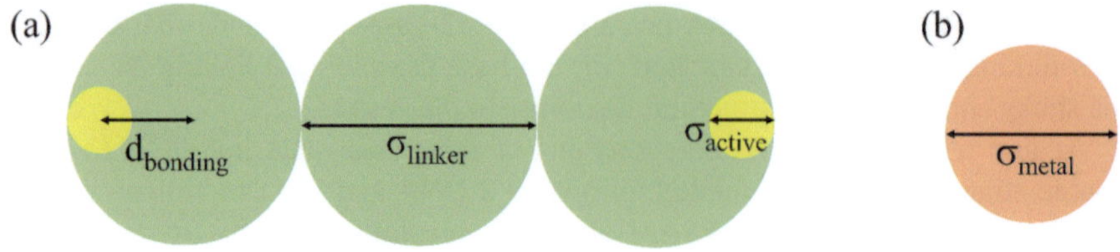

Fig. 5 **a** A linear linker molecule model consisting of bonding atoms, a linear molecule, and an active site and **b** Metal atom

$$U_{active-linker} = -k_{active-linker}(r-d)^2 \quad (9)$$

Angular harmonic potential for all the angles can be written as,

$$U_{linker-linker}(\theta_{linker-linker}) = -k_\theta\left(\theta_{linker-linker} - \theta_{0,linker-linker}\right)^2 \quad (10)$$

$$U_{active-linker}(\theta_{active-linker}) = -k_\theta\left(\theta_{active-linker} - \theta_{0,active-linker}\right)^2 \quad (11)$$

The interparticle interactions employed a shifted (12, 6) Lennard–Jones potential, ensuring continuity of potential $U_{LJ}(r) = 4 \in_{ij} \left[\left(\frac{\sigma_{ij}}{r}\right)^{12} - \left(\frac{\sigma_{ij}}{r}\right)^6\right]$ and $U_{LJ}\prime(r_{cut})$ its first derivative at $r = r_{cut}$ can be written as

$$U_{LJ} = \begin{cases} U_{LJ}(r) - U_{LJ}(r_{cut}) + U_{LJ}\prime(r_{cut})(r - r_{cut}) & r < r_{cut} \\ 0 & Otherwise \end{cases} \quad (12)$$

The LJ parameters $\sigma_{\text{linker}} = \sigma$ and $\in_{\text{linker-linker}} = \in$ are set as the length and energy units, respectively.

Reduced units were used for time $\tau = t\sqrt{\in/m\sigma^2}$ and $T = kT/\in_{linker-linker}$, and density $\rho = \frac{R\sigma_c^2 + N_{metal}\sigma_{metal}^2}{L_x L_y}$, where N_{metal} are the metal atoms, and R is the segment number in the linear linker. The composition $\chi = N_{linker}/N_{total}$, where N_{linker} is the number of linker molecules, $N_{\text{total}} = N_{\text{linker}} + N_{\text{metal}}$. Linker-linker and linker-active site interaction energies were considered as $\in_{\text{linker-linker}} = \in_{\text{active-active}} = \in \ominus$, while metal-active site interactions were stronger, $\in_{metal-active} = 5.0 \in$. Linker-site interaction energy and the diameter are given by $\in_{\text{active-linker}} = \in$ and $\sigma_{active-linker} = (\sigma_{active} + \sigma_{linker})/2$ respectively. Interaction cutoffs between the metal and active sites are given as $r_{cut,metal-active} = 2\sigma_{metal-active}$, and $r_{cut,ij} = \sigma_{ij}$ for other pairs where ij = active-active, active-linker, linker-linker, linker-metal, metal–metal. Solvent effects were implicitly modeled by screening non-active interactions. High harmonic constants $k_{active-linker} = k_{linker-linker}$ set to $1000 \in /\sigma^2$ and $k_{theta} = 1000 \in /(rad)^2$ ensured rigidity.

Simulations were performed using LAMMPS in the NVT ensemble with a velocity-Verlet integration scheme and a timestep of $\Delta t = 0.001\tau$. Initial NPT runs established densities, followed by equilibration runs using the Berendsen thermostat $\tau_B = 10\tau$ and production runs with the Nosé–Hoover chain algorithm $\tau_H = 10\tau$. Temperature grids $\Delta T = 0.01$ facilitated cooling from disordered to ordered self-assembled states.

4.8 Inert Tetrapod Model (3D)

A bead spring model to represent a coarse-grained modeling of a tetrapod has been implemented [22]. Each arm is symmetrically connected to a central core, forming an equidistant arrangement in 3D (Fig. 6a). All the neighboring beads [7] are connected by the harmonic spring using the equation.

$$\mathrm{V_h(l)} = \mathrm{k_r(l - l_0)^2} \tag{13}$$

where the spring constant is given by $\mathrm{k_r}$, the equilibrium bond length is given by $\mathrm{l_0}$. To ensure the structural integrity of the tetrapod and to allow only small fluctuations, $\mathrm{k_r}$ is $300\mathrm{k_B T}$.

To replicate the building block geometry, four beads are symmetrically arranged around a central bead with bond angles of 109.5°.

The excluded volume interaction among the tetrapods is modeled using truncated Lennard Jones (LJ) potential, commonly known as Weeks-Chandler-Andersen (WCA) potential as.

$$U_{LJ}(r_{ij}) = \begin{cases} 4 \in \left[\left(\frac{\sigma_{ij}}{r_{ij}} \right)^{12} - \left(\frac{\sigma_{ij}}{r_{ij}}^{6} \right) \right] + \varepsilon, \ for\ r_{ij} \le 2^{\frac{1}{6}} \sigma_{ij} \\ 0, \qquad Otherwise \end{cases} \tag{14}$$

Here, $\mathrm{r_{ij}} = |\mathrm{r_i} - \mathrm{r_j}|$ represents the interparticle distance, $\sigma_{ij} = (\sigma_i + \sigma_j)/2$ is the overall size of the interaction domain of the interacting particle of size σ_i, σ_j, and ϵ is the strength of the interaction. To simulate good solvent conditions, only the repulsive component of the Lennard–Jones potential was employed by truncating the attractive term. An experimental (Scanning Electron Microscope) image showing the bead-like ZnO-T structure equivalent to the theoretical model, synthesized using a modified catalyst-free thermal sublimation technique, followed by chemical treatment, is shown in Fig. 8.

4.9 Non-Inert Model of Tetrapod (3D)

To introduce functionality, small patches designated as patch-I and patch-II are attached to the end of each arm (Fig. 6b and c). Patches of the same type interact repulsively, whereas patches of opposite types exhibit attractive interactions [22].

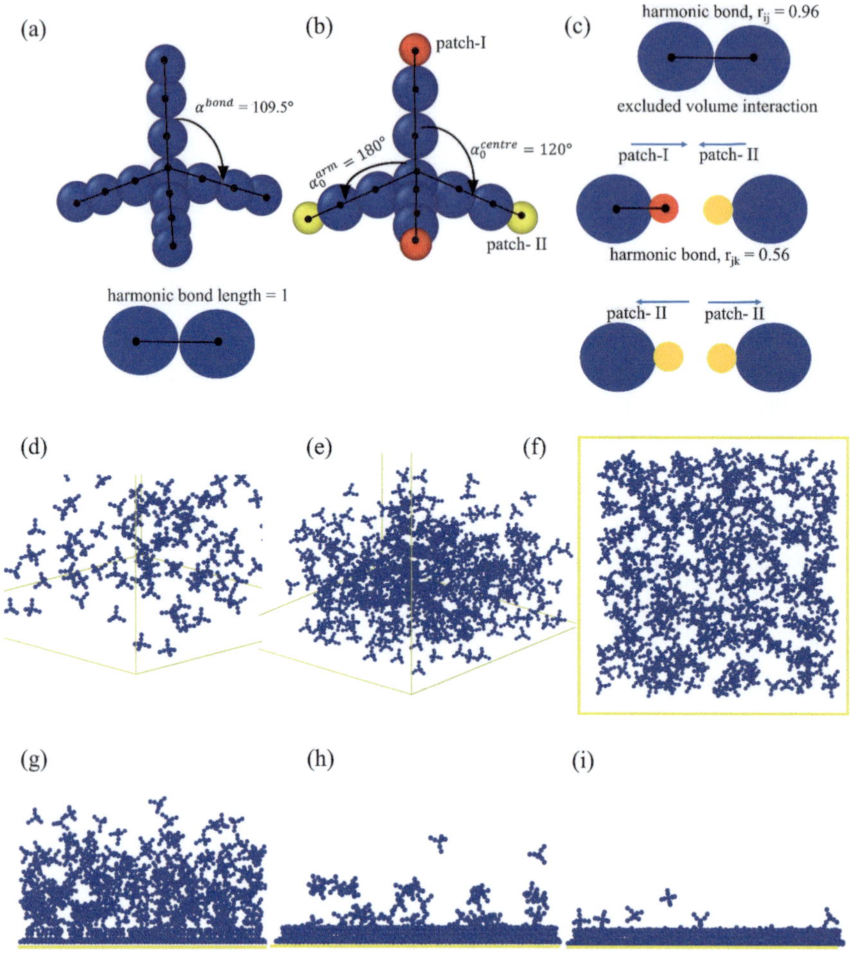

Fig. 6 Schematic representation of the bead spring model of a tetrapod. **(a)** 3D inert model. **(b)** 3D non-inert model. **(c)** Excluded volume interaction and attractive interaction between patch I and patch II. **(d)** and **(e)** Different numbers of tetrapods falling under gravity from the top surface. **(f)** The top view of the depositing surface showcases the self-assembly of tetrapods. **(g)**, **(h)** and **(i)** are the stages of tetrapods depositing and settling down on the depositing surface

The excluded volume effect among the individual beads is modeled using the truncated LJ potential, referred to as the WCA potential, using Eq. 14. In our simulation, to prevent overlap between the tetrapod arms, a repulsive interaction is implemented between the large beads using $\in=\in_{LJ}$, $\sigma = \sigma_{LJ}$, $r_{\text{cutoff}} = 1.12\ \sigma_{LJ}$. Furthermore, for the small beads in the patches, a repulsive interaction is applied between identical patches using parameters $\in=\in_{LJ}$, $\sigma = 0.67\sigma_{LJ}$, $r_{\text{cutoff}} = 0.67\ \sigma_{LJ}$, thereby preventing attractive interactions between different patches.

Attractive interactions among complementary patches are modeled using the full Lennard–Jones potential truncated at a farther distance,

$$U_{LJ}(r_{ij}) = \begin{cases} 4 \in \left[\left(\frac{\sigma_{ij}}{r_{ij}} \right)^{12} - \left(\frac{\sigma_{ij}}{r_{ij}}^{6} \right) \right] & for\ r_{ij} \leq 5\sigma_{ij} \\ 0 & otherwise \end{cases} \quad (15)$$

The interaction among patch-I and patch-II is given by, $\in = 4.5 \in_{LJ}, \sigma = 0.2\sigma_{LJ}$. The harmonic spring links all neighboring large beads via Eq. 13. The equilibrium bond length between two large beads is taken to be 0.96 σ_{LJ} and between the large bead and the patch is 0.56 σ_{LJ}. Angle constant is taken by the harmonic potential $V_{angle}(\theta) = K_\theta(\theta - \theta_0)^2$, where the spring constant is given by K_θ and the bond angle is θ_0 at equilibrium bond angle. K_θ is used for maintaining the bead-patch rigidity. $\alpha_0^{arm} = 180°$ with each arm, and the three arms are symmetrically arranged around the central core, each positioned at a minima of $\alpha_0^{center} = 120°$ is provided.

Our simulation parameters are defined using the characteristic energy ϵ, mass m, and length σ scales. The fundamental time unit is given by $t = \sqrt{m\sigma_{LJ}^2/\in_{LJ}}$, while the corresponding force scale is $f = \epsilon/\sigma$, expressed in ps and pN, respectively.

Initially, tetrapods are positioned with uniform spacing at varying concentrations on the top surface of a simulation box with a height of 60σ. The bottom (deposition) surface exhibits an attractive interaction characterized by a cutoff distance of 1.5σ. Tetrapods experience a gravitational force of 0.1 LJ units directed along the negative z-axis (Fig. 7). We performed this coarse-grained simulation using the LAMMPS simulation package [13]. The dynamics of each bead is governed using the equation of motion and is described by the Langevin relation, which can be written as

$$m\frac{d^2r(t)}{dt^2} = -\xi\frac{dr(t)}{dt}(t) + F^R(t) - \nabla(U_{LJ} + U_{harmonic}) + mg \quad (16)$$

where position is defined by r(t), mass by m, gravity as g, friction coefficient by ξ, and F^R refers to the random force pertaining to thermal fluctuation fulfilling the fluctuation–dissipation theorem,

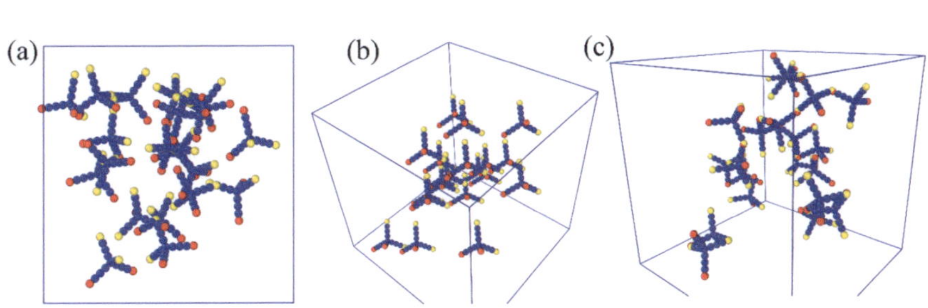

Fig. 7 Non-inert model of tetrapods. **(a–c)** are the distinct representations of tetrapods falling under gravity and exhibiting self-assembly with different patches at the tips from the top and side view

Fig. 8 Experimental images of ZnO-T. Scanning Electron Microscope (SEM) images **(a-d)** showing the bead-like ZnO-T structure equivalent to the theoretical model, synthesized using a modified catalyst-free thermal sublimation technique followed by chemical treatment. The enlarged SEM images show the tetrapod arms as composed of bead-like ZnO particles. (Image Courtesy: Dr. D. Paul Joseph, Department of Physics, NIT-Warangal, Unpublished data)

$$\langle F_i^R(t)\rangle = 0, \left\langle F_i^R(t) \cdot F_j^R(t')\right\rangle = 2dk_BT\ \xi\ \delta_{ij}\delta(t-t') \tag{17}$$

Here, D is the friction coefficient defined as $k_B T/\xi$, relating it to the diffusivity of the particle. Here, the dimension of the system is given by d, in the present model, d = 3, the Boltzmann constant is k_B, and temperature by T. In the simulation, the dimensionless parameters are temperature T = 1.2 and friction $\xi = 0.7$ in LJ units. The Langevin thermostat was employed to perform the simulations, beginning with a random initial configuration giving the same results as the ordered initial configuration. The equations of motion are integrated at each time step using the velocity-Verlet algorithm. Following an initial equilibration period, the tetrapods are allowed to move downward from the top surface. To ensure satisfactory statistical accuracy, data were averaged over 100 independent realizations, using a simulation time step of $\Delta t = 0.001$.

5 Results

5.1 Effect of Shape and Size

Material properties are governed by characteristics such as shape, size, crystal structure, dispersion behavior, and phase composition. Optimizing these parameters enhances performance in biomedical applications, including biosensors and targeted drug delivery, by improving surface interactions and functional properties [23].

Nanoparticles possess unique physicochemical properties, such as a high surface-to-volume ratio and biomolecule-scale dimensions, enabling efficient transport and penetration through biological membranes. Among metal oxides, ZnO is a cost-effective semiconductor widely used in biosensing, catalysis, solar cells, and opto-electronics. Its excellent biocompatibility, minimal toxicity, and strong antimicrobial properties further expand its applications in medicine, where it serves as a biomarker, drug carrier, and therapeutic agent. The biological interactions of ZnO structures are highly size- and shape-dependent. Smaller ZnO nanoparticles and nanorods can readily penetrate bacterial cells, while larger structures, such as ZnO-T and heterostructures, predominantly interact with cell membranes (Fig. 8). These interactions facilitate ion diffusion and free radical generation, leading to oxidative damage in cellular components, including proteins, lipids, and DNA.

5.2 Effect of Morphology

Material morphology plays a crucial role in optimizing the performance of devices in optics, photonics, sensors, and solar cells. The surface area and shape of nanostructures significantly affect their antibacterial activity, making synthesizing materials with controlled morphologies and enhanced adsorption sites essential. Tailoring these properties enhances bio applications by enabling tunable and improved performance [24]. Warheit et al. [25] investigated the impact of ZnO morphology on cytotoxicity in lung cells, reporting low toxicity in rat L2 lung epithelial cells and primary alveolar macrophages exposed to ZnO nanoparticles.

Like other metal oxides, ZnO exhibits tunable physicochemical properties depending on its environment, but controlling the release of toxic Zn ions in solution is essential. The toxicity of ZnO nanoparticles is primarily influenced by zinc ion concentration and interactions with sulfur, phosphate, and ferric oxide. Phosphate, in particular, mitigates ZnO toxicity by rapidly encapsulating nanoparticles, thereby inhibiting zinc ion release [26].

ZnO-T synthesized by Agne et al. (JOPC, 2021) [26] via combustion offered a cost-effective, high-yield approach. Size-based separation via centrifugation resulted in small tetrapods (STs) with leg diameters of ~ 5 nm and lengths of ~ 50 nm, while larger tetrapods (LTs) exhibited leg diameters of ~ 20 nm and lengths of ~ 150 nm.

High-resolution TEM (HRTEM) confirmed highly crystalline structures with dominant prismatic facets. Compared to ZnO nanorods (NRs) and nanoparticles (NPs), ZnO-T demonstrated unique surface alignment, minimizing percolation junctions and enhancing electron transport efficiency.

Surface area and porosity analyses of ZnO-T synthesized via chemical vapor deposition revealed that ZnO-T exhibited the highest surface area and mesoporosity, facilitating superior adsorption and capillary condensation compared to commercial ZnO NRs and NPs. The distinct structural and morphological attributes of ZnO-T make them highly promising for applications requiring high surface area, efficient electron transport, and enhanced adsorption properties.

5.3 *Effect of Hydrophobic Ratio on Self-Assembly and Structure*

The effect of the hydrophobic ratio of tetrapod nanoparticles was studied by Yusuke et al., 2019 by calculating the mean-tip connecting number $\langle N_{tip} \rangle$ in the tip-connecting sites. The self-assembly of ZnO-T with varying hydrophobic ratios (HRs) was analyzed by tracking the mean tip-connecting number, $\langle N_{tip} \rangle$, over time. Distinct assembly behaviors emerged across low, middle, and high HR systems [17]. At low HR, $\langle N_{tip} \rangle$ increased linearly until equilibrium (t $\sim$ 900), stabilizing at an average of five connections. In middle HR systems, a two-stage process was observed: rapid nucleation led to an initial $\langle N_{tip} \rangle$ increase to $\sim$4.0, followed by a slower diffusion-driven growth phase reaching $\sim$4.8 before equilibrium. This behavior, analogous to micelle formation, suggests that tip connection dynamics follow a nucleation-growth mechanism. High HR systems also exhibited two-stage assembly but reached equilibrium faster (t $\sim$ 200), stabilizing at a lower $\langle N_{tip} \rangle \sim 3.6$. The increased hydrophobic surface area enhanced collision rates and stabilized bonding, reducing recombination and overall connectivity.

Cluster analysis revealed that high HR systems formed percolated networks, restricting tetrapod diffusion and limiting large-scale tip connections. Equilibrium studies indicated that $\langle N_{tip} \rangle$ peaked at an HR of 27.3% before declining. Size distribution analysis showed that monomers dominated low HR systems, while middle and high HR systems exhibited peak connectivity at six and three tips, respectively. These findings underscore the intricate relationship between HR, self-assembly dynamics, and equilibrium morphology in ZnO-T, providing a basis for optimizing their structural and functional properties in various applications.

5.4 Effect of Arm Length on Self-Assembled Structure

The influence of ZnO-T arm length (AL) on self-assembly was examined using DPD simulations for ALs ranging from 34 to 70 nm, categorized as short (34 nm), middle (52 nm), and long (70 nm). Two distinct self-assembly behaviors emerged depending on AL. Short ALs exhibited a nearly constant average number of tip connections ($\langle N_{tip} \rangle \approx 2.2$) due to the limited surface area of hydrophobic tips, restricting bonding opportunities. In contrast, middle and long ALs followed a two-stage growth mechanism: an initial rapid nucleation phase ($t \lesssim 100$) marked by an increase in $\langle N_{tip} \rangle$, followed by a slower, diffusion-controlled growth phase. Long ALs demonstrated greater stability with reduced fluctuations in $\langle N_{tip} \rangle$, attributed to stronger bonding from larger tip surface areas, which minimized recombination events. At equilibrium, $\langle N_{tip} \rangle$ increased with AL up to 49 nm, beyond which it plateaued [17].

Size distribution analysis revealed that short ALs predominantly formed monomers, with their numbers exceeding dimers and trimers by over an order of magnitude. In contrast, middle and long ALs shifted toward larger structures, decreasing monomers and increasing medium-sized aggregates (3–8 tips). While increasing AL enhanced tip stabilization and bonding strength, the overall size distribution remained unchanged beyond a threshold length. These findings highlight the critical role of hydrophobic tip surface area in governing ZnO-T self-assembly, influencing dynamic growth and final equilibrium structures.

5.5 Diffusion Dynamics of Tetrapods

To elucidate the dynamics of tetrapods, we investigate MSD as a function of time. The MSD for the tetrapod's center of mass is given by,

$$MSD(t) = \langle r^2(t) \rangle = \frac{1}{N} \sum_{i=1}^{N} \left\langle \left[r_i(t + t') - r_i(t) \right]^2 \right\rangle \tag{18}$$

where the tetrapods number is given by N, and t' is time difference between steps. MSD of center of mass of tetrapods are calculated until the deposition for various concentrations have been observed. MSD follows power law form: $\langle \Delta r^2(t) \rangle \sim t^{\alpha}$, where the exponent α is defined as the anomalous diffusion exponent.

For the inert tetrapod model, at short time scales, when time t is much less compared to the relaxation time τ_R ($t < \tau_R$), the dynamics exhibit a scaling exponent of $\alpha = 1.5$. The intermediate time scale corresponds to the relaxation time for tetrapods initially confined in a crowded environment, during which they reorganize and expand to occupy the available space. At longer time scales ($t > \tau_R$), the MSD exhibits a scaling behavior with an exponent $\alpha = 1.7$ (Fig. 9a).

A similar trend is observed in the MSD behavior for the non-inert model. At short timescales as compared to the relaxation time τ_R, tetrapods exhibit diffusive motion

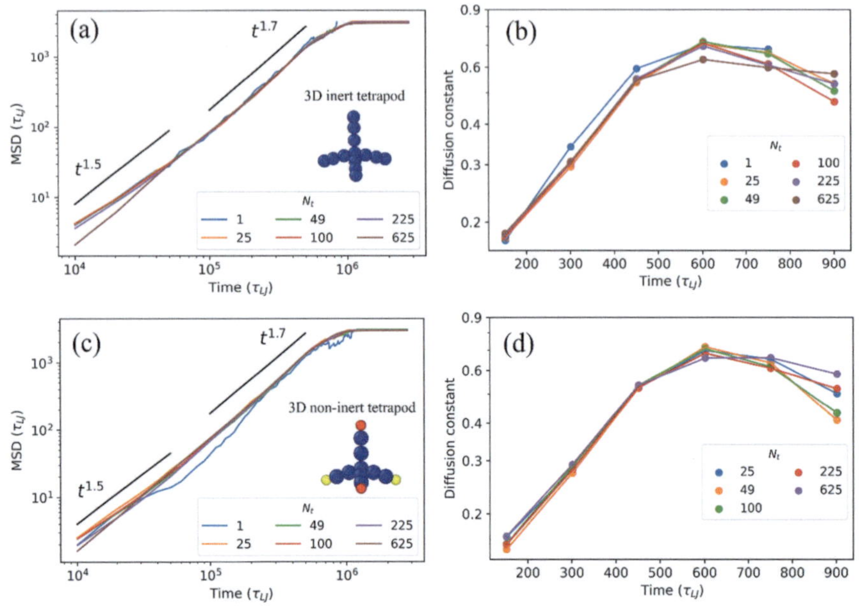

Fig. 9 The dynamics of the tetrapods falling under the gravity from the top surface. **(a)** and **(b)** MSD of center of mass of the inert and non-inert tetrapods with time for various initial concentrations. At small time scales, MSD exhibits the variation as ~ $t^{1.5}$, whereas tetrapods show faster diffusion with a scaling of ~ $t^{1.7}$ for longer times **(c)** and **(d)** Diffusion constant D for the inert and non-inert tetrapods as a function of simulation time τ_{LJ} [22]

with a scaling exponent $\alpha = 1.5$, whereas at longer times, the motion becomes super diffusive with $\alpha = 1.7$ (Fig. 9c) [22].

Diffusion constant D is derived from the MSD as,

$$\langle r^2(t) \rangle = 2dDt \tag{19}$$

where $r^2(t)$ is the MSD over time t, d is system's dimension, and the diffusion constant as D. For d = 3, D can be calculated as the rate of variation of MSD as,

$$D = \frac{\langle r^2(t) \rangle}{6t} \tag{20}$$

At short times, the diffusion constant is low due to slower individual tetrapod motion. However, at intermediate times, increased density leads to interlocking and intertwining of tetrapods, exhibiting collective motion and resulting in a higher diffusion constant over time. Furthermore, with increasing density, the increasing number of tetrapods induces competition between excluded volume effects and entropy, resulting in reduced diffusion [22]. This trend in the diffusion constant is consistent across both inert and non-inert tetrapod models (Fig. 9b and d).

5.6 *Self-Assembly*

5.6.1 Geometric Cluster Formation

The tetrapodal shape has gained significant interest in self-assembled structures due to its unique geometry and tenability [6]. Functionalization can be achieved either across the entire surface or selectively at the tips, enabling precise control over interactions. By adjusting geometry, size, and surface properties, tetrapods can be tailored for applications in biosensors, controlled drug delivery, electronic components, and lasers.

Controlled self-assembly of tetrapods follows two primary approaches. In single-component suspensions, assembly is regulated by surface charge modification, where functionalization dictates pairwise interactions and aggregate structures. Alternatively, introducing charged nanoparticles as a secondary component modifies the effective tetrapod interactions, leading to a broader design space. In binary systems, size asymmetry drives entropic effects, leading to diverse aggregation behaviors, including depletion attraction and oscillatory potentials. Particle-based computational approaches, such as molecular dynamics and Monte Carlo simulations, provide insights into self-assembly in size-asymmetric mixtures by explicitly capturing all degrees of freedom, including fluctuation phenomena. However, large-size asymmetry imposes computational limitations, necessitating advanced algorithms. The Geometric Cluster Algorithm (GCA) improves efficiency by moving entire particle clusters via symmetry operations like point reflections, creating a rejection-free Monte Carlo scheme. While effective for translational degrees of freedom, it struggles with orientational constraints in anisotropic particles. This limitation is addressed by extending the GCA with arbitrary plane and line reflections, although periodic boundary conditions introduce distortions in interparticle distances (Fig. 10).

The hyperspherical GCA (h-GCA) employs isotropic hyperspherical geometry, integrating symmetry operations through arbitrarily oriented planes passing through the hypersphere's center to mitigate these challenges. This approach maintains rejection-free configurations, ensuring computational efficiency. Validation against conventional Metropolis Monte Carlo data using asymmetric Lennard–Jones dimers

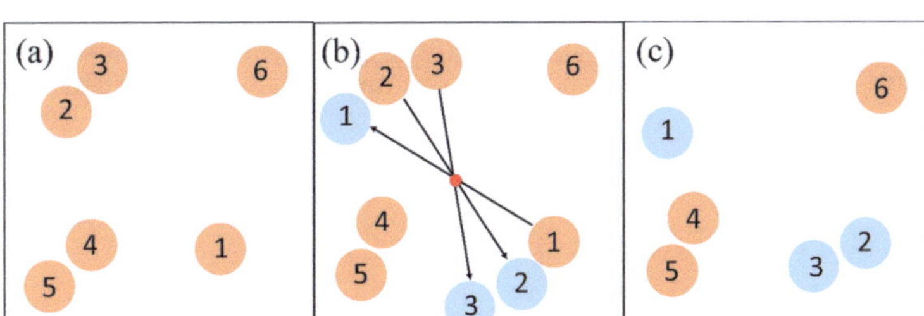

Fig. 10. 2D Geometric Cluster Algorithm. **(a)** Initial configuration. **(b)** Formation of a new cluster via point reflection of 1–3 about the pivot. **(c)** Final configuration

confirms its accuracy. The h-GCA framework offers a powerful tool for simulating self-assembly in anisotropic tetrapodal systems with extreme size asymmetry, expanding the scope of computational studies in nanomaterials.

Geometric Cluster Formation in Self-Assembled Structures

The Geometric Cluster Algorithm (GCA) provides an efficient framework for simulating self-assembly in complex systems by enabling collective particle movement through cluster formation [27]. Starting with a random pivot, particles are moved via point reflection, and neighbors are included in the cluster based on an energy-dependent probability. For systems with pair potentials V(r), the GCA is generalized as follows: after a particle m is reflected to a new position r'_m, particles interacting with m at its original or new positions are considered for inclusion in the cluster. Each particle n is added with a probability:

$$p_{mn} = \max\left[1 - \exp\left\{-\beta\left(V\left(\left|r'_m - r_n\right|\right) - V\left(\left|r_m - r_n\right|\right)\right)\right\}, 0\right] \tag{21}$$

where $\beta = 1/(k_B T)$. This approach ensures detailed balance and facilitates the exploration of cooperative dynamics, which are central to self-assembly processes. The algorithm is particularly useful for systems with large size asymmetries, high packing fractions, and complex interaction potentials. For example, in the limit of hard-core interactions, the GCA recovers its original formulation as a special case. Suppressing critical slowing down near phase transitions allows efficient sampling of configurations critical for studying nucleation and growth in self-assembled systems.

The GCA is also adaptable to non-spherical particles, electrostatic interactions, and systems with varying degrees of polydispersity. Its efficiency is illustrated in systems where the energy autocorrelation time exhibits only a weak dependence on system size or interaction strength, providing a significant advantage over conventional Monte Carlo methods. Moreover, the algorithm's ability to simulate mixtures with size ratios up to 100, involving millions of particles, makes it a robust tool for exploring self-assembly in binary fluids, patchy particles, and charged colloids. The collective motion captured by the GCA provides direct insights into pathways and mechanisms of self-assembly, particularly under conditions of high density and strong interactions. This framework accelerates the study of equilibrium structures and sheds light on kinetic pathways, offering a comprehensive tool for understanding self-assembly in diverse systems.

Hyperspherical Geometric Cluster Formation

The hyperspherical Geometric Cluster Algorithm (h-GCA) is a sophisticated simulation technique designed to address the challenges of modeling anisotropic particles and systems with significant size asymmetries [6]. It introduces two key innovations and builds upon the conventional Geometric Cluster Algorithm (GCA). First, simulations are performed on the three-dimensional surface of a hypersphere, an isotropic finite-volume space. This geometry eliminates distortions in interparticle distances that arise from reflections under periodic boundary conditions. Second, the algorithm incorporates arbitrary reflections about planes passing through the center of the hypersphere, enabling full relaxation of both translational and rotational degrees of freedom. These advancements overcome the orientation limitations of the original GCA, making h-GCA suitable for studying complex systems.

Clusters in the h-GCA are constructed using reflection-based symmetry operations, ensuring a rejection-free Monte Carlo scheme while maintaining detailed balance. The algorithm's validity was confirmed through comparisons with conventional Metropolis Monte Carlo simulations of asymmetric Lennard–Jones dimers, yielding consistent results. To evaluate efficiency, simulations of rod–sphere mixtures with increasing size asymmetry $= \sigma_{rod}/\sigma_{sphere}$, where σ_{rod} and σ_{sphere} represent rod and sphere diameters, demonstrated that the h-GCA significantly reduces autocorrelation times. Efficiency gains exceeded four orders of magnitude for size ratios as large as S = 9.

The h-GCA employs an orientation order parameter based on the second Legendre polynomial to analyze simulation outcomes. This parameter is defined as:

$$\frac{1}{N(N-1)}\sum_{i}\sum_{j\neq i}\left(\frac{3}{2}\left[t_i\cdot t_j-\frac{(t_i.n_j)(t_j\cdot n_i)}{1+n_i\cdot n_j}\right]^2-\frac{1}{2}\right) \tag{22}$$

where n_i is the four-dimensional unit vector representing particle position on the hypersphere and t_i is the tangential director vector.

The h-GCA's utility was demonstrated in studies of tetrapod self-assembly, particularly in functionalized ZnO-T–nanoparticle mixtures. These simulations revealed that the algorithm efficiently handles large particle asymmetries and enables detailed investigations of aggregation mechanisms. For instance, tip-functionalized ZnO-T exhibited aggregation behavior controlled by nanoparticle concentration and charge. The algorithm also allowed the identification of specific bonding configurations, such as unidirectional and bidirectional bonds, and their dependence on tetrapod tip charge.

Moreover, the addition of nanoparticles introduced novel aggregation mechanisms, including nanoparticle-mediated tip-to-tip and center-to-center bonding. These bonding tendencies were systematically analyzed as functions of nanoparticle charge and concentration. The h-GCA's ability to efficiently simulate such interactions highlights its potential for exploring ZnO-T self-assembly in engineered nanostructures.

The hyperspherical GCA is a transformative tool for simulating complex systems with anisotropic particles and large-size asymmetries. Its rejection-free approach and adaptability to hyperspherical geometry provide unparalleled efficiency, making it indispensable for studying self-assembly and designing functional nanostructures of tetrapod structures.

5.6.2 Linear Linker Molecule

The study demonstrates how adjusting the composition of a binary mixture drives self-assembly into ordered networks [21]. At an equal ratio of linkers and metal atoms, metal clustering due to soft-core interactions, i.e., short-range repulsive forces preventing complete overlap, leads to an imperfect square network. Increasing the linker fraction from the equal ratio to 75% suppresses clustering, forming a nearly perfect square lattice.

Radial distribution functions (RDFs) reveal the transition from clustered metal atoms $r \approx 0.5$ at $\chi = 0.5$ to separated atoms $r \approx 3.5$ at $\chi = 0.75$. Bond-orientation order parameters (2D BOOP) quantify this self-organization, showing significant improvement in order and symmetry at higher linker concentrations. BOOP, with respect to the metal atom, can be written as,

$$B_l^{2D} = \frac{1}{N_b}\left|\sum_{m}\sum_{m \neq n} \exp(lm\theta_{mn})\right| \tag{23}$$

where m runs over all metal atoms, n for all the neighbors of m, θ_{mn} is the angle from the axis between the bond connecting with m and n, N_b is the number of bonds in the system, l = 2,3,4,5,6. B_l^{2D} is zero for disordered and 1 for ordered structures.

The self-assembly of binary mixtures was further examined for specific conditions. At $\chi = 0.25$ with threefold more metal atoms than linkers, disordered "spaghetti-like" structures formed. Metal atoms clustered into dimers, confirmed by a prominent radial distribution function (RDF) peak $r \approx 0.5$.

Both linker-metal ratios and particle size critically influence self-assembly. Increasing linker concentration consistently promotes the stabilization of ordered networks. These findings provide valuable insights for designing functional nanostructures, particularly in materials science and molecular assembly applications.

5.6.3 Tip-To-Tip Connection of Tetrapod

Gel-phase formation (self-assembly of DNA nanostars) is observed across various temperatures and densities. At T > 0.5, nanostars remain in a disordered, gas-like state, while at T < 0.3, they self-assemble into percolating networks spanning the simulation box [21]. The structural properties of these networks are characterized using the radial distribution function (RDF):

$$S(l) = \frac{1}{4\pi l^2 \rho N}\sum_{m=1}^{N}\sum_{n \neq m}^{N}\langle \delta |l_{ij} - l| \rangle \tag{24}$$

The RDF profile is flat at high temperatures, indicating a lack of structure. At lower temperatures, distinct peaks emerge, signifying network formation. The first peak corresponds to the bonded distance between nanostars, with additional peaks reflecting higher-order arrangements. An increase in density reduces peak heights due to proportional increases in nanostar pairs and system density.

The amorphous nature of these networks is confirmed by the structure factor (SF):

$$SF(q) = \left\langle \frac{1}{N} \sum_{m=1}^{N} \sum_{m \neq n}^{N} \exp\left[iq \cdot (r_m - r_n)\right] \right\rangle \tag{25}$$

At high temperatures, the SF is flat, consistent with fluid-like behavior. Structural organization appears without crystallization at lower temperatures, as evidenced by the absence of Bragg peaks. Distinct network motifs emerge at very low temperatures: trihedral structures for Y-DNA, tetrahedral networks for X-DNA, and pentagonal arrangements for 5WJ DNA. These motifs become more pronounced at higher densities. At low densities, the increase in SF at $q \to 0$ reflects high compressibility and liquid-like behavior, while at higher densities, reduced compressibility indicates gel-like characteristics.

5.6.4 Aggregation of Colloidal Tetrapods

The simulation model is based on the droplet method given by Gaurav Modi et al. [28]. Each droplet features a spherical particle of radius r at the middle core (type P) with Q binders dispersed on its surface. Each binder consists of two types of particles: one on the outside (type R or S), accountable for binding between the complementary particles, and the excluded volume interaction is considered by an inside particle (type Q). The binders are initialized in a Fibonacci arrangement to minimize overlap in the initial configuration. The setup resembles the amalgamation of single-stranded sticky ends and double-stranded tethers, analogous to experimental DNA systems (Fig. 11).

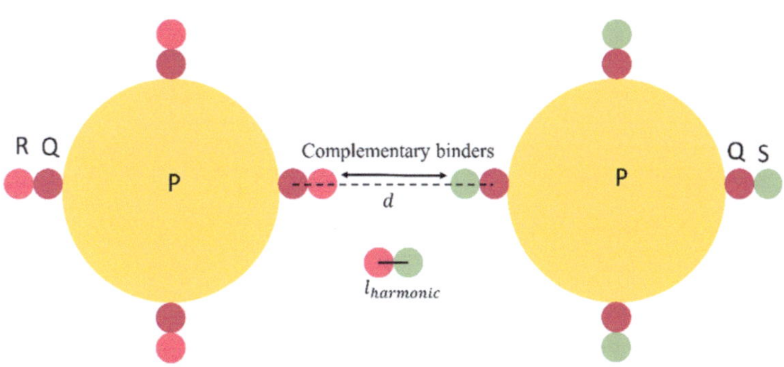

Fig. 11 Schematic representation of the bonding model exhibiting the dynamic binding between two droplets with outer binder particles. P is the central core with Q binders dispersed on its surface. R and S represent the pair of complementary binders

The vertical orientation of the binders is controlled by an angular term. Binders diffuse on the surface via a harmonic bond to the droplet center, characterized by the spring constant k_{PQ} and rest length l^0_{PQ}. Similarly, harmonic bonds hold the inner and outer particles together, with spring constant k_{QR} and $l_{QR} = r_Q + r_R$, the sum of their radii. To ensure outward orientation, binder particles are aligned in the direction of the radial vector from the center of the droplet using a harmonic angular potential with parameters k_{PQR} and rest angle $\theta_{PQR} = 180°$.

Spring constants are taken in such a way that the thermal standard deviation $\sigma = \sqrt{k_B T / k}$ lies in the range of the rest length or angle's small fraction, balancing rigid bonds and manageable MD time steps.

Non-Binding Interactions

To prevent particle overlap, a soft repulsive interaction is used:

$$U_s(l) = \begin{cases} \in_s \left[1 - \left(\frac{l}{l_{cut}}\right)^4\right] & if l < l_{cut} \\ 0 & if l \geq l_{cut}, \end{cases} \tag{26}$$

This interaction is used between all types of particles, except between complementary types of pairs of outer binder particles, where the opposite interaction could affect the binding. Here, $\in_s$ represents the interaction energy (in units of $k_B T$) and l_{cut} is the threshold distance. Tuning the effective diameter of Q particles modulates steric repulsion between binders, analogous to adjusting electrostatic screening between surface-bound DNA strands.

The study of droplet interaction is done by a coarse-grained molecular dynamics model in three dimensions. Additionally, the required confining potential is incorporated to replicate experimental conditions where droplets are observed in a plane due to gravitational effects. A force-shifted Lennard–Jones wall potential enforces a quasi-2D arrangement by confining the droplet center at a fixed z-position.

5.6.5 Colloidal Self-Assembly via Explicit Mobile Binders

Binder interactions are modeled via covalent bonds, with adhesive bonds forming only between complementary outer binders of different droplets. The model accommodates droplets with mixtures of binder types, with the potential for multiple types beyond just two. Harmonic bonds are employed having spring constant $k_{harmonic}$ and bond length $l_{harmonic} = r_R + r_S$, representing the sum of the radii of bonding particles. Each binder engages in a single binding event simultaneously, ensuring independent interactions. Each binding event follows a two-state model, with the effective free energy difference between bound and unbound states given by:

$$\Delta F = k_B T \ln\left(\frac{k_b}{k_{ub}}\right) \equiv \in \tag{27}$$

Here, k_b and k_{ub} represent the rate constants for binding and unbinding, respectively. The parameter $\in$ is tuned to control the affinity between binders.

Simulation Details

Molecular dynamics simulations of droplets with mobile binders were conducted using HOOMD-blue. Particle dynamics are integrated using the Langevin integrator moving forward in time. Two distinct drag coefficients were used: one for droplets γ_P and another for binders γ_b.

The Langevin dynamics equation of motion for ith particle is:

$$m_i r_i''(t) = -\nabla U_i - \gamma_i r_i'(t) + \sqrt{2\gamma k_B T}\eta(t) \tag{28}$$

where m_i is the particle's mass, Boltzmann constant is represented by k_B, the drag coefficient is given by γ, $r'(t)$ is the particle's velocity, U is total potential energy, and $\eta(t)$ is delta-correlated random white noise having zero mean and unit variance. Non-bonded calculations and bond pair construction were accelerated using a tree neighbor list.

5.6.6 Qualitative Representation of Self-Assembly

The formation of chains begins with droplets interacting through binding interactions. Monomers initially form dimers, and amalgamation of them with other monomers leads to the formation of trimers or combination with dimers forms tetramers (Fig. 6d–i) and (Fig. 12). This hierarchical assembly process reflects the self-organizing nature of the system, where smaller building blocks self-assemble into larger, more complex structures. The simulations reveal that patch formation progresses through two distinct stages: an initial rapid recruitment phase and a slower saturation phase. These stages illustrate the emergent order driven by binder interactions, highlighting the self-assembly process where local interactions form larger, stable structures.

5.7 *Effect of Confinement (Free Standing Structure)*

Freestanding, slender, and load-bearing structures can be engineered in confined spaces using tetrapod-like particles. Unlike traditional granular materials, which form squat, sloped heaps, these structures exhibit verticality akin to masonry walls or columns [29]. By pouring specially designed non-convex particles into molds and

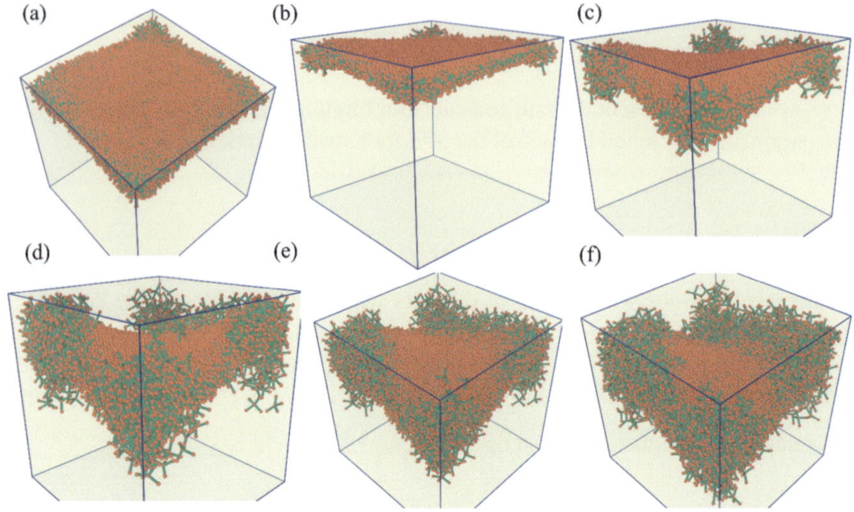

Fig. 12 Self-assembly visualization of a large concentration of inert tetrapods. **(a)–(f)** shows the various stages of tetrapods falling under the effect of gravity and exhibiting self-assembly or gel formation-like structures without attractive interaction among themselves

subsequently removing confinement, stable, freestanding columns with high aspect ratios (>10) are formed, capable of bearing compressive loads. A combination of experiments and simulations highlights the critical role of particle geometry in structural stability, with a particular focus on Z-shaped particles. These particles interlock to form rigid aggregates, exhibiting pronounced anisotropic stress propagation and strain-stiffening behavior. Mechanical testing under uniaxial compression, bending, and vibration reveals that Z-shaped particles outperform other geometries, such as U-shaped particles and rods, in structural integrity [30].

Z-shaped particles uniquely combine flowability in the unjammed state with the ability to entangle into self-confining aggregates, forming stable columns without external confinement. These structures exhibit near-zero Poisson's ratio, enhancing resistance to axial loads. Mechanical tests demonstrate that pre-straining increases rigidity and resistance to disintegration. Compared to U and Z90 particles, Z-shaped particles exhibit superior strain-stiffening and overall structural performance. Furthermore, lightweight structures can be fabricated using minimal material and low pre-strain, allowing for customizable architectural designs.

This study explores a novel class of cohesionless granular materials that transition between two typically incompatible states: a freely flowing, unjammed state and a rigidly jammed configuration. Aggregates of Z-family particles can flow into arbitrary spaces and subsequently form unconfined vertical structures capable of withstanding external stresses. Their rigidity, tunable via axial pre-strain, enables the design of recyclable architectural elements, including arches and slender columns.

Within the Z family, particle geometry dictates mechanical behavior. Z90 particles form rigid packings without pre-strain, U-shaped particles resist dissociation through chain formation, and Z-shaped particles exhibit the most significant rigidity increase under pre-strain while maintaining high flowability. Simulations show that these particles preferentially align horizontally in the presence of gravity, promoting axial stress propagation. The bent arms of Z-family particles inhibit sliding and rotation, fostering self-confinement and structural stability. Future optimization using evolutionary algorithms could further enhance these properties.

6 AI-ML Based Models

Artificial intelligence (AI) and machine learning (ML) are revolutionizing the study of formation, compatibility, and self-assembly in complex materials, particularly in tetrapod-based systems. Neural networks have outperformed traditional predictive modeling techniques, offering more profound insights into the mechanisms governing the self-organization of these non-convex particles. ML has been employed to design nanoparticle assemblies by training algorithms on polymers, gels, and colloidal systems, optimizing subcomponent configurations through perturbation theory and predictive analytics [31].

The ML tools can be powerful for designing patchy particles in simulations, enabling researchers to explore how different tuning strategies influence the particles' final properties before moving to experimental synthesis. Tetrapods with patchy particles having carboxylated tips are believed to result from amide bond formation. These structures exemplify "patchy particles"—nanoparticles coated with or covalently bound to polymers for diverse applications. In such systems, several critical studies have examined patchy interactions, highlighting the versatility of this approach. Within the patchy particle category, it becomes evident that even a single nanoparticle—just a basic unit of soft matter—can exhibit an extraordinary range of structural and functional variants, as reported throughout the literature [31–33]. In tetrapod-based self-assembly, ML has been instrumental in predicting structural arrangements, optimizing particle interactions, and fine-tuning suspension parameters to achieve desired morphologies. For example, Bayesian regression models have linked environmental conditions, such as pH and ionic strength, to the formation of stable colloidal assemblies, enabling precise control over the assembly process. These advancements underscore the growing role of AI-driven techniques in characterizing and designing tetrapod-based self-assembling materials [31].

ML methodologies provide a systematic approach to understanding and optimizing self-assembly. Supervised learning, leveraging labeled datasets, enhances predictive modeling of tetrapod packing and network formation, allowing for accurate regression and classification of assembly outcomes. In contrast, unsupervised learning uncovers emergent structural correlations in unlabeled data, revealing hidden assembly pathways. Neural networks, including feedforward architectures and long short-term memory (LSTM) models, facilitate studying dynamic self-assembly processes, capturing temporal evolution and metastable states. Additionally, Gaussian process modeling and ML-driven interatomic potentials enable accurate predictions of key properties, such as mechanical strength, band gaps, and thermal conductivity, further guiding the design of tetrapod-based materials. Despite these advancements, a key challenge remains in identifying critical system features and selecting optimal statistical approaches to establish precise structure–property relationships.

Beyond theoretical modeling, ML accelerates the experimental realization of tetrapod-based self-assembled architectures. Deep learning enhances the detection and arrangement of nanostructures, enabling automated fabrication of stable, free-standing tetrapod assemblies. Furthermore, ML-driven clustering techniques, such as k-means, improve dispersion analysis, a crucial factor in ensuring uniform self-assembly. These innovations position ML as a transformative tool for designing and optimizing tetrapod-based materials with tunable mechanical and functional properties.

By providing deeper insights into the structure–property relationships of tetrapod assemblies, ML-driven approaches contribute to the development of advanced self-assembled materials, paving the way for precise control over mechanical stability, defect tolerance, and structural integrity. Integrating AI in tetrapod self-assembly enhances fundamental understanding and accelerates the development of novel materials for applications in photonics, structural engineering, and nanotechnology.

7 Discussion and Conclusion

Recent advancements in the statistical modeling of tetrapods have provided significant insights into their self-assembly behaviors. This chapter presented an in-depth analysis of various coarse-grained models designed to capture the structural and dynamic properties of ZnO-T and DNA-based hydrogels. The combination of molecular geometry, functionalization, and interaction potentials governs the self-assembly mechanisms, enabling applications across nanotechnology, biomedicine, and optoelectronics.

The self-assembly of tetrapods is highly dependent on hydrophobic ratios, arm lengths, morphology, and interaction forces. A balance between attractive and repulsive interactions dictates the formation of ordered structures. Various modeling techniques, including Langevin dynamics and DPD, have successfully captured the role of thermal fluctuations and solvent-mediated interactions in self-assembly. These models reveal that hydrophobic tip interactions and patchy colloids influence the final assembled morphology. Tetrapods exhibit structural similarities with DNA hydrogels, where complementary sticky ends facilitate predictable self-assembly. The principles governing DNA-based networks can be extended to ZnO-T aggregation, bridging soft matter physics and nanomaterial science.

Studies revealed that the hydrophobic ratio significantly affects self-assembly. Tetrapods formed small, isolated clusters at lower hydrophobic ratios, whereas higher ratios led to larger, percolated networks. Similarly, the arm length played a crucial role in defining equilibrium structures, with longer arms promoting more stable, interconnected assemblies. The study also demonstrated the impact of interaction potentials, where stronger attractive forces facilitated tip-to-tip binding, enhancing the stability of the final structures.

The morphology of tetrapods was found to play a key role in determining their self-assembly behavior. Variations in arm geometry and core connectivity influenced the overall stability and packing of the assembled structures. Complex morphologies facilitated hierarchical organization, optimizing tetrapods for applications in optoelectronics and sensing technologies. Additionally, shape-dependent electron transport efficiency was observed, suggesting that tailored tetrapod geometries could enhance functional device performance.

External constraints, such as gravitational effects and spatial confinement, were also investigated in the simulations. Freestanding tetrapods exhibited different aggregation patterns than those influenced by confinement, highlighting the importance of environmental factors in self-assembly. The interplay between directional forces and constraints provides valuable insights into designing controlled self-assembly systems for targeted applications.

Furthermore, studies indicated that tetrapod self-assembly shares similarities with patchy colloidal aggregation. Selective attractive patches introduced directional bonding, influencing the formation of well-defined superstructures. The ability to program self-assembly through engineered patches offers a promising avenue for developing advanced nanostructured materials.

The study of tetrapod self-assembly offers new perspectives for material science, particularly in fabricating programmable nanostructures. ZnO-T are excellent candidates for hierarchical assembly due to their stability and multifunctionality. Future research can explore the integration of experimental validation with computational models to refine predictive accuracy, the development of hybrid organic–inorganic systems combining ZnO-T with bio-inspired scaffolds, and the influence of external fields, such as electric or magnetic forces, on the directed assembly of tetrapods. By advancing statistical modeling techniques and leveraging AI-driven simulations, the self-assembly of tetrapods can be fine-tuned for next-generation applications in photonics, drug delivery, and energy storage systems.

Acknowledgements The author, SKG, would like to thank MATRICS, ANRF (Project No. MTR/2022/000833) for financial support and acknowledge NIT Warangal for research support.

References

1. Sun, K., Qi, J., Zhang, Q., Yang, Y., Zhang, Y.: A novel logic switch based on individual ZnO nanotetrapods. Nanoscale **3**, 2166 (2011). https://doi.org/10.1039/c1nr10115c

2. Sinkovits, D.W., Luijten, E.: Nanoparticle-controlled aggregation of colloidal tetrapods. Nano Lett. **12**, 1743–1748 (2011). https://doi.org/10.1021/nl200961z
3. Roy, S.M., Garg, V., Barman, S., Ghosh, C., Maity, A.R., Ghosh, S.K.: Kinetics of nanomedicine in tumor spheroid as an in vitro model system for efficient tumor-targeted drug delivery with insights from mathematical models. Front. Bioeng. Biotechnol. **9**, 785937 (2021). https://doi.org/10.3389/fbioe.2021.785937
4. Pandey, R.K., Dutta, J., Brahma, S., Rao, B., Liu, C.-P.: Review on ZnO-based piezotronics and piezoelectric nanogenerators: aspects of piezopotential and screening effect. J. Phys. Mater. **4**, 044011 (2021). https://doi.org/10.1088/2515-7639/ac130a
5. Ghosh, S.K., Kundu, T., Sain, A.: From chemosensing in microorganisms to practical biosensors. Phys. Rev. E **86** (2012). https://doi.org/10.1103/physreve.86.051910
6. Deng, K., Luo, Z., Tan, L., Quan, Z.: Self-assembly of anisotropic nanoparticles into functional superstructures. Chem. Soc. Rev. **49**, 6002–6038 (2020). https://doi.org/10.1039/d0cs00541j
7. Roy, S.M., Garg, V., Sivaraman, S.P., Barman, S., Ghosh, C., et al.: Overcoming the barriers of nuclear-targeted drug delivery using nanomedicine-based strategies for enhanced anticancer therapy. J. Drug Deliv. Sci. Technol. **83**, 104408 (2023). https://doi.org/10.1016/j.jddst.2023.104408
8. Newton, M.C., Warburton, P.A.: ZnO tetrapod nanocrystals. Mater. Today **10**, 50–54 (2007). https://doi.org/10.1016/s1369-7021(07)70079-2
9. Mishra, Y.K., Adelung, R.: ZnO tetrapod materials for functional applications. Mater. Today **21**, 631–651 (2018). https://doi.org/10.1016/j.mattod.2017.11.003
10. Yan, L., Uddin, A., Wang, H.: ZnO tetrapods: synthesis and applications in solar cells. Nanomater. Nanotechnol. **5**, 19 (2015). https://doi.org/10.5772/60939
11. Roy, S.M., Barman, S., Kishore, P., Chatterjee, B., Bag, P., et al.: Acidic pH-triggered release of doxorubicin from ligand-decorated polymeric micelles potentiates efficacy against cancer cells. ACS Appl. Nano Mater. **6**, 18988–18998 (2023). https://doi.org/10.1021/acsanm.3c03481
12. Garg, V., Mathew, R., Ibrahim, R., Singh, K., Ghosh, S.K.: Crowding induced switching of polymer translocation by the amalgamation of entropy and osmotic pressure. iScience **27**, 109348 (2024). https://doi.org/10.1016/j.isci.2024.109348
13. Baran, Ł: Coarse-grained modeling of on-surface self-assembly of mixtures comprising di-substituted polyphenyl-like compounds and metal atoms of different sizes. ACS Omega **6**, 25193–25200 (2021). https://doi.org/10.1021/acsomega.1c02857
14. Li, X., Li, Z., Bian, X., Deng, M., Kim, C., et al.: Dissipative particle dynamics, overview. In: Encyclopedia of Nanotechnology, pp. 1–8 (2015). https://doi.org/10.1007/978-94-007-6178-0_100954-1
15. Xing, Z., Ness, C., Frenkel, D., Eiser, E.: Structural and linear elastic properties of DNA hydrogels by coarse-grained simulation. Macromolecules **52**, 504–512 (2019). https://doi.org/10.1021/acs.macromol.8b01948
16. Naskar, S., Bhatia, D., Lin, S.-T., Maiti, P.K.: Mechanistic insight into the structure, thermodynamics and dynamics of equilibrium gels of multi-armed DNA nanostars. Phys. Chem. Chem. Phys. **25**, 7847–7858 (2023). https://doi.org/10.1039/d2cp04683k
17. Araki, Y., Kobayashi, Y., Arai, N.: Programmed self-assembly of tetrapod nanoparticles with an amphiphilic surface pattern: the effect of arm length and hydrophobic ratio. Mol. Syst. Des. Eng. **4**, 1095–1102 (2019). https://doi.org/10.1039/c9me00112c
18. Garg, V., Mishra, M., Jayavelu, Y., Kumar, K.S., Joseph, D.P., Ghosh, S.K.: Coarse grain modeling of ZnO Tetrapods and generic DNA hydrogels. In: Proceedings of the International Conference on Emerging Multifunctional Materials and Devices for Sustainable Technologies, IEMDST Springer (2025) pp. 216–224. https://doi.org/10.1007/978-981-96-5863-3_28
19. Garg, V., Mathew, R., Chatterjee, A., Ghosh, S.K.: Protocols for translocation processes of flexible polymers through a pore using LAMMPS. STAR Protoc. **6**, 103568 (2025). https://doi.org/10.1016/j.xpro.2024.103568
20. Yao, G., Zhang, F., Wang, F., Peng, T., Liu, H., et al.: Meta-DNA structures. Nat. Chem. **12**, 1067–1075 (2020). https://doi.org/10.1038/s41557-020-0539-8

21. Fosado, Y.A.G.: Nanostars planarity modulates the rheology of DNA hydrogels. Soft Matter **19**, 4820–4828 (2023). https://doi.org/10.1039/d2sm00221c
22. Henrich, O., Gutiérrez Fosado, Y.A., Curk, T., Ouldridge, T.E.: Coarse-grained simulation of DNA using LAMMPS. Eur. Phys. J. E **41** (2018). https://doi.org/10.1140/epje/i2018-11669-8
23. Babayevska, N., Przysiecka, Ł., Iatsunskyi, I., Nowaczyk, G., Jarek, M., Janiszewska, E., Jurga, S.: ZnO size and shape effect on antibacterial activity and cytotoxicity profile. Sci. Rep. **12** (2022). https://doi.org/10.1038/s41598-022-12134-3
24. Modi, G.: Zinc oxide tetrapod: a morphology with multifunctional applications. Adv. Nat. Sci: Nanosci. Nanotechnol. **6**, 033002 (2015). https://doi.org/10.1088/2043-6262/6/3/033002
25. Warheit, D.B., Sayes, C.M., Reed, K.L.: Nanoscale and fine zinc oxide particles: can in vitro assays accurately forecast lung hazards following inhalation exposures? Environ. Sci. Technol. **43**, 7939–7945 (2009). https://doi.org/10.1021/es901453p
26. Sulciute, A., Nishimura, K., Gilshtein, E., Cesano, F., Viscardi, G., et al.: ZnO nanostructures application in electrochemistry: influence of morphology. J. Phys. Chem. C **125**, 1472–1482 (2021). https://doi.org/10.1021/acs.jpcc.0c08459
27. Liu, J., Luijten, E.: Rejection-free geometric cluster algorithm for complex fluids. Phys. Rev. Lett. **92** (2004). https://doi.org/10.1103/physrevlett.92.035504
28. Mitra, G., Chang, C., McMullen, A., Puchall, D., Brujic, J., Hocky, G.M.: A coarse-grained simulation model for colloidal self-assembly via explicit mobile binders. Soft Matter **19**, 4223–4236 (2023). https://doi.org/10.1039/d3sm00196b
29. Mishra, Y.K., Modi, G., Cretu, V., Postica, V., Lupan, O., et al.: Direct growth of freestanding ZnO tetrapod networks for multifunctional applications in photocatalysis, UV photodetection, and gas sensing. ACS Appl. Mater. Interfaces **7**, 14303–14316 (2015). https://doi.org/10.1021/acsami.5b02816
30. Murphy, K.A., Reiser, N., Choksy, D., Singer, C.E., Jaeger, H.M.: Freestanding loadbearing structures with Z-shaped particles. Granul. Matter **18** (2016). https://doi.org/10.1007/s10035-015-0600-2
31. Moud, A.A.: Recent advances in utility of artificial intelligence towards multiscale colloidal based materials design. Colloid Interface Sci. Commun. **47**, 100595 (2022). https://doi.org/10.1016/j.colcom.2022.100595
32. Kress, R.N., Jones, M.R.: Colloidal interactions get patchy and directional. Proc. Natl. Acad. Sci. U.S.A. **117**, 15382–15384 (2020). https://doi.org/10.1073/pnas.2006330117
33. Staneva, I., Frenkel, D.: The role of non-specific interactions in a patchy model of protein crystallization. J. Chem. Phys. **143** (2015). https://doi.org/10.1063/1.4935369

FSC
www.fsc.org
MIX
Papier aus verantwortungsvollen Quellen
Paper from responsible sources
FSC® C105338

If you have any concerns about our products,
you can contact us on
ProductSafety@springernature.com

In case Publisher is established outside the EU,
the EU authorized representative is:
Springer Nature Customer Service Center GmbH
Europaplatz 3, 69115 Heidelberg, Germany

Printed by Libri Plureos GmbH
in Hamburg, Germany